Novel Approaches in Mycotoxins Research: Detection, Prevention and Mode of Action

Novel Approaches in Mycotoxins Research: Detection, Prevention and Mode of Action

Guest Editors

Daniela Jakšić

Ana Juan-García

MDPI

Basel • Beijing • Wuhan • Barcelona • Belgrade • Novi Sad • Cluj • Manchester

Guest Editors

Daniela Jakšić
Department of Microbiology
Faculty of Pharmacy and Biochemistry
University of Zagreb
Zagreb
Croatia

Ana Juan-García
Laboratory of Food Chemistry and Toxicology
Faculty of Pharmacy and Food Science
University of Valencia
Valencia
Spain

Editorial Office
MDPI AG
Grosspeteranlage 5
4052 Basel, Switzerland

This is a reprint of the Special Issue, published open access by the journal *Toxins* (ISSN 2072-6651), freely accessible at: https://www.mdpi.com/journal/toxins/special_issues/WK7285W406.

For citation purposes, cite each article independently as indicated on the article page online and as indicated below:

Lastname, A.A.; Lastname, B.B. Article Title. *Journal Name* **Year**, *Volume Number*, Page Range.

ISBN 978-3-7258-4233-9 (Hbk)
ISBN 978-3-7258-4234-6 (PDF)
https://doi.org/10.3390/books978-3-7258-4234-6

Contents

Ana Juan-García and Daniela Jakšić
Novel Approaches in Mycotoxins Research: Detection, Prevention and Mode of Action
Reprinted from: *Toxins* **2025**, *17*, 161, https://doi.org/10.3390/toxins17040161 1

Luna Bridgeman, Cristina Juan, Houda Berrada and Ana Juan-García
Effect of Acrylamide and Mycotoxins in SH-SY5Y Cells: A Review
Reprinted from: *Toxins* **2024**, *16*, 87, https://doi.org/10.3390/toxins16020087 4

Alan Inglis, Andrew C. Parnell, Natarajan Subramani and Fiona M. Doohan
Machine Learning Applied to the Detection of Mycotoxin in Food: A Systematic Review
Reprinted from: *Toxins* **2024**, *16*, 268, https://doi.org/10.3390/toxins16060268 34

Ivana Dodlek Šarkanj, Nada Vahčić, Ksenija Markov, Josip Haramija, Natalija Uršulin-Trstenjak, Krunoslav Hajdek, et al.
First Report on Mycotoxin Contamination of Hops (*Humulus lupulus* L.)
Reprinted from: *Toxins* **2024**, *16*, 293, https://doi.org/10.3390/toxins16070293 64

Tamara Krska, Krisztian Twaruschek, Gerlinde Wiesenberger, Franz Berthiller and Gerhard Adam
Mechanism of Fumonisin Self-Resistance: *Fusarium verticillioides* Contains Four Fumonisin B_1-Insensitive-Ceramide Synthases
Reprinted from: *Toxins* **2024**, *16*, 235, https://doi.org/10.3390/toxins16060235 78

Laura F. Cadenillas, Guillaume Billerach, Christopher Hernandez, Vanessa Durrieu and Jean-Denis Bailly
Inhibition of Aflatoxin B1 Production by Procyanidins Present in *Annona muricata* and *Uncaria tomentosa* Aqueous Extracts
Reprinted from: *Toxins* **2024**, *16*, 454, https://doi.org/10.3390/toxins16110454 90

Elba Beraza, Maria Serrano-Civantos, Maria Izco, Lydia Alvarez-Erviti, Elena Gonzalez-Peñas and Ariane Vettorazzi
High-Performance Liquid Chromatography–Fluorescence Detection Method for Ochratoxin A Quantification in Small Mice Sample Volumes: Versatile Application across Diverse Matrices Relevant for Neurodegeneration Research
Reprinted from: *Toxins* **2024**, *16*, 213, https://doi.org/10.3390/toxins16050213 105

Soraia V. M. de Sá, Miguel A. Faria, José O. Fernandes and Sara C. Cunha
In Vitro Digestion and Intestinal Absorption of Mycotoxins Due to Exposure from Breakfast Cereals: Implications for Children's Health
Reprinted from: *Toxins* **2024**, *16*, 205, https://doi.org/10.3390/toxins16050205 122

Marianna Pauletto, Mery Giantin, Roberta Tolosi, Irene Bassan, Anisa Bardhi, Andrea Barbarossa, et al.
Discovering the Protective Effects of Quercetin on Aflatoxin B1-Induced Toxicity in Bovine Foetal Hepatocyte-Derived Cells (BFH12)
Reprinted from: *Toxins* **2023**, *15*, 555, https://doi.org/10.3390/toxins15090555 140

Jensen E. Cherewyk, Taylor J. Grusie-Ogilvie, Sarah E. Parker, Barry R. Blakley and Ahmad N. Al-Dissi
The Impact of Storage Temperature and Time on Ergot Alkaloid Concentrations
Reprinted from: *Toxins* **2023**, *15*, 497, https://doi.org/10.3390/toxins15080497 166

Silvia Trombetti, Alessandra Cimbalo, Michela Grosso, Pilar Vila-Donat, Jordi Mañes and Lara Manyes
Proteomic Analysis of the Murine Liver Response to Oral Exposure to Aflatoxin B1 and Ochratoxin A: The Protective Role to Bioactive Compounds
Reprinted from: *Toxins* **2025**, *17*, 29, https://doi.org/10.3390/toxins17010029 . . . . **177**

toxins

MDPI

Editorial

Novel Approaches in Mycotoxins Research: Detection, Prevention and Mode of Action

Ana Juan-García [1,*] and Daniela Jakšić [2]

1 Department of Preventive Medicine and Public Health, Food Sciences, Forensic Medicine and Toxicology, University of Valencia, 46100 Burjassot, Valencia, Spain
2 Department of Microbiology, Faculty of Pharmacy and Biochemistry, University of Zagreb, 10000 Zagreb, Croatia; djaksic@pharma.unizg.hr
* Correspondence: ana.juan@uv.es

Mycotoxins, which are toxic secondary metabolites produced by fungi, pose significant risks to food safety and public health. The research included in this Special Issue primarily focused on the regulated mycotoxin aflatoxin B1 (AFB1), fumonisins, and ochratoxin A (OTA), addressing their toxic effects, detection methods, and potential mitigation strategies. The highlights of this research include the genetic basis of fumonisin resistance in *Fusarium verticillioides* (Krska et al.), the development of a validated HPLC-FLD method for the quantification of ochratoxin A in models of mice neurotoxicity (Beraza et al.), and the assessment of mycotoxin contamination in hops when used in brewing (Sarkanj). Advances in machine learning for the rapid detection of mycotoxins (Inglis et al.) and protective compounds such as quercetin (Pauletto et al.) and functional food ingredients, e.g., fermented whey (Trombetti et al.), provide novel insights into the mitigation of toxins. Studies on polyphenolic plant extracts (Cadenillas et al.) also highlight eco-friendly antifungal alternatives.

As described in Krska et al. (2024), *Fusarium verticillioides* produces fumonisins, which inhibit the biosynthesis of sphingolipids in various organisms. These mycotoxins function as virulence factors in plant pathogens and influence interactions between competing fungi. It is reported that fumonisin-producing *Fusarium verticillioides* exhibits higher resistance to fumonisin B1 (FB1) compared to non-producing *F. graminearum*. By investigating the genetic basis of this resistance, Krska and colleagues found that the overexpression of certain ceramide synthases, particularly FUM18, conferred a high level of resistance, suggesting that *F. verticillioides* possesses a redundant self-resistance mechanism.

Similarly, Beraza et al. (2024) underscore the dangers of ochratoxin A (OTA), a mycotoxin that is commonly found in food products and has been increasingly linked to neurodegeneration. However, despite its relevance, no fully validated HPLC analytical methods are currently available for the quantification of OTA in mice, which is used as an animal model in neurotoxicity research. In order to address this gap, Beraza's team developed a highly sensitive and robust HPLC-FLD method that was validated according to FDA and EMA guidelines. This methodology enables the precise quantification of OTA in crucial tissues such as the brain and intestine, enabling more accurate neurotoxicity studies.

Meanwhile, Sarkanj provides insights into mycotoxin contamination in hops, a crucial ingredient in brewing. Analyzing 62 hop samples from Croatian craft breweries, the study confirmed that the hops had been contaminated with *Alternaria* and *Fusarium* toxins. Tenuazonic acid was detected in all samples, while deoxynivalenol appeared in 98% of them. However, the absence of *Aspergillus* and *Penicillium* toxins indicated that appropriate storage conditions were being employed. Additionally, regional differences in toxin levels

https://doi.org/10.3390/toxins17040161

highlighted the necessity for targeted monitoring to ensure the safety of hops in the brewing and pharmaceutical industries.

In another study, Inglis et al. (2024) explore the application of machine learning (ML) in mycotoxin detection. Traditional laboratory analyses, while effective, are time-consuming and impractical for large-scale screenings. ML applications have emerged as promising alternatives, offering accuracy and efficiency. Inglis reviews recent ML models used for mycotoxin detection, noting that neural networks, particularly convolutional ones, dominate the field. However, challenges remain, including the lack of detailed reporting on hyperparameters and open-source code, which hampers reproducibility and the optimization of models.

Expanding on mycotoxin-induced toxicity, Pauletto et al. (2023) investigate the protective role of quercetin (QUE) against aflatoxin B1 (AFB1) in bovine fetal hepatocyte-derived cells. AFB1 induces oxidative stress, inflammation, and transcriptional changes linked to carcinogenesis. Encouragingly, QUE reduced AFB1-induced cytotoxicity and the peroxidation of lipids, exerting broader transcriptional modifications than AFB1 alone. Notably, QUE reversed AFB1-induced alterations in the enzymatic activity of CYP3A, further supporting its protective role. These findings pave the way for in vivo studies that explore the potential role of QUE in mitigating aflatoxicosis.

Additionally, Trombetti et al. (2025) examine the hepatotoxic effects of exposure to AFB1 and OTA in Wistar rats and evaluate the mitigating properties of fermented whey (FW) and pumpkin (P). Proteomic analysis revealed the significant downregulation of differentially expressed proteins (DEPs) in the presence of AFB1 and OTA, suggesting a synergistic toxic effect. However, FW and P supplementation helped counteract these harmful effects, underscoring their potential application as functional ingredients in mitigating mycotoxin-induced damage.

Exploring alternative protective strategies, Cadenillas et al. (2024) investigate the antifungal effects of plant extracts against AFB1. Extracts from *Annona muricata* and *Uncaria tomentosa* inhibited the synthesis of AFB1 in a dose-dependent manner, correlating with their polyphenol content. More specifically, catechin and epicatechin played crucial roles in the inhibition of AFB1, with catechin reducing the production of toxins by 45% at concentrations comparable to the extracts. These findings highlight the potential use of plant-derived compounds as eco-friendly alternatives to synthetic fungicides.

Cherewyk et al. (2023) and Sá et al. (2024) address concerns regarding the stability and absorption of mycotoxins in food products. Cherewyk's research on ergot alkaloids in wheat reveals that the total ergot concentration fluctuates over time, emphasizing the need for proper storage and timely quantification. Meanwhile, Faria examines the bioaccessibility of AFB1, enniatin B, and sterigmatocystin in breakfast cereals. Notably, milk significantly influenced the bioaccessibility of mycotoxins, with variations depending on the milk type. These insights stress the importance of studying co-occurrence and dietary interactions to assess the exposure of humans to mycotoxins comprehensively.

Lastly, in a review article, Bridgeman et al. (2024) examine the neurological effects of acrylamide (AA) and mycotoxins, which frequently co-occur in food matrices such as cereals and coffee. With a focus on the SH-SY5Y neuroblastoma cell model, this study analyzed the cytotoxicity, apoptosis, oxidative stress, and axonopathy induced by AA and mycotoxins over the past decade. Bridgeman highlighted the growing scientific interest in studying their combined effects, as well as the need for further research into mitigation strategies. Notably, while bioactive compounds in food have been shown to counteract the toxicity of mycotoxins, there remains a significant gap in the knowledge regarding AA, underscoring the need for continued investigation.

While these findings provide crucial insights into the detection of mycotoxins, their resistance mechanisms, and mitigation strategies, several questions remain. Standardized ML models and the long-term effects of food contaminants combined with mycotoxins require further exploration. Additionally, the stability of mycotoxins in stored food and the potential for bioactive compounds to counteract toxicity warrant deeper investigation.

The most critical gap in mycotoxin research is the lack of comprehensive data regarding human exposure and validated biomarkers for risk assessment. While analytical advancements, such as the validated HPLC-FLD method for OTA in mice (Beraza et al.), provide a foundation, there remains a significant need for standardized, sensitive detection methods that can be employed in human biological matrices (e.g., blood, urine, tissues). Without robust biomonitoring tools, the assessment of real-world exposure levels and long-term health effects remains challenging. Additionally, the interplay between mycotoxins and food components, as well as their cumulative and synergistic toxic effects (e.g., with acrylamide, as highlighted by Bridgeman et al.), are underexplored. The limited use of advanced human-relevant models, such as organoids and co-culture systems, also hinders a deeper mechanistic understanding of mycotoxin toxicity in humans. Addressing these gaps through interdisciplinary research that combines analytical chemistry, toxicology, and computational modeling will be crucial for improving risk assessments and developing targeted mitigation strategies.

Conflicts of Interest: The authors declare no conflict of interest.

List of Contributions:

1. Beraza, E.; Serrano-Civantos, M.; Izco, M.; Alvarez-Erviti, L.; Gonzalez-Peñas, E.; Vettorazzi, A. High-Performance Liquid Chromatography–Fluorescence Detection Method for Ochratoxin A Quantification in Small Mice Sample Volumes: Versatile Application across Diverse Matrices Relevant for Neurodegeneration Research. *Toxins* **2024**, *16*, 213.
2. Bridgeman, L.; Juan, C.; Berrada, H.; Juan-García, A. Effect of Acrylamide and Mycotoxins in SH-SY5Y Cells: A Review. *Toxins* **2024**, *16*, 87.
3. Cadenillas, L.F.; Billerach, G.; Hernandez, C.; Durrieu, V.; Bailly, J.-D. Inhibition of Aflatoxin B1 Production by Procyanidins Present in Annona muricata and Uncaria tomentosa Aqueous Extracts. *Toxins* **2024**, *16*, 4544.
4. Cherewyk, J.E.; Grusie-Ogilvie, T.J.; Parker, S.E.; Blakley, B.R.; Al-Dissi, A.N. The Impact of Storage Temperature and Time on Ergot Alkaloid Concentrations. *Toxins* **2023**, *15*, 497.
5. Inglis, A.; Parnell, A.C.; Subramani, N.; Doohan, F.M. Machine Learning Applied to the Detection of Mycotoxin in Food: A Systematic Review. *Toxins* **2024**, *16*, 268.
6. Krska, T.; Twaruschek, K.; Wiesenberger, G.; Berthiller, F.; Adam, G. Mechanism of Fumonisin Self-Resistance: Fusarium verticillioides Contains Four Fumonisin B1-Insensitive-Ceramide Synthases. *Toxins* **2024**, *16*, 235.
7. Pauletto, M.; Giantin, M.; Tolosi, R.; Bassan, I.; Bardhi, A.; Barbarossa, A.; Montanucci, L.; Zaghini, A.; Dacasto, M. Discovering the Protective Effects of Quercetin on Aflatoxin B1-Induced Toxicity in Bovine Foetal Hepatocyte-Derived Cells (BFH12). *Toxins* **2023**, *15*, 555.
8. Dodlek Šarkanj, I.; Vahčić, N.; Markov, K.; Haramija, J.; Uršulin-Trstenjak, N.; Hajdek, K.; Šarkanj, B. First Report on Mycotoxin Contamination of Hops (*Humulus lupulus* L.). *Toxins* **2024**, *16*, 293.
9. Sá, S.V.M.d.; Faria, M.A.; Fernandes, J.O.; Cunha, S.C. In Vitro Digestion and Intestinal Absorption of Mycotoxins Due to Exposure from Breakfast Cereals: Implications for Children's Health. *Toxins* **2024**, *16*, 205.
10. Trombetti, S.; Cimbalo, A.; Grosso, M.; Vila-Donat, P.; Mañes, J.; Manyes, L. Proteomic Analysis of the Murine Liver Response to Oral Exposure to Aflatoxin B1 and Ochratoxin A: The Protective Role to Bioactive Compounds. *Toxins* **2025**, *17*, 29.

toxins

MDPI

Review

Effect of Acrylamide and Mycotoxins in SH-SY5Y Cells: A Review

Luna Bridgeman, Cristina Juan, Houda Berrada and Ana Juan-García *

Laboratory of Food Chemistry and Toxicology, Faculty of Pharmacy and Food Science, University of Valencia, Av. Vicent Andrés Estellés s/n, Burjassot, 46100 València, Spain; luna.birdgeman@uv.es (L.B.); cristina.juan@uv.es (C.J.); houda.berrada@uv.es (H.B.)
* Correspondence: ana.juan@uv.es

Abstract: Thermal processes induce the formation of undesired toxic components, such as acrylamide (AA), which has been shown to induce brain toxicity in humans and classified as Group 2A by the International Agency of Research in Cancer (IARC), as well as some mycotoxins. AA and mycotoxins' toxicity is studied in several in vitro models, including the neuroblastoma cell line model SH-SY5Y cells. Both AA and mycotoxins occur together in the same food matrix cereal base (bread, pasta, potatoes, coffee roasting, etc.). Therefore, the goal of this review is to deepen the knowledge about the neurological effects that AA and mycotoxins can induce on the in vitro model SH-SY5Y and its mechanism of action (MoA) focusing on the experimental assays reported in publications of the last 10 years. The analysis of the latest publications shows that most of them are focused on cytotoxicity, apoptosis, and alteration in protein expression, while others are interested in oxidative stress, axonopathy, and the disruption of neurite outgrowth. While both AA and mycotoxins have been studied in SH-SY5Y cells separately, the mixture of them is starting to draw the interest of the scientific community. This highlights a new and interesting field to explore due to the findings reported in several publications that can be compared and the implications in human health that both could cause. In relation to the assays used, the most employed were the MTT, axonopathy, and qPCR assays. The concentration dose range studied was 0.1–10 mM for AA and 2 fM to 200 μM depending on the toxicity and time of exposure for mycotoxins. A healthy and varied diet allows the incorporation of a large family of bioactive compounds that can mitigate the toxic effects associated with contaminants present in food. Although this has been reported in some publications for mycotoxins, there is still a big gap for AA which evidences that more investigations are needed to better explore the risks for human health when exposed to AA and mycotoxins.

Keywords: acrylamide; mycotoxins; food processed contaminants; neurotoxicity; in vitro

Key Contribution: Acrylamide (AA) (0.1 to 10 mM) and mycotoxins (2 fM to 200 μM) causes toxic effects in neuroblastoma cells, SH-SY5Y. This review highlights the importance of performing studies based in the mixture of mycotoxins and food processed contaminants such as acrylamide.

1. Introduction

The food baking process is the most important sub-process responsible for the main chemical, physical, and sensory properties of the final product, as well as the development of bioactive and antioxidant compounds [1–3]. However, this thermal process induces the formation of unwanted toxic components, including acrylamide (AA), due to the Maillard reaction occurring at high temperatures [1,4–6]. In detail, the Maillard reaction involves three major steps: (i) condensation of free amino groups (such as asparagine) with reducing sugars (glucose and fructose) to form acrolein; (ii) Strecker degradation of amino acids to aldehydes and ammonia; and (iii) brown nitrogenous compounds combining with acrylic acid to form AA (Figure 1) [7].

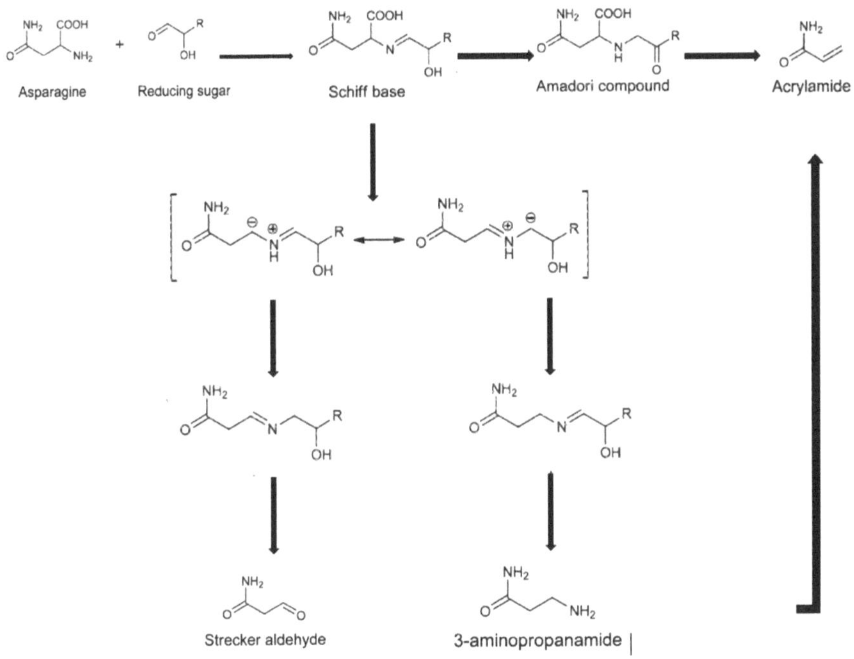

Figure 1. Proposed mechanism for AA formation as a side reaction of the Maillard reaction.

The compound AA is a colorless and odorless, highly reactive, and water-soluble crystalline compound with a low molecular weight of 71.08 kDa. AA is used in many different industrial areas, such as water treatment and the production of paper, fabrics, or cosmetics, and is widely employed in laboratories for gel chromatography [8]. The widespread use of AA causes a high level of occupational exposure by inhalation and skin contact [9]. Furthermore, in 2002, Swedish researchers reported that several heat-treated carbohydrate-rich foods contained significantly higher levels of AA than other known food carcinogens [6]. However, several factors can influence the level of AA formation in foods, such as the type of raw materials, product composition, pH, and moisture. In addition, the highest levels of AA are detected in foods derived from thermal processes involving the frying and browning of potatoes, cocoa beans, and coffee roasting, and the baking of cereals and cakes [10]. It is interesting to highlight that AA was classified as a compound "possibly carcinogenic to humans" by the International Agency for Research on Cancer [11] and classified in Group 2A. Likewise, the European Commission (2017) [12] classifies AA as a Class 1B carcinogen and mutagenic substance and Class 2 for reproductive toxicity, while the European Chemicals Agency (2022) [13] has included AA in the list of substances of very high concern. Therefore, it is of great interest, and further studies of AA toxicity are necessarily required to assess the potential risk for human health.

The toxicodynamic characteristics associated with AA are based on rapid absorption in the blood and wide distribution to tissues due to its low molecular weight and high solubility properties [14]. AA is mainly metabolized in the liver and brain by the glutathione-S-transferase enzyme, forming N-acetyl-S-(3-amino-3-oxypropyl)-cysteine [15]. Moreover, AA can be converted by cytochrome P450 2E1 (CYP2E1) to glycidamide (GA), a more reactive epoxide that can react with hemoglobin and DNA [16,17]. Thus, GA is responsible for the mutagenic and carcinogenic effects of AA in vivo [18]. AA has been shown to induce systemic side effects in animals, while brain toxicity has been detected in humans [19,20]. It has also been reported that AA neurotoxicity is associated with the imbalance between oxidation and antioxidant function brought on by lipid peroxidation and a rise in intracellular reactive oxygen species (ROS) [8].

Mycotoxins are toxic compounds produced by various *Aspergillus*, *Penicillium*, *Fusarium*, and *Trichoderma* fungal species that are health-hazardous poisons frequently present in various agricultural products, including grains, nuts, and spices [21]. These toxins pose significant health risks to humans and animals when consumed [22]. Understanding the effects of mycotoxins on different cell types is crucial for assessing their toxicological impact [23–25].

Studying mycotoxins in vitro using SH-SY5Y cells allows researchers to assess their potential neurotoxic effects and elucidate the underlying mechanisms of toxicity. These studies involve exposing the cells to various concentrations of specific mycotoxins, such as ochratoxin A (OTA) or fumonisins, and evaluating their impact on cell viability, morphology, proliferation, apoptosis, oxidative stress, neurotransmitter function, and other relevant endpoints [26,27].

Likewise, the European Commission has reported more than 400 alerts for fungal contamination in food and feed, with OTA being one of the most reported toxins, accounting for 10% of the notifications [28].

Nowadays, there is a strong effort toward studying the combination of compounds that can occur in the same food matrix, environment, or are daily ingested through the diet, especially regarding the effects that mixed compounds can cause. AA and mycotoxins neurotoxicity is reviewed in several in vitro models, including undifferentiated and differentiated human neuroblastoma cell line (SH-SY5Y) models. SH-SY5Y cells are used in toxicological research to assess neuronal differentiation, metabolism, and function, as well as in neuroadaptive, neurodegenerative, and neuroprotection processes [29]. The human SH-SY5Y cells are a subclone of SK-NSH, which, once stimulated with retinoic acid (RA), differentiates into dopaminergic neuron-like cells, acquiring many biochemical and functional properties of neurons [30,31].

Moreover, despite numerous reviews offering scientific insights into the chronic exposure [32], toxicity [33,34], presence in food [35], and various extraction procedures [36,37] related to AA and mycotoxins individually, there remains an unaddressed gap in the literature: a review uniting the toxic effects of AA and mycotoxins within an in vitro model, such as SH-SY5Y cells. In addition, a new study establishes the combined effects of these compounds in SH-SY5Y cells, emphasizing the importance of collecting data on AA and mycotoxins individually to understand their mechanisms of action [38]. Furthermore, the interest in the shared toxic effects documented in both acrylamide (AA) and multiple mycotoxins is not just due to their occurrence in our diets but also attributable to the potential combination of these effects, raising important implications for human health, particularly concerning neuronal impacts. Therefore, this review aimed to evaluate the effects of AA and mycotoxins as well as their mechanism of action (MoA) in the human neuroblastoma cell line SH-SY5Y, as reported in studies of the last 10 years.

The results have been organized according to findings of AA and mycotoxins that might coexist together in food and cereal food-based products. Thus, the section has been divided into the following sub-sections: cytotoxicity, apoptosis, oxidative stress, network degeneration, signaling pathways, protein expression, and natural compounds against the adverse effects of AA.

2. Cytotoxicity

2.1. Acrylamide (AA)

Non-viable or dead cells are associated with a loss of membrane integrity, which is accomplished by alterations in the movement of molecules either into or out of cells across membranes that have become leaky [39]. Cytotoxicity assays are convenient for understanding the dose- and time-dependent toxicity of chemicals and their reversibility and impacts on the cell cycle. Multiple methods are known, depending on their endpoints; some are based on a visual morphological scoring process (e.g., elution test, direct and indirect contact), while other methods use colorimetric and numbering rapid analysis techniques [i.e., trypan blue exclusion, propidium iodide uptake, 3-(4,5-dimethylthiazol-2-yl)-2,5-diphenyl-2H-tetrazolium bromide (MTT), etc.] [40].

In this sense, to find out the cytotoxicity of AA at the neuronal level, several authors investigated the cytotoxicity by MTT assay in SH-SY5Y cells when incubated with different concentrations of AA for 24 h [41]. Reductions in cell viability were detected from 15% to 40% in SH-SY5Y cells treated with AA (from 2 to 5 mM) in a dose-dependent manner [41–43]. Furthermore, changes in the shape of cells were detected after exposure to 2.5 or 5 mM of AA, for 24 h, as most cells shrank, and the cell body became round [41]. In the same line of research but one step forwards, the same group of investigators studied the protective effect of curcumin on AA-induced cytotoxicity for which SH-SY5Y cells AA-treated from 2.5 to 10 mM obtained a decrease in the number of living cells from 14% to 61% with respect to the control by a CCK8 assay [42], while the treatment of SH-SY5Y cells with 100 μM of α-lipoic acid (LA) suppressed AA apoptosis induction and the loss of cell viability.

Similarly, comparing the cytotoxicity of SH-SY5Y and a human glioblastoma cell line (U-1240 MG), in different time- and dose-dependent assays, it was shown that upon treatment with AA from 0.1 to 2 mM up to 72 h, SH-SY5Y viability was significantly reduced up to 40% upon exposure to 2 mM AA; while for U-1240 MG cells, viabilities were significantly reduced up to 92% for that same concentration of AA (2 mM) [44]. Coinciding with the results for SH-SY5Y cells in reducing viability, significance was also reported at 100 μM and after 72 h of exposure [45].

In addition to the common cytotoxic assay of MTT, viability alterations produced by AA have been carried out by SF assay [46], and trypan blue and lactate dehydrogenase leakage (LDH) [47]. The effects reported for SH-SY5Y cells by SF assay revealed that viability was altered at high concentrations of AA (10 mM) causing definite cellular damage and cell death at short times of exposure (8 h) [46]; while cytoxicity evaluation by trypan blue exclusion and LDH leakage was reported at the same doses of AA (10 mM) but, in this case, at 6 h with a viability decreased by 38% [47]. Compared with U-1240 MG cells, at 24 h, a reduction in viability of 20% and 35% at 5 and 10 mM was shown, respectively, concluding that the cell viability reduction in U-1240 MG cells was less than that reported for SH-SY5Y cells [47].

Another study carried by Okuno et al. (2006) [48], also in SH-SY5Y cells, showed that AA cytotoxicity was dose (0.5–5 mM) and time (1–24 h) dependent, by the fact that trypan blue exclusion decreased and LDH leakage increased [48]. The WST-8 assay is another assay for cell viability and it was compared with LDH leakage on SH-SY5Y exposed to a concentration range of 1–5 mM AA [49]; the results revealed a WST-8 decrease and LDH increase according to the AA dose, but it helped to confirm the previous results [49].

On the other hand, the possibility of SH-SY5Y cells to be differentiated as neurons allowed researchers to study the effect of AA by the LDH assay, revealing that it was largely non-cytotoxic at 1 h of exposure, except at higher doses (10 mM); this was similar to the results when undifferentiated SH-SY5Y cells were tested [50].

Furthermore, Frimat et al. (2010) [51] demonstrated through the viability assay with CellTiter-Blue that an AA concentration of 5 mM reached a 50% inhibition concentration (IC_{50}) for SH-SY5Y cells, while 0.26 mM caused a 20% reduction in the network formation equivalent to the control at 24 h of exposure.

In another study, the basal cytotoxicity was determined by measuring the total cellular protein content, and the total protein from the 50 S subunit (TP_{50}) and TP_{20} values (1.34 mmol/L and 0.61 mmol/L, respectively) reflected severe and moderate cytotoxicity, respectively, in SH-SY5Y cells exposed to AA [52].

However, other authors showed that 0.5–2.0 mM of AA did not produce any cytotoxicity (assessed by fluorescence microscopy) at 24 h, but concentrations higher than 4 mM caused a significant loss of HuD-positive neurons (58% of control) and a decrease in axon number by 21% with respect to the control [53]. Similarly, as proved by Calcein-AM assay, (a fluorescence-based cell viability assay), 1 mM of AA had no statistically significant toxicity at 48 h in SH-SY5Y cells or the embryonic carcinoma cell line (P19) derived from an embryo-derived teratocarcinoma in mice; however, it was cytotoxic for pheochromocytoma of the rat adrenal medulla (PC12) cells [54]. A study of cytotoxicity through fluorescence microscope observation by cytotoxicity tests using the Live/Dead Cell Staining Kit II following exposure to different concentrations of AA (0.01, 0.28, 7 mM) in the SH-SY5Y cell line at 24 h reported that the ratios significantly decreased at 0.28 mM and 7 mM, respectively [55].

In order to obtain more accurate results, the cytotoxicity of AA on SH-SY5Y cells was studied by MTS assay and the toxicity after 24 h of incubation was evaluated [56]. The amount obtained in the abluminal concentration after 1 h was also measured in the blood–brain barrier (BBB) model, to finally compare the results of cytotoxicity between AA assayed directly in SH-SY5Y after 24 h with the AA obtained after passing through an in vitro BBB (4 d/24 w) model for 1 h. It was proved that AA was cytotoxic at 100 μM in SH-SY5Y at 24 h, but not the abluminal AA concentration after BBB transport [56].

2.2. *Mycotoxins*

2.2.1. Beauvericin (BEA)

Studies on the impact of BEA on SH-SY5Y cells have illustrated varied results, predominantly through the utilization of the MTT assay. Viability was shown to be reduced by 50% during a 72 h exposure at 2.5 μM, with some studies citing a 43% decrease in viability at the highest concentration assayed (2.5 μM) after 48 h of exposure [57–60]. Diverse findings have been noted regarding IC_{50} values at different exposure times, such as 1.9 μM at 6 h, 1.7 μM at 24 h, and 1.5 μM at 48 h [61,62]. LDH leakage was also found to be induced by 1 μM of BEA, corroborating results obtained by the MTT assay [61].

2.2.2. Deoxynivalenol (DON)

Regarding the effects of DON in SH-SY5Y cells, the MTT assay was carried out for both studies, but with different concentrations and results. The IC_{50} for Pérez-Fuentes N. et al. 2021 [61] was 2.25 μM at 24 h, but 120 μM for Kalagatur et al. [63] at the same time. Pérez-Fuentes N. et al., 2021 [61] did not find any increase in the extracellular LDH level, but Kalagatur et al. [63] reported a directly proportionate extracellular LDH level in comparison to the MTT assay.

2.2.3. Enniatin A and Enniatin B (ENN A and ENN B)

Six hours of ENN A exposure resulted in a 41.0 ± 8.5% drop in cell viability at 2.5 μM and 94.3 ± 1.6% at 5 μM, respectively, with total inhibition at 10 μM. After incubation for 24 and 48 h, cell viability at 2.5 and 5 μM showed a decrease greater than 74.5%, and at the highest tested dose, a complete reduction was observed once more. Concerning ENN B, cell viability did not reach total inhibition at the highest dose at 6 h [61]. Nevertheless, at 24 h, ENN B had an IC_{50} of 0.43 μM. Also, ENN A increased LDH release at 5 and 10 μM at 24 and 48 h. After treatment for 24 h, ENN B increased LDH release by roughly 20% at values higher than 0.25 μM. After 6 h of incubation, at 5 and 10 μM, ENN A showed substantial differences from control cells (79.0 ± 8.9% and 67.7 ± 14.5%, respectively) [61].

2.2.4. Fumonisin B1 (FB1)

Concerning the effects of FB1, several studies investigated the toxic potential of this mycotoxin in SH-SY5Y cells. In total, 100 μM of FB1 decreased the cell viability of SH-SY5Y cells after incubation for 48–144 h [64]. In the same line but at lower exposure times, after 48 h of incubation, FB1 reduced the cell viability at the highest concentration (30 μM), and it appeared to promote cell proliferation at the lowest doses (0.1 μM) after 24 and 48 h. Furthermore, no IC_{50} was found, and in the LDH assay no cytotoxic effects were shown following FB1 treatment [62]. Similarly, in another study, 50 μM treatment of FB1 was compared to the control, and LDH was released at 12 h, but this release decreased at 24 h and at 48 h of treatment and there was no difference from the control [65]. Then, Domijan et al. (2011) [66] investigated how FB1 (at concentrations ranging from 0.5 to 200 M) affected the viability of cell cultures after 24 h of exposure. None of the FB1 concentrations utilized in the experiment caused cell death in neuroblastoma cells. Neuroblastoma cells treated for 24 h with the maximum FB1 dose (200 μM) showed 98.0 ± 1.85% cell survival.

2.2.5. Ochratoxin A (OTA)

Concerning the viability of SH-SY5Y cells treated with OTA, the IC_{50} values observed for 24 h and 48 h were 9.1 μM and 5.8 μM, respectively [67]. A significant decline in viability was noted beyond concentrations exceeding 3.12 μM or 6.25 μM for 24 h and 48 h, resulting in a decrease from 74% to 25% and from 80% to 49% at 24 h and 48 h, respectively [67]. Similarly, less dramatic decreases in mitochondrial activity were observed under different concentrations of OTA treatment (74 ± 12 and 74 ± 44% control at 10 and 100 μM OTA, respectively) at 24 h, and the LDH activities of OTA at 1, 10, or 100 μM were 128 ± 1, 125 ± 2, and 200 ± 1% of the untreated control [68].

2.2.6. T-2 Toxin

Regarding the exposure of SH-SY5Y cells to the T-2 toxin, this resulted in a significant decrease in cell viability, with percentages of 81.9%, 40.8%, and 35.5% observed for cells exposed to 5, 10, and 20 ng/mL of the toxin for 48 h, respectively. Additionally, LDH levels were significantly elevated by 1.8, 2.9, and 3.2 times compared to the control group [69].

2.2.7. Zearalenone (ZEA) and Its Metabolites

Referring to ZEA effects on SH-SY5Y cells, the MTT assay stated that at 6 h of incubation, ZEA did not cause any cell damage. However, it significantly reduced cell viability when the concentration exceeded 20 μM at 24 and 48 h, resulting in a reduction of more than 50% compared to the control cells. The inhibitory concentration (IC_{50}) of ZEA was determined to be 17.4 μM [61]. However, after 24 h of ZEA treatment, in a previous study [70] similar results were obtained in SHSY5Y cells with a significant decrease in cell viability even at a lower concentration (25 μM). The viability of SH-SY5Y cells decreased by as much as 86% at 200 μM ZEA [70]. Regarding the cell viability in ZEA metabolites (α-ZEL and β-ZEL), the following IC_{50} values were obtained for α-ZEL: 20.8 at 48 h, 14.0 at 72 h and for β-ZEL: 94.3 at 24 h, 9.1 at 48 h, and 7.5 at 72 h [59].

In summary, we can state that AA produces cytotoxicity in SH-SY5Y cells, at high concentrations (5 mM or higher), and that in lower concentrations it reduces cell viability in a time- and dose-dependent manner (Table 1). On the other hand, mycotoxins produce cytotoxicity at low concentrations of 200 μM or less and this is time- and dose-dependent in some cases. The most cytotoxic compound was ENN B, followed by DON and ENN A. (Table 2). Therefore, AA and mycotoxins directly affect cell viability in all cell models studied; the most sensitive cells exposed to AA were PC12, followed by SH-SY5Y, and U-1240 MG.

Table 1. Cytotoxicity in SH-SY5Y cells. AA exposure conditions, assays, and effects.

Dose	Exposure Time	Assays	Effects	References
1 to 10 mM	8 h	Cell Count Reagent SF	At 10 mM AA	[46]
1–5 mM	16 h, 20 h	Caspase-3 activity	At 1–5 mM	[45]
2.5, 5, 7.5, and 10 mM	24 h	CCK8	Reduction of 14%, 35%, 48%, and 61%, respectively	[42]
0.8, 20 and 500 μg/mL	24 h	MTT	Reduction of 60% and 70% at the higher concentrations	[55]
2, 2.5, 3, 4 and 5 mM	24 h	MTT	Reduction of 10%, 18%, 20%, 25%, and 30%	[41]

Table 1. *Cont.*

Dose	Exposure Time	Assays	Effects	References
1.25, 2.5, 5 mM	24 h	MTT	Significant loss of cell viability	[43]
0–500 mM	24 h	CellTiter-Blue Cell Viability Assay	IC_{50} 5 mM NI_{20} 0.26 mM	[52]
1, 10 and 100 μM	24 h	MTT, ATP, Caspase	100 μM reduced cell viability	[56]
0.01 mM to 12 mM.	24 h	Propidium iodide and calcein-AM	0.5–2.0 mM of AA was not cytotoxic; 4 mM, produced a significant loss of HuD-positive neurons (58 ± 7%), and the number of axons decreased to 21 ± 5%	[42]
0.5–5 mM	24 h	Trypan blue	Cytotoxic at 5 mM	[48]
0.1–10 mM	24 h	LDH assays	AA was cytotoxic at 10 mM	[49]
0.1–1000 μM	48 h	Calcein-AM, βIII-tubulin, and LDH assay	10% reduction at 1 mM	[54]
0–1000 μM	72 h	Total protein content	>1000 μM IC_{50}	[52]
0, 0.5, 1, 2, 5, and 10 mM	0, 6, 12, 24, 48, and 72 h	Trypan blue and LDH	10 mM reduced 38% cell viability at 6 h	[47]
0, 0.1, 0.5, 1, and 2 mM	24, 48, 72 h	MTT	Reduction of 20%, 30%, and 50%	[44]
1–1000 μM.	24, 48, and 72 h; 3, 6, and 10 days.	BradUrd, Sub-G_1, ^{3}H-thimidine, and MTT	Reduction of 30% (48 h) and 90% (72 h) At 100 μM or higher viability decreased at Day 3 and 6.	[45]

AA: acrylamide, ATP: adenosine triphosphate, CCK8: Cell counting kit-8, IC_{50}: 50% inhibition concentration, LDH: lactate dehydrogenase, MTT: (3-[4,5-dimethylthiazol-2-yl]-2,5 diphenyl tetrazolium bromide), NI_{20}: 20% reduction in network formation.

Table 2. Cytotoxicity in SH-SY5Y cells. Mycotoxin, exposure conditions, assays, and effects.

Mycotoxin	Dose	Exposure Time	Assays	Effects	References
DON	0–10 μM	6 h and 24 h	MTT LDH	IC_{50}: 2.25 μM at 24 h; 1.5 at 48 h Not cytotoxic	[61]
	0–200 μM	6 h and 24 h	MTT LDH	IC_{50}: 120 μM at 24 h	[63]
BEA	0.009–25 μM	24 h, 48 h, and 72 h	MTT assay	BEA IC_{50}: 2.5 at 72 h	[59]
	0.08–2.5 μM	24 h and 48 h	MTT assay	Decrease of 43% in cell viability at 2.5 μM after 48 h	[60]
	0.08–2.5 μM	24 h and 48 h	MTT assay	IC_{50}: 2.5 μM at 72 h	[58]
	0.08–25 μM	24 h and 48 h	MTT assay	IC_{50}: 3.2 μM at 24 h; 5 μM at 48 h	[62]
	0.1–30 μM	6 h, 24 h, and 48 h	MTT assay LDH assay	IC_{50}: 1.9 μM for 6 h, 1.7 μM for 24 h, and 1.5 μM for 48 h Inducing LDH leakage at 1 μM	[61]
	0–2.5 μM	24 h, 48 h, and 72 h	MTT assay	IC_{50} at 72 h was 2.5 μM	[57]

Table 2. *Cont.*

Mycotoxin	Dose	Exposure Time	Assays	Effects	References
ZEA α-ZEL and β-ZEL	0.39–100 μM	24 h, 48 h, and 72 h	MTT assay	IC_{50}: α-ZEL: 20.8 at 48 h, 14.0 at 72 h β-ZEL: 94.3 at 24 h, 9.1 at 48 h, and 7.5 at 72 h	[59]
	0.4–12.5 μM	24 h and 48 h	MTT assay	β-ZEL: 6.25, 12.5, and 25 μM from 31% to 82%	[60]
	0.1–30 μM	6 h, 24 h, and 48 h	MTT assay LDH assay	ZEA presented a IC_{50} of 17.4 μM LDH leakage at 20 μM (9.1 ± 8.9%) and 30 μM (19.5 ± 4.3%) at 24 h	[61]
	25, 50, 75, 100 and 200 μM		MTT assay LDH	Decreased by 86% at 200 μM	[70]
FB1	0.1–30 μM	6 h, 24 h, and 48 h	MTT assay LDH assay	At 30 μM after 48 h LDH not cytotoxic	[61]
	50 μM	12 h, 24 h, and 48 h	LDH assay	FB1 led to LDH release at 12 h	[65]
ENN's	0.1–30 μM	6 h, 24 h, and 48 h	MTT assay LDH assay	ENN A IC_{50} of 2.4 μM at 6 h and 2.25 μM at 24 h ENN B IC_{50} of 0.43 μM at 24 h ENN A increased LDH at 5 and 10 μM at 6 h and 24 h ENN B showed LDH release of 20% at 0.25 μM after 24 h	[61]
OTA	1, 10, or 100 μM	30, 60 min, 24 h,	LDH	LDH at 1, 10, or 100 μM were 128% ± 1%, 125% ± 2%, and 200% ± 1%	[68]
	0.2–50 μM	24 h and 48 h	MTT assay	IC_{50}: 9.1 μM at 24 h; 5.8 μM at 48 h	[67]
	130–0.20 nM	24 h, 48 h, and 72 h 7 days	MTT assay	No IC_{50} was reached	[71]
T-2	5–20 ng/mL T-2	1–48 h	CCK-8 assay LDH assay	Viability reduced 81.9%, at 5 ng/mL, 40.8% at 10 ng/mL, and 35.5% at 20 ng/mL LDH levels were increased by 1.8, 2.9, and 3.2 times	[69]

BEA: Beauvericin, CCK8: Cell counting kit-8, DON: Deoxynivalenol, ENN's: enniatin, FB1: Fumonisin B1, IC_{50}: 50%inhibition concentration, LDH: lactate dehydrogenase, MTT: (3-[4,5-dimethylthiazol-2-yl]-2,5 diphenyl tetrazolium bromide), OTA: Ochratoxin A, ZEA: Zearalenone, α-ZEL: alpha-zearalenol, β-ZEL: beta-zearalenol.

3. Apoptosis

3.1. Acrylamide (AA)

In order to assess cellular apoptosis, the levels of mono- and oligonucleosomes were monitored in SH-SY5Y and U-1240 MG cells treated with AA. This revealed an increase in the enrichment factor of both mono- and oligonucleosomal fragments in both cell lines when exposed to concentrations ranging from 0.5 to 10 mM of AA over a period of 0 to 72 h. In SH-SY5Y cells, the levels of DNA fragments showed a significant increase with longer exposure times and higher concentrations of AA. On the other hand, U-1240 MG cells exhibited apoptotic responses only after 48 h of exposure to all concentrations of AA, except at 0.5 mM [47]. Moreover, when SH-SY5Y cells were treated with 2.5 mM of AA for 24 h, there was an approximately four-fold increase in the total apoptotic rate compared to the control group [38]. However, inhibiting c-Jun N-terminal kinase (JNK) using a nuclear factor κB (NF-κB) inhibitor led to a reduction in both early and late apoptosis in AA-treated cells [41].

To investigate the potential role and underlying mechanism of Sphingosine Kinase 1 (SphK1) in AA-induced nerve injury in SH-SY5Y cells, the expression of SphK1 was examined. It was observed that SphK1 levels decreased in correlation with increasing concentrations of AA. Specifically, after exposure to AA, the apoptosis rates and activation of SphK1 increased proportionally with the AA concentration [72]. At concentrations of 1.25 mM and 2.5 mM, the apoptosis rates were 3.06% ± 0.13% and 6.86% ± 0.67, respectively, while in the SphK1 activator group, the rates at the same concentrations were 2.12% ± 0.33% and 3.53% ± 0.17, respectively [72]. Furthermore, in a related study, Ning et al., 2021 [68], took the investigation a step further by using AA as a model to induce apoptosis in zebrafish, exposing them to 10 mM of AA. This resulted in the observation of AA-induced neuroapoptosis through fluorescent assays in the zebrafish [73].

3.2. *Mycotoxins*

3.2.1. Beauvericin (BEA)

Regarding cell cycle alterations, BEA significantly reduced the percentage of viable cells, producing an increase in apoptotic cell death of 55.9% ± 8.6% [61]. Similar results were obtained by Agahi et al., 2021 [74], who stated that apoptotic/necrotic cells increased for both times of exposure, which was up to 89% for 24 h and up to 38.8% for 48 h. Furthermore, Agahi et al., 2021 [74] observed that after 48 h of exposure, there was a notable increase in necrotic cells at 0.39 and 0.78 μM by almost two times compared to control cells.

3.2.2. Deoxynivalenol (DON)

In the case of DON, Kalagatur et al. [63] reported that a dose of 120 μM induced DNA damage and led to the formation of apoptotic nuclei.

3.2.3. Enniatin A and Enniatin B (ENN A and ENN B)

ENN A and ENN B reduced cell viability, producing an increase in apoptotic cell death of 49.2 ± 8.2% and 46.0 ± 9.3%, respectively [61].

3.2.4. Zearalenone (ZEA) and Its Metabolites

ZEA, on the other hand, did not produce a significant increase in cell death at the concentrations tested [61]. Nevertheless, α-ZEL treated cells increased significantly in apoptotic and apoptotic/necrotic cells at 24 h. And β-ZEL showed a significant increase in apoptotic cells exposed to the highest concentration assayed (12.5 μM) after 24 h and 48 h [74].

AA increases the apoptosis rate in neuronal cells (SH-SY5Y and U-1240 MG cells) proportional to the AA concentration, as well as to the time of exposure. Activating SphK1 could improve the survival rate of SH-SY5Y cells and reduce the apoptotic rate. Also, it has been shown to produce neuroapoptosis in zebrafish (Table 3). Mycotoxins like BEA, ENNs, and ZEA metabolites (α-ZEL and β-ZEL) could reduce the cell viability in SH-SY5Y cells and produce an increase in apoptotic cell death (Table 4).

Table 3. Apoptosis in SH-SY5Y cells. AA exposure conditions, assays, and effects.

Dose	Time	Assays	Effects	References
2.5 mM	24 h	Annexin V-FITC/PI apoptosis	The early apoptotic rate was 5.5%, the late apoptotic rate was 11.8%	[42]
2.5 mM	24 h	Annexin V-FITC/PI apoptosis	Induced apoptosis	[41]
0.5–10 mM	0–72 h	DNA fragmentation detection (mono and oilgonucleosomes)	DNA fragments increased	[47]
1.25–2.5 mM	24 h	Annexin V-FITC/PI apoptosis	Increased with increasing AA concentrations	[72]
10 mM	24 h	ELISA and cytometry	AA induced neuroapoptosis in zebrafishes and produced free radicals	[73]

AA: acrylamide, DNA: deoxyribonucleic Acid, JNK: c-Jun N-terminal kinase, NF-κB: nuclear factor κB, SH-SY5Y: neuroblastoma cell line.

Table 4. Apoptosis in SH-SY5Y cells. Mycotoxins exposure conditions, assays, and effects.

Mycotoxin	Dose	Exposure Time	Assays	Effects	References
DON	0–10 μM	6 h and 24 h	Cell cycle, FC	Did not produce cell death	[61]
	0–200 μM	6 h and 24 h	Apoptosis by FC	Induced apoptotic nuclei formation	[63]
BEA	2.5, 1.25, 0.78 and 0.39 μM	24 h and 48 h	Cell cycle Apoptosis/necrosis	Cell cycle arrest at G0/G1; Apoptotic/necrotic cells increased up to 89% for 24 h and up to 38.8% for 48 h	[74]
	0.1–30 μM	6 h, 24 h, and 48 h	FC analysis (Apoptosis)	Increase of 55.9 ± 8.6%. in apoptotic cells	[61]
ZEA α-ZEL and β-ZEL	12.5, 6.25, 3.12, and 1.56 μM	24 h and 48 h	Cell cycle Apoptosis/necrosis	Concentration-dependent increase in G0/G1 phase at 24 h; α-ZEL increased apoptotic and apoptotic/necrotic cells at 24 h; β-ZEL increase apoptotic cells exposed to 12.5 μM after 24 and 48 h	[73]
	0.1–30 μM	6 h, 24 h, and 48 h	FC analysis (Apoptosis)	Did not produce cell death	[61]
FB1	0.1–100 μM	48 h, 72 h, and 144 h	Cell viability (propidium Iodide)	Viability decreases at 100 μM at 48–144 h	[64]
	0.1–30 μM	6 h, 24 h, and 48 h	FC analysis (Apoptosis)	Did not produce cell death	[61]
	50 μM	12 h, 24 h, and 48 h	Apoptosis assay (FC)	Induced apoptosis	[65]
ENN's	0.1–30 μM	6 h, 24 h, and 48 h	FC analysis (Apoptosis)	Increase in apoptotic cell death of 49.2 ± 8.2% at 24 h and 46.0 ± 9.3% at 48 h	[61]
OTA	130–0.20 nM	24 h, 48 h, and 72 h 7 days	Cell cycle (FC)	Low OTA doses altered neuronal differentiation and cell cycle	[71]

BEA: Beauvericin, DON: Deoxynivalenol, ENN's: enniatin, FB1: Fumonisin B1, FC: flow cytometry, OTA: Ochratoxin A, ZEA: Zearalenone, α-ZEL: alpha-zearalenol, β-ZEL: beta-zearalenol.

4. Oxidative Stress

4.1. Acrylamide (AA)

The study of ROS in the SH-SY5Y cell line revealed an accumulation after AA exposure [46]. These results are supported by those reported indicating an intracellular decrease in glutathione (GSH) production in a dose-dependent manner and increased malondialdehyde (MDA) and ROS generation when SH-SY5Y cells were treated with 2.5 and 5 mM of AA [41]. Furthermore, when SH-SY5Y cells were exposed to AA at 2.5 mM for 24 h, it caused oxidative stress as revealed by the distinct increase in cellular ROS and the MDA level and a significant decrease in GSH content [42].

4.2. Mycotoxins

4.2.1. Beauvericin (BEA)

In view of the effects on cytotoxicity and apoptosis that mycotoxins had, oxidative stress was analyzed by several authors. Agahi et al. [75] studied the ROS production and the GSH/GSSG ratio in which BEA produced a slight decrease at 1.25 and 2.5 μM, from 45 to 120 min compared to the control, and GSH/GSSG ratio increased after 24 h in cells exposed to BEA from 103% to 142%. Subsequently, in the same line of investigation, they studied the activity of GPx, GST, CAT, and SOD [76]. The results showed an increase ranging from 9- to 17-fold and from 2- to 9-fold after 24 h and 48 h, respectively. Also, BEA increased GST activity at doses above 0.78 μM for 48 h of exposure by 4–32%. CAT activity was not altered when cells were exposed to the BEA mycotoxin. Finally, SOD activity increased significantly after being exposed to BEA after 48 h by up to 1-fold. In other study [61], the mitochondrial membrane potential measurement was analyzed, concluding that BEA had the capacity to depolarize the mitochondrial membrane at concentrations ranging from 2.5 to 10 μM at 6 h and 24 h.

4.2.2. Deoxynivalenol (DON)

Also, Kalagatur et al. [63] reported that at 120 μM, DON induced ROS and oxidative stress over the induction of LPO and exhaustion of antioxidant enzymes (GSH, CAT, and SOD). Also, DON induced MMP loss [63].

4.2.3. Enniatin B (ENN B)

At 24 h, this emerging toxin affected the $\Delta\Psi m$ value at all of the concentrations tested. ENN B treatment altered $\Delta\Psi m$ at the highest doses employed (5 and 10 μM) after 6 h of incubation and caused differences at 24 h [61].

4.2.4. Fumonisin B1 (FB1)

In terms of oxidative stress, FB1 treatment resulted in a markedly higher production of ROS as compared to control cells. Additionally, they measured the amount of ROS accumulation in the mitochondria and found that FB1 therapy, independent of treatment duration, enhanced ROS accumulation in the mitochondria [65]. It is interesting to note that adding 0.5 μM of FB1 to neuroblastoma cells increased the rate of ROS by 1.2 times. In neuroblastoma cells, increasing the FB1 concentration to 5 and 50 μM did not activate the rate of HEt fluorescence in a dose-dependent manner [66]. On the other hand, at all of the time points (24, 48, 72, 96, 120, and 144 h) and fumonisin concentrations (0.1 to 100 μM) employed, treatment with FB1 had no effect on the formation of ROS [64].

When free radical levels are high, ROS can be harmful to cells because they significantly reduce the quantity of endogenous antioxidants. One of the main CNS antioxidant pathways is supplied by GSH. Oxidative stress causes GSH to be oxidized, and CNS disease results from GSH deficiency [67]. Reduced GSH levels (61% of controls) were seen in SH-SY5Y cells following a 144 h incubation period with 100 μM FB1, but not at earlier times or with lower FB1 doses [60].

After incubating with 100 μM FB1 for 24 h, SH-SY5Y cells already showed increased MDA concentrations [61]. Following a 72 h incubation period, cells exposed to 10 μM FB1 exhibited elevated MDA levels. MDA levels at the other time periods did not statistically differ from those of controls [61].

In a different study, the neuroblastoma cells' cellular Ca^{2+} level considerably increased following FB1 exposure in comparison to the control group. Following ER stress, cellular Ca^{2+} is released, which causes the loss of mitochondrial membrane potential and, in neuroblastoma cells, results in cell death [65]. Nevertheless, Pérez-Fuentes N et al., 2021 [61], found that 30 µM FB1 treatment did not result in appreciable alterations in the $\Delta\Psi m$ of SH-SY5Y cells during either of the two incubation periods. While MitoSOX florescence rose at all doses, only the lowest concentration (0.5 µM) had a discernible impact [66].

4.2.5. Ochratoxin A (OTA)

When it came to ROS, 100 µM OTA for 30 and 60 min, followed by loading cells with 2 mM DCF-DA for 30 min, showed that OTA increased the intensity of DCF-DA fluorescence in comparison to the corresponding untreated control cells. This suggests that OTA-treated SH-SY5Y cells may have been exposed to oxidative stress [68].

4.2.6. T-2 Toxin

In this investigation, ROS levels dramatically increased to 3.8 and 5.0 times, respectively, in cells treated with 5 and 10 ng/mL of the T-2 toxin. As a result, for cells treated with 5 and 10 ng/mL of the T-2 toxin, the ratio of GSH to GSSG was drastically reduced to 79.8% and 60.7%, respectively. The ATP content dropped to 66.7% and 51.5% at the 5 and 10 ng/mL concentrations, respectively, whereas the mitochondrial membrane potential decreased to 60.7% and 41.5% at the same concentrations [69].

4.2.7. Zearalenone (ZEA) and Its Metabolites

ZEN administration dramatically boosted the formation of ROS in SH-SY5Y cells in relation to oxidative stress [70]. Additionally, ZEN administration led to a dose-dependent increase in MMP loss and lipid peroxidation [60,61]. Regarding the ZEA metabolites, α-ZEL at 25 µM elevated ROS from 5 to 60 min and moderately decreased from 90 to 120 min in comparison to their control. Conversely, β-ZEL dropped for 12.5 µM over a 90 min period [75]. After 24 h, the GSH/GSSG ratio in cells exposed to mycotoxins in fresh media increased dramatically from 111% to 148%, and for α-ZEL and β-ZEL, from 68% to 131%, respectively [75]. Deepening on our understanding of the harmful consequences of ZEA metabolites, the study conducted by Agahi et al. [76] examined the effects of α-ZEL and β-ZEL on the activity of enzymes in SH-SY5Y cells. The findings showed that α-ZEL and β-ZEL exposure increased the activity of GPx, CAT, and GST at all concentrations examined, and that SOD and α-ZEL increased the activity of SOD and GST after 48 h.

Summarizing, AA produces an increase in cellular ROS and a reduction in intracellular GSH at concentrations higher than 2.5 mM for more than 24 h (Table 5). Mycotoxins (DON, FB1, OTA, T-2 toxin, and ZEA) produced an increase in cellular ROS and in mitochondrial ROS in the case of FB1, and a reduction in the GSH ratio at concentrations of 120 µM for DON, 0.5 µM for FB1, 100 µM for OTA, 5 and 10 ng/mL for T-2 toxin, and 12.5 µM for α-ZEL and β-ZEL (Table 6).

Table 5. Oxidative stress. AA dose, exposure time, assays, effects, and references.

Dose	Time	Assays	Effects	References
10 mM	6 h	ROS	Induces ROS	[46]
2.5 mM and 5 mM	24 h	GSH, MDA, and ROS	2.5 and 5 mM decreased intracellular GSH production and increased MDA and ROS generation	[41]
2.5 mM	24 h	ROS (DCFH-DA) and LPO (MDA, GSH)	Maximal ROS production; 2.5 mM (+40%) MDA content: +100% 2.5 GSH content: −50% 2.5 mM	[42]

DCFH-DA: Dichloro-dihydro-fluorescein diacetate, **GSH:** glutatión, **LPO:** lipid peroxidation, **MDA:** malondialdehyde, **ROS:** reactive oxygen species.

Table 6. Oxidative stress. Mycotoxins dose, exposure time, assays, effects, and references.

Mycotoxin	Dose	Exposure Time	Assays	Effects	References
DON	0–200 μM	6 h and 24 h	ROS generation by fluorometry LPO and antioxidant enzyme levels MMP	ROS generation for 6 and 24 h Induction of LPO and exhaustion of antioxidant enzymes Induction of MMP loss	[63]
BEA	2.5, 1.25, 0.78, and 0.39 μM	2 h, 24 h, and 48 h	ROS GSH	ROS decrease at 1.25 and 2.5 μM, from 45 to 120 min; GSH/GSSG ratio increased from 103% to 142% after 24 h	[75]
	2.5, 1.25, 0.78, and 0.39 μM	24 h and 48 h	Glutathione peroxidase activity (GPx) Glutathione S-transferase (GST) activity Catalase (CAT) activity Superoxide dismutase (SOD) activity	GPx activity, BEA promoted a 9- to 17-fold increase after 24 h and a 2- to 9-fold for 48 h; GST activity, a slight increase (4–32% at 48 h) was obtained, while for SOD, a 2.5-fold increase at 48 h was shown; No significant differences for CAT were observed	[76]
	0.1–30 μM	6 h, 24 h, and 48 h	MMP	Alteration of mitochondrial membrane at 2.5–10 μM during 6–24 h	[61]
ZEA α-ZEL and β-ZEL	25, 12.5, 6.25, and 3.12 μM	120 min, 24 h, and 48 h	ROS GSH	α-ZEL: at 25 μM increase from 5 to 60 min; β-ZEL: decrease at 12.5 μM from 5 to 90 min and at 25 μM from 15 to 30 min and from 60 to 120 min; GSH/GSSG ratio increased after 24 h at all concentrations from 111% to 148%, for α-ZEL and from 68% to 131% for β-ZEL	[75]
	12.5, 6.25, 3.12 and 1.56 μM	24 h and 48 h	Glutathione peroxidase activity (GPx) Glutathione S-transferase (GST) activity Catalase (CAT) activity Superoxide dismutase (SOD) activity	GPx activity increased by 13.5- to 23-fold for α-ZEL and 9- to 17-fold for β-ZEL. For 24 h and 48 h; GST increased after 48 h of exposure to β-ZEL at 3.12 and 12.5 μM by 22% and 102%, respectively; CAT activity increased from 0.4- to 1.4-fold for α-ZEL and 1–4.2-fold for β-ZEL exposure after 48 h at all concentrations; SOD activity increased for α-ZEL up to 1.4-fold, and for β-ZEL up to 2.5-fold after 48 h	[76]
	0.1–30 μM	6 h, 24 h, and 48 h	MMP	ZEA produced a slight decrease at 30 μM after 24 h of incubation	[61]
	25, 50, 75, 100 and 200 μM		ROS Lipid peroxidation assay MMP	ZEN increased ROS generation, LPO, and loss of MMP in a dose-dependent manner	[70]

Table 6. *Cont.*

Mycotoxin	Dose	Exposure Time	Assays	Effects	References
FB1	0.1–100 μM	48 h, 72 h, and 144 h	ROS Lipid Peroxidation GSH	Did not affect production of ROS; 100 μM increased MDA concentrations after a 24 h, 10 μM increased MDA levels at 72 h, 100 mM decreased GSH levels (61% of controls) at 144 h	[64]
	0.5–200 μM	24 h and 48 h	ROS MitoSOX	0.5, 5, and 50 μM, increased ROS production; All doses increased MitoSOX florescence	[66]
	0.1–30 μM	6 h, 24 h, and 48 h	MMP	No significant changes in ΔΨm	[61]
	50 μM	12 h, 24 h, and 48 h	ROS Mitochondrial superoxide level Ca^{2+} measurement	Increase in ROS production, mitochondrial ROS accumulation, and cellular Ca^{2+} level	[65]
ENN's	0.1–30 μM	6 h, 24 h, and 48 h	MMP	ENN A affected at 5 (79.0 ± 8.9%) and 10 μM (67.7 ± 14.5%,) after 6 h. At 24 h, ENN A affected to ΔΨm at all the concentrations tested; ENN B altered ΔΨm at 5 and 10 μM after 6 h and 24 h	[61]
OTA	1, 10, or 100 μM OTA	30, 60 min, 24 h,	ROS	100 μM for 30 and 60 min, indicated the evocation of oxidative stress	[68]
T-2	5–20 ng/mL T-2	1–48 h	ROS level, GSH/GSSH ratio, ATP content, MMP	ROS levels significantly increased to 3.8 times at 5 ng/mL and 5.0 times at 10 ng/mL; The ratio of GSH to GSSG decreased to 79.8% at 5 ng/mL and 60.7% at 10 ng/mL; 5 and 10 ng/mL decreased the ΔΨm to 60.7% and 41.5%, while the ATP content decreased to 66.7% and 51.5%	[69]

ATP: adenosine triphosphate, BEA: Beauvericin, DON: Deoxynivalenol, ENN's: enniatin, FB1: Fumonisin B1, GSH: glutatión, LPO: lipid peroxidation, MDA: malondialdehyde, MMP: Mitochondrial membrane potential, ROS: reactive oxygen species, OTA: Ochratoxin A, ZEA: Zearalenone, α-ZEL: alpha-zearalenol, β-ZEL: beta-zearalenol.

5. Network Degeneration

Another way to assess how AA exerts effects at a neuronal level is by explaining whether it degrades or impedes the development of the neural network during the process of differentiation; a study found that degeneration was not noticeable for the first 48 h after receiving 1 mM AA treatments, but that the network levels had considerably decreased by the 72 h mark [77]. In this instance, network deterioration could not be stopped by co-treatment with BDNF or calpeptin. Similar results were later reported proving that the number of neurites was significantly reduced after 100 nM following 3 days of exposure and after 6 days of exposure to 10 pM of AA [45], showing an impairment in the neurite outgrowth in a time- and dose-dependent manner in future studies [55].

Furthermore, 0.5 mM caused a shortening in SH-SY5Y neurite morphology, while 1 and 2 mM of AA resulted in cells with no neurite morphological extensions [44]. Similar results were found when cells were treated with 0.25 mM of AA, causing a 50% neurite degeneration at 24 h [78]. At 96 h with 4 mM of AA, SH-SY5Y neurites were reduced by up to 67.7% [53]. Thus, AA induces network degeneration in a dose- and time- dependent manner [79].

In order to obtain more accurate results, Nordin-Andersson et al., 1998 [79], exposed differentiated human neuroblastoma (SH-SY5Y) cells to a series of contaminants, among them AA, studying general cytotoxicity (IC_{20}) and the neurite degenerative effect (ND_{20}), concluding after 72 h that the IC_{20} was 670 mM and the ND_{20} 250 mM. Thus, the ND_{20} values for AA were significantly (65%) lower than the IC_{20}, concluding the induction of axonopathy. Later on in 2003, during the AA treatment, there were observations of basal cytotoxicity, morphological modifications, and changes in cell physiological and neurochemical activities in differentiated human neuroblastoma (SH-SY5Y) cells [52]. After 72 h of exposure, AA caused a 20% drop in the number of neurites per cell at 0.21–0.25 mM and a 20% decrease in the rate of protein synthesis at 0.17 mmol/L. Moreover, there was a 49% and 38% rise in the baseline intracellular calcium concentration ($[Ca^{2+}]i$) fluxes, measuring 0.25 mmol/L and 0.5 mmol/L, respectively. The SH-SY5Y cells recovered 48 h after the AA exposure was stopped; that is, their basal level of $[Ca^{2+}]$, rate of protein synthesis, and number of neurites per cell were all similar to those of control cells.

Taken together, AA impairs neurite outgrowth in the SH-SY5Y cell line during differentiation in a time- and concentration-dependent manner, especially after 72 h of exposure. However, at 48 h after cessation of AA exposure, SH-SY5Y cell recover (Table 7).

Table 7. Axonopathy in SH-SY5Y cells. AA exposure conditions, assays, and effects.

Dose	Time	Assays	Effects	References
0.5 mM or 1 mM	0, 24, 48, and 72 h	Neuronal network integrity (neurites connections)	1 mM AA, at 72 h network levels reduction	[77]
100 nm and 10 pm	3 and 6 days	Phase-contrast microscope	Impairs neurite outgrowth in a time- and dose-dependent manner	[45]
0.8, 20 and 500 μg/mL	24 h	Phase-contrast microscope	Number and the length of processes showed a gradually decreasing tendency	[55]
0.5–2 mM	24, 48, 72 h	Biochemical indicators for morphologic changes;	1 and 2 mM resulted in cells with no neurite morphological extensions	[44]
0.1–1 mmol/L	72 h	Neurite quantification;	ND_{20} 0.21 mmol/L	[52]
0.25 mM	24 h	Axonopathy (neurite degeneration assay	0,25 mM 50% neurite degeneration	[78]
4 mM	24 h and 96 h	Neuronal and axon number	4 mM reduced neuron number by 96 h, to 67.7%	[53]
6 mM	8 h	Axonopathy	Dose-dependent from 2.5 to 10 mM	[79]
250 μM 670 μM	72 h	Axonopathy	Concentration-dependent decrease in the number of neurites/cells. ND_{20}: 250 IC_{20}: 670	[80]
0.1–10 mM	4 h, 24 h, and 48 h	Morphology assay	Differentiated cells were more sensitive to 1 mM compared with undifferentiated cells	[50]

AA: acrylamide, BDNF: brain-derived neurotrophic factor, IC_{20}: concentration causing 20% protein reduction, ND_{20}: concentration causing 20% neurite degeneration.

6. Signaling Pathways

To fully understand how AA interferes at the neuronal level, various studies scrutinized the involvement of different transcription factors crucial in neurodevelopment, cell differentiation, apoptosis, and inflammatory responses. These included cAMP response element-binding (CREB), mitogen-activated protein kinases (MAPKS), JNK, PERK-eIF2α, and C/EBP homologous protein (CHOP), along with the nuclear factor NF-κB.

AA was studied with the CREB protein signaling during neuronal differentiation, concluding that in the CREB signaling pathway the expression of 17 genes was significantly changed after exposures to 1 or 70 μM, reporting that AA interferes with important cholinergic and dopaminergic neuronal markers during differentiation, which were downregulated after exposure to 70 μM AA [80].

In other studies [41], the intracellular levels of interleukin 6 (IL-6) and tumor necrosis factor alpha (TNF-α) in SH-SY5Y cells were assessed in order to look into the potential pro-inflammatory effects of AA. The findings showed that, in comparison to the control value, the TNF-α content was significantly raised following treatment with 2.5 and 5 mM AA; however, the level of IL-6 was only significantly increased in the 5 mM AA group when compared to the control group. In addition, nuclear NF-κB and the mitogen activated protein kinases (MAPK) signaling pathway with JNK and p38 were activated by AA.

The same research group studied the induction of AA in phosphorylated tau aggregation, phosphorylated CREB protein reduction, and Bax/Bcl-2 ratio upregulation in SH-SY5Y cells at the same concentrations reported above. Furthermore, AA induced glycogen synthase kinase-3β (GSK-3β) and upregulated activating transcription factor 4 (ATF4) and CHOP in SH-SY5Y cells. It also activated the protein kinase RNA-like endoplasmic reticulum kinase (PERK)-Eukaryotic Initiation Factor 2 alpha (eIF2α) signaling [42].

However, 1.25, 2.5, and 5 mM of AA were found to increase the cleavage of caspase-3 and the poly ADP ribose polymerase (PARP) enzyme (the initiator and effector caspases in the intrinsic apoptotic pathway, respectively). They also decreased Akt phosphorylation in SH-SY5Y cells, which prevented the activation of the PI3K/Akt pathway [43]. GSH levels were restored and H_2O_2 stimulation was inhibited by LA pretreatment, which counteracted AA-induced alterations in GSH loss and H_2O_2 production. The downregulation of AMP-activated protein kinase (AMPK) and glycogen synthase kinase-3 beta (GSK3β) phosphorylation were induced by AA in a dose-dependent manner; AA elicited energy deficits in the SH-SY5Y cells by regulating the AMPK/GSK3β cascade, resulting in Ca^{2+} disturbance, ATP depletion, and CREB/BDNF signaling impairment. In addition, Sirtuin 1 (Sirt1) and peroxisome proliferator-activated receptor γ co-activator 1 α (PGC-1α) were downregulated by AA; however, AA elicited cellular autophagy in terms of the conversion of microtubule-associated protein light chain 3 (LC3)-I to LC3-II and beclin-1 protein expression. Also, AA induced the overexpression of prostaglandin-endoperoxide synthase 2 (COX-2) and inducible nitric oxide synthase (iNOS), and AA reduced the phosphorylation of extracellular signal-regulated kinases (ERK) while intensifying the phosphorylation of JNK and p38 in MAPKs. In addition, AA mediated a reduction in nuclear factor erythroid 2–related factor 2 (Nrf2) nuclear translocation and cytoplasmic Kelch-like ECH-associated protein 1 (Keap1) expression, resulting in the downregulation of the expression of phase II enzymes, namely HO-1 and NAD(P)H Quinone Dehydrogenase 1 (NQO1) [43].

In brief, AA modulated several signaling pathways related to differentiation, inflammation, and apoptosis, disturbing transcriptomic factors such as nuclear NF-κB and MAPKs signaling pathway with JNK and p38 activation, COX-2 and iNOS overexpression, and AMPK and GSK3β downregulation. AA also inhibited the activation of the PI3K/Akt pathway and downregulated Sirt1 and PGC-1α. Furthermore, it interfered with the CREB signaling pathway, an important cholinergic and dopaminergic neuronal marker during SH-SY5Y differentiation (Table 8).

Table 8. Signaling pathways in SH-SY5Y cells. AA exposure conditions, assays, and effects.

Dose	Time	Assays	Effects	References
1–70 μM	9 days	qPRC	CREB signaling was downregulated	[80]
0, 1.25, 2.5 and 5 mM	24 h	Cytokine assessment	NF-κB, MAPKs, JNK, and p38 signaling pathway were activated;	[41]
2.5 mM	24 h	Immunofluorescence (tau hyperphosphorylation (pS262))	AA: +50% pS262 Activated the PERK-eIF2α signaling, triggered the activation of GSK-3β, and upregulated ATF4 and CHOP; Also, inhibited the activation of the PI3K/Akt pathway	[42]
1.25, 2.5, 5 mM	24 h	Western blot	Inhibited the activation of the PI3K/Akt pathway; Downregulation of AMPK, Sirt1, PGC-1α, and GSK3β; Phosphorylation by AA in a dose-dependent manner; Overexpression of COX-2 and iNOS	[43]

AA: acrylamide, AMPK: AMP-activated protein kinase, COX: Prostaglandin-endoperoxide synthase, CREB: cAMP response element binding, CHOP: C/EBP homologous protein, GSK-3β: glycogen synthase kinase-3β, iNOS: inducible nitric oxide synthase, JNK: c-Jun N-terminal kinase, MAPKS: mitogen-activated protein kinases, PERK/eIF2α: protein kinase RNA-like endoplasmic reticulum kinase—Eukaryotic Initiation Factor 2 alpha, Pyk2: proline-rich tyrosine kinase 2 RNA (PKR)-like/Pancreatic ER Kinase, PI3K: phosphatidylinositol 3-kinase, PGC-1α: Peroxisome proliferator-activated receptor-gamma coactivator, qPCR: quantitative polymerase chain reaction, Sirt 1: NAD-dependent deacetylase sirtuin-1.

7. Gene Expression

7.1. Acrylamide (AA)

The impact of AA on gene expression, as elucidated by recent research, highlights a complex and nuanced response within cellular systems. A comprehensive analysis of differentially expressed genes (DEGs) following exposure to AA revealed distinct patterns of gene dysregulation.

Following exposure to 70 μM AA over a six-day differentiation period, a notable dysregulation was observed in a set of eight genes. Among these genes, NTRK2 (Neurotrophic Receptor Tyrosine Kinase 2, a receptor of BDNF), FGF1, RASD2, BMP7, SEMA3F (Semaphorin 3F), and SEMA5A (Semaphorin 5A) exhibited significant alterations, indicating potential pathways affected by AA-induced toxicity during cellular differentiation [81]. In the same study, they also studied other compounds; therefore, the shared dysregulation of CNR1 (encoding cannabinoid receptor 1) and SEMA5A between AA and valproic acid exposures, as well as OPRD1 (encoding opioid receptor delta 1) between AA and rotenone, underscores the intriguing overlap in genetic response among these neurotoxic compounds [81].

In addition to examining markers linked with CREB signaling pathways, Attoff et al., 2020 [81] opted to incorporate indicators associated with cholinergic (CHAT) and dopaminergic (DRD2 and MAOA) neurons in consideration of their crucial role in the neuronal differentiation process of SH-SY5Y cells. They also included synaptotagmin 1 (SYT1), known for its protein adduct formation with AA, along with genes previously impacted by AA exposure (such as MAOA and FGF1). Additionally, BDNF and tropomyosin receptor kinase B (TrkB), closely associated with the CREB signaling pathway, were part of the selection. The analysis revealed a notable decrease in the expression of 6 out of 16 genes following exposure to 70 μM of AA, with 5 of them demonstrating decreased expression even after exposure to 1 μM of AA. Interestingly, the expression of CHAT, TGFB1, STXBP2, BDNF, and DRD2 showed an increase during differentiation compared to undifferentiated cells. However, these genes were all downregulated upon exposure to AA. Notably, while MAOA demonstrated an increased expression compared to undifferentiated cells, it exhibited significant reduction only after exposure to 70 μM of AA [80].

7.2. Deoxynivalenol (DON)

For DON, Kalagatur et al. [63] studied the ability of these mycotoxins to alter the expression of significant neuronal markers was analyzed, reporting a significant downregulation at 50 and 100 μM of BDNF, AADC, and TH mRNA expression.

7.3. Ochratoxin A (OTA)

Similarly, but at higher concentrations of OTA (130–0.20 nM) and at 7 days of exposure, an alteration in several cell cycle (p21, p53, cyclin B and D) and neuronal differentiation (GAP43, Wnt5a and TUBB3) genes was found [71]. Concerning the gene and protein expression, several genes and proteins were studied, all of them implicated in the cell proliferation (BAX, P53, MAPT, TPPP p21, cyclin B and D, GAP43, Wnt5a, and TUBB3). At the highest dose studied, 1 μM OTA exposure for two days reduced the expression of P53, BAX, and MAPT mRNA at Days 1 and 2. In comparison to the control exposure groups, TPPP mRNA expression rose on Day 2 after exposure to 1 μM OTA but decreased on Day 1 [82]. The duration of exposure and the relationship between the duration of exposure and the 1 μM OTA dose were indicated by the expression of BDNF mRNA. For 11 days, cells exposed to lesser doses (2 fM, 20 fM) showed no change in gene expression [71].

7.4. T-2 Toxin

To investigate the impact of the T-2 toxin on mitochondrial biogenesis in SH-SY5Y cells, researchers measured the mtDNA copy number and the expression levels of NRF2, KEAP1, PGC-1α, NRF1, and TFAM [69]. For cells treated with 5 and 10 ng/mL, the mtDNA copy number decreased to 80.3% and 60.9%, respectively. The expression of NRF2 mRNA increased to 1.7 and 2.8 times, respectively [69]. Conversely, KEAP1's mRNA expression dropped to 45.4% and 77.0%. The T-2 toxin significantly reduced the expressions of PGC-1α and its downstream targets NRF1 and TFAM. The most notable reductions were observed in the gene levels, which were lowered by 5 and 10 ng/mL of T-2 toxin, respectively, to 56.3% and 21.6% [69].

7.5. Zearalenone (ZEA)

When the concentration of ZEA increased in the comet assay, the tail dispersion increased in a dose-dependent manner. Additionally, a substantial increase in apoptotic nuclei was observed in treated cells compared to control cells at 12 and 24 h after exposure. This "ladder" pattern is indicative of oligonucleosomal DNA fragmentation. Additionally, BDNF and TH mRNA expression was not changed and gradually declined with an increase in toxin concentration for 12 and 24 h after ZEN exposure when gene expression was examined at the lower dose of ZEA (25 μM). Conversely, at 25 μM, AADC mRNA expression was inhibited. Further suppression of AADC mRNA expression was achieved by increasing the concentration of ZEA toxin (50 and 100 μM) [70]. In the case of the metabolites, other genes were studied and similar results were obtained; in α-ZEL treated cells, the mRNA of CASP3 and BAX mRNA were overexpressed at 12.5 and 25 μM. And in the case of β-ZEL, it upregulated Erβ mRNA at 12.5 μM while it downregulated the expression of CASP3 and BCL2 [75].

7.6. Fumonisin B1 (FB1)

Finally, for FB1, the ability to produce DNA damage was studied, through the comet assay, TUNEL assay, and DNA fragmentation assay. In the comet assay and when compared to other treatments, DNA damage was the highest 48 h after FB1 treatment, as indicated by an increased tail length and a decreased head diameter. Furthermore, compared to the control, considerable DNA damage resulted from a 24 h toxin treatment. Additionally, TUNEL assay results showed that FB1 treatment for 12, 24, and 48 h increased TUNEL-positive cells relative to the control [65]. In concordance, DNA ladders were seen after treatment with 30 and 100 μM FB1 for 48 h [77].

To sum up, mycotoxins have demonstrated the capacity to modulate the gene expression in SH-SY5Y cells, as well as to produce DNA damage (Table 9).

Table 9. Gene expression and genotoxicity in SH-SY5Y cells. Mycotoxin dose, exposure conditions, assays, and effects.

Mycotoxin	Dose	Exposure Time	Assays	Effects	References
DON	0–200 μM	6 h and 24 h	Comet assay; Gene expression by qRT-PCR	Damage in DNA at 24 h; BDNF and TH mRNA expression were downregulated at 50 and 100 μM; AADC gene expression was downregulated at all concentrations	[63]
BEA	2.5, 1.25, 0.78, and 0.39 μM	2 h, 24 h, and 48 h	Gene expression assay by RT-PCR	Upregulated BCL2 mRNA	[73]
ZEA α-ZEL and β-ZEL	25, 12.5, 6.25, and 3.12 μM	120 min, 24 h, and 48 h	Gene expression assay by RT-PCR	α-ZEL: mRNA of CASP3 and BAX mRNA were overexpressed at 12.5 and 25 μM; β-ZEL: upregulated ERβ mRNA at 12.5 μM, while downregulated expression of CASP3 and BCL2	[75]
	25, 50, 75, 100, and 200 μM	3 h, 24 h	COMET assay; DAPI staining plasmid breakage assay; Gene expression by qRT-PCR	25 μM increased the tail length by 31% ± 2.6%, and at 100 μM 83% ± 5.3%; Showed DNA fragmentation at 12 h and 24 h; 25 μM of ZEA: BDNF and TH mRNA expression decreased with increased concentration for 12 h and 24 h; AADC mRNA expression was suppressed at 25 μM; The mRNA expression of AADC was further abrogated by 50 and 100 μM	[70]
FB1	1, 10, and 100 μM	12 h, 24 h, 48 h, 72 h, and 144 h	DNA fragmentation	DNA ladders were obtained at 30 and 100 μM for 48 h	[76]
	50 μM	12 h, 24 h and 48 h	Comet assay; TUNEL assay	Increased tail length and decreased head diameter time-dependently; At 24 and 48 h, increased TUNEL-positive cells	[65]

Table 9. *Cont.*

Mycotoxin	Dose	Exposure Time	Assays	Effects	References
OTA	2 fM, 20 fM and 2 pM 0.001 μM, 0.01 μM, 0.05 μM, 0.1 μM, 0.5 μM, and 1 μM	2 and 11 days	Gene expression by RT-qPCR	Expression of P53, BAX, and MAPT mRNA were reduced at Days 1 and 2; TPPP mRNA expression decreased at Day 1 at 1 μM	[82]
	0.1, 0.25, 0.5, 1.0 and 2.5 μM	24 h and 48 h	DNA fragmentation assay	Concentration-dependent DNA laddering	[83]
T-2	5–20 ng/mL T-2	1–48 h	Quantification of mitochondrial DNA copy number quantification; Gene expression by RT-qPCR	Reduced mtDNA copy number of 80.3% at 5 ng/mL and 60.9%, at 10 ng/mL; NRF2 mRNA expression increased to 1.7-fold at 5 ng/mL and 2.8-fold by 10 ng/mL; The expression of KEAP1 mRNA decreased to 77.0% and 45.4%; The expressions of PGC-1α and its downstream targets NRF1 and TFAM were distinctly inhibited	[69]

BAX: Bcl-2 Associated X-protein, BCL2: B-cell leukemia/lymphoma 2 protein, BDNF: Brain-derived neurotrophic factor, BEA: Beauvericin, CASP3: cysteine–aspartic acid protease, DON: Deoxynivalenol, ENN's: enniatin, Erβ: Estrogen receptor beta, FB1: Fumonisin B1, KEAP1: Kelch-like ECH-associated protein 1, NRF1: Nuclear respiratory factor-1, NRF2: nuclear factor erythroid 2–related factor 2, OTA: Ochratoxin A, P53: tumor protein p53, PGC-1α: Peroxisome proliferator-activated receptor-gamma coactivator, qPCR: quantitative polymerase chain reaction, TFAM: Mitochondrial transcription factor A, ZEA: Zearalenone, α-ZEL: alpha-zearalenol, β-ZEL: beta-zearalenol.

8. Protein Expression

8.1. Acrylamide (AA)

Although the study of transcription factors is important to know which signaling pathways can be altered, the importance lies in the final expression or suppression; this is why the protein content and how the presence of AA modulates it was considered.

According to the signaling pathways and transcriptomic factors altered as mentioned above, the proteins involved in processes of stress, neuronal differentiation, inflammation, and cell death were analyzed, such as heat shock proteins (HSPs), microtubule-associated proteins (MAPs), focal adhesion kinase (FAK), and CHOP.

In this sense, on one hand, it has been proved that the treatment of cells with AA increased the cell death through the caspase-3 activity and an increase in cells in the sub-G1 phase. Both caspase-3 activation and the sub-G1 cell population peaked when the cells were exposed to 3 mM AA. Interestingly, a higher dose of AA (4–5 mM) resulted in less caspase-3 activity, even though the cytotoxicity increased. This indicates that the decrease in caspase-3 activity at 5 mM of AA is not solely due to heightened cytotoxicity [48].

One year later, the same research team found that, after exposing human neuroblastoma cells (SH-SY5Y) to 0.5–5 mM AA for 18 h, the levels of HSPs of 90, 70, and 27 kDa (Hsp90, Hsp70, and Hsp27, respectively) were elevated in the incubation media depending on the dose of AA, while only the Hsp70 level increased within cells [83].

According to other work, exposure to AA predominantly causes eukaryotic translation initiation factor 2α (eIF2α) to be phosphorylated. This is followed by the buildup of ATF4, the protein that binds to eIF2α. Additionally, it was demonstrated that CHOP mRNA expression dramatically increased at all investigated time points and doses, with a notable increase of up to 100-fold at 7.5 or 10 mM AA following an 8 h exposure. Conversely, there was no alteration on the protein levels of GRP94, glucose-regulated protein (GRP78), or endoplasmic reticulum (ER) chaperones. These results imply that the AA exposure of SH-SY5Y cells causes the production of the pro-apoptotic CHOP protein but not of the ER chaperones that provide cytoprotection [46].

The effects of AA on the p53 protein and intracellular signal transduction pathways were examined using human neuroblastoma SH-SY5Y cells. p53, phosphorylated p53, and p53-associated protein murine double minute 2 (MDM2) were all upregulated by AA. The Ser15 position was the only location where p53 was phosphorylated. Extracellular signal-regulated protein kinase (ERK) and p38 were phosphorylated by MAPKs at increasing AA doses (0.5–5 mM), but not c-Jun NH2-terminal kinase [48].

Numerous proteins are impacted by AA-induced adduct formation, including SNAP-25 (synaptosome-associated protein) and other proteins found in synaptic vesicles [8]. In order to confirm the decrease in axon quantity when AA was present, cocultures were labelled with antibodies against both III-tubulin and the pan-neuronal marker PGP 9.5 [49]. The number of axon profiles in response to AA decreased in both instances, and the results matched the information obtained from SNAP-25 immunocytochemistry.

According to the findings of other research groups' in vitro studies, activating SphK1 in SH-SY5Y cells also controlled MAPK signaling. This included raising the phosphorylation of extracellular signal-regulated protein kinases (ERK) and lowering that of JNK and p38 [71]. These findings imply that SphK1 activation can provide nerve cell protection against AA-induced damage.

MAPs are responsible for the polymerization, stabilization, and dynamics of the microtubule network; it has been detected that the expressions of MAP1b and MAP2c in SH-SY5Y cells were reduced by 39% and 64%, respectively, when they treated the cells with 2 mM of AA [44], while the expressions of MAP1b and MAP2c in U-1240 MG cells were reduced by 52% and 57%, respectively. In terms of Janus kinase (JAK)-signal transducer and activator of transcription (STAT) signaling, JAK1 expression was increased by 93% in 10 μM retinoic acid (RA)-stimulated SH-SY5Y cells and by 108% in 10 μM butyric acid (BA)-stimulated U-1240 MG cells. Moreover, JAK1 expression in U-1240 MG cells was reduced by 74% when treated in combination with 10 μM BA and 2 mM AA, whereas JAK1 expression in SH-SY5Y cells was reduced by 68% when treated in combination with 2 mM AA and 10 μM RA. These findings demonstrated that in SH-SY5Y and U-1240 MG cells, AA-inhibited differentiation was mediated through MAPs expression and JAK-STAT signaling, arguing that exposure to AA inhibits cellular differentiation in human neuroblastoma and glioblastoma cells and that downregulation of the JAK-STAT signal pathway may contribute to the understanding of AA-induced neurodegeneration [44].

The differentiating process of neuroblastoma SH-SY5Y cells has also been studied with AA inducing time-dependent tyrosine phosphorylation of FAK [pY^{397}] and Pyk2[pY^{402}] [78].

Shortly, we can conclude that the expressions of MAP1b and MAP2c, typically involved in microtubule networks, were reduced in SH-SY5Y and U-1240 MG cells which indicated a clear impairment of cell shutdown upon AA exposure. In addition, a significant pro-inflammatory response with JAK1 overexpression was observed. Similarly, in vitro transcriptomic analysis showed how AA exposure (2.5–5 mM) induced apoptosis (increased levels of p53, MDM2, and caspase-3 expression), oxidative stress (ATP4 and CHOP overexpression), and inflammation with high levels of IL-6 and TNF- α. It is also necessary to underline the downregulation of SphK1 which is associated with inflammation and antiapoptotic processes (Table 10).

Table 10. Protein expression in SH-SY5Y cells. AA exposure conditions, assays, and effects.

Dose	Time	Assays	Effects	References
1–5 mM	16 h and 20 h		Increased the caspase-3 activity	[48]
0.5–5 mM	18 h	Protein analysis	Hsp90, 70, and 27 levels increased, dose-depending; HSPs increases as the AA grows	[84]
1–10 mM	8 h	qPCR	Induction of the pro-apoptotic CHOP protein expression	[47]
0.5–5 mM	1–24 h	Immunoprecipitation	p53 protein accumulated	[49]
0.01–12 mM.	24, 96, or 144 h	Immunoblotting	30% reduction in SNAP-25-positive axon profiles	[53]
0, 0.1, 0.5, 1, and 2 mM	24, 48, 72 h	Western blot	MAPs expression and JAK-STAT signaling were involved	[45]
6 mM	8 h	Western blot analysis and PCR	FAK and Pyk2 time-dependent tyrosine phosphorylation	[78]

AA: acrylamide, CHOP: C/EBP homologous protein, FAK: focal adhesion kinase, HSPs: Heat shock protein, IC_{20}: concentration causing 20% protein reduction, IC_{50}: 50% inhibition concentration, JAK-STAT: Janus kinase–signal transducer and activator of transcription, PERK: protein kinase, Pyk2: proline-rich tyrosine kinase 2 RNA (PKR)-like/Pancreatic ER Kinase, qPCR: quantitative polymerase chain reaction, SNAP: synaptic vesicle protein S-nitroso-N-acetyl penicillamine.

8.2. Mycotoxins

8.2.1. Fumonisin B1 (FB1)

Other studies have evaluated the capacity of this mycotoxin to induce apoptosis; therefore, some authors investigated the protein expression alterations (caspase 9 and 3, p53, Bax, Bcl-2, Bcl-XL, and Mcl-1) when FB1 was applied to SH-SY5Y cells. Initially, exposure to FB1 did not result in significant alterations in caspase-3-like protease activity. However, after a full day, 100 µM FB1 caused a slight but statistically insignificant increase in caspase-3 activity. During either FB1 treatment period, there was no caspase-3 or caspase-9 cleavage. Exposure to 100 µM FB1 for 72 h also did not affect the amounts of p53 in nuclear or cytoplasmic extracts. These findings are consistent with earlier research that did not discover a statistically significant rise in cell mortality at the concentrations examined (0.1–30 µM) [61,65,77].

On the other hand, in comparison to the control cells, FB1 had no effect on the expression of the pro-apoptotic protein Bax at any of the doses or time points. Furthermore, in SH-SY5Y cells, FB1 had no effect on the investigated anti-apoptotic proteins (Bcl-2, Bcl-XL, and Mcl-1) [85]. Another study suggested that mitochondria may be involved in FB1-mediated cell death because Bax expression rose 48 h after FB1 treatment in comparison to the control, whereas cytochrome C expression increased 24 h after FB1 treatment [61]. After exposure of neuroblastoma cells to FB1, there was a significant increase in JNK phosphorylation, which peaked at 48 h. Additionally, there was a significant increase in the expression of IRE1-α and PERK. Lastly, there was a significant activation of CHOP after 24 h of FB1 exposure [65].

8.2.2. Ochratoxin A (OTA)

In contrast, OTA treatment in SH-SY5Y cells activated caspase-9 and caspase-3 in one study, while another study did not detect activated caspase-3 or changes in p53 phosphorylation. This suggests that OTA-induced cytotoxicity may have different mechanisms at play [68,78].

8.2.3. T-2 Toxin

As for the T-2 toxin, it led to an increase in NRF2 protein expression, potentially indicating an adaptive response to oxidative stress induced by the toxin [69].

Regarding mycotoxins, FB1 significantly increased the phosphorylation of JNK in neuroblastoma cells, as well as significantly increasing the expression of CHOP, IRE1-α, and PERK. OTA treatment in SH-SY5Y activated caspase-9 and caspase-3 in one study but not in another. Finally, the T-2 toxin increased NRF2 protein expression. Therefore, more studies are needed in order to state that FB1, OTA, and the T-2 toxin lead to a modulation in protein expression in SH-SY5Y cells (Table 11).

Table 11. Protein expression in SH-SY5Y cells. Mycotoxin dose, exposure conditions, assays, and effects.

Mycotoxin	Dose	Exposure Time	Assays	Effects	References
FB1	1, 10, and 100 μM	12 h, 24 h, 48 h, 72 h, and 144 h	Caspase-3-like protease activity Western blot	No significant changes in caspase-3-like protease activity; 100 μM for 72 h had no effects on p53 levels; Bax Bcl-2, Bcl-XL, and Mcl-1 were not affected.	[85]
OTA	1, 10, and 100 μM	30, 60 min, and 24 h	Western blot	Activated caspase-3 was not detected; p53 phosphorylation was not detected	[68]
	0.1, 0.25, 0.5, 1.0, and 2.5 μM	24 h, 48 h	Western blot	Activation of caspase-9 and caspase-3	[78]
T-2	5–20 ng/mL	1–48 h	Western blot	NRF2 protein expression increased 1.2-fold by 5 ng/mL and 1.4-fold by 10 ng/mL. The mRNA protein expression decreased to 95.6% and 85.6%	[69]

BAX: Bcl-2 Associated X-protein, BEA: Beauvericin, DON: Deoxynivalenol, ENN's: enniatin, FB1: Fumonisin B1, NRF2: nuclear factor erythroid 2–related factor 2, OTA: Ochratoxin A, P53: tumor protein p53, ZEA: Zearalenone, Z-VAD-fmk: carbobenzoxy-valyl-alanyl-aspartyl-[O-methyl]- fluoromethylketone, Z-DEVD-fmk: Z-D(OMe)E(Ome)VD(OMe)-FMK. Z-DEVD-fluoromethylketone. Z-DEVD-fluoromethyl ketone. specific inhibitor of caspase-3, α-ZEL: alpha-zearalenol, β-ZEL: beta-zearalenol.

9. Natural Bioactive Compounds

For all of the above, we can conclude that AA produces toxicity in several neuronal cell lines in a dose- and time- dependent manner; nevertheless, some authors have shown that besides AA toxicity, some natural compounds may reduce this effect on neuronal cells. Pre-treatment with 6 μM curcumin was found to considerably reduce the neuronal toxicity caused by 2.5 mM AA. This was demonstrated by improved cell viability, reduced levels of intracellular ROS and MDA, and increased levels of GSH. Additionally, curcumin pretreatment eliminated aberrant tau phosphorylation, P-CREB decrease, and CHOP-induced apoptosis in SH-SY5Y cells, further lowered GSK-3β and ATF4 activity, and blocked PERK-dependent eIF2α phosphorylation [42]. In this same line, but with another compound, Chen and Chou 2015 [44] demonstrated that 0.25, 0.5, and 1 mM of caffeine attenuated 2 mM AA-inhibited phosphorylation of MAPKs in U-1240 MG cells. In addition, 50 μM of Z-VAD-fmk, a pan-caspase inhibitor, could be a protective compound against 0 to 5 mM of AA by abolishing caspase-3 activities in cells exposed to AA, lowering LDH leakage and increasing cell viability [50].

One step further, and in order to obtain more accurate results, Ning et al., 2021 [73], showed that 0.014 to 10 μg/mL of triphala, a herbal mixture from India, suppressed 10 mM AA-induced neurotoxicity and scavenged free radicals in a zebrafish model, suggesting that triphala could be a potential agent to treat neurodegenerative diseases associated with oxidative stress (Table 12).

Table 12. Natural bioactive compounds dose, time of pretreatment, compound, and effect.

	Dose	Time	Compound	Effects	References
AA	6 μM	2 h pretreatment	Curcumin	Increased cell viability; Alleviated oxidative stress; GRP78, P-PERK, and P-eIF2α expression decreased; Blocked PERK-eIF2α signaling activation; Suppressed tau hyperphosphorylation; Mitigated neuronal apoptosis.	[42]
	0, 0.25, 0.5, and 1 mM	30 min pretreatment	Caffeine	Attenuated AA-inhibited phosphorylation of MAPK;	[44]
	50 μM	1 h	Z-VAD-fmk	Abolished caspase-3 activities, lowered LDH leakage, and increased cell viability	[50]
	0.123, 0.370, 1.11, 3.33, and 10.0 μg/mL	20 h	Triphala	Decreased the level of free radicals; Neuroprotective	[73]
BEA, α-ZEL and β-ZEL		24 h and 48 h	*Allium sativum* L. garlic	Cell viability increased	[60]
BEA		24 h and 48 h	Coffee by-products (coffee silverskin and spent coffee)	Cytoprotection	[58]
FB1		24 h	NAC	ROS generation at 24 h	[65]
BEA			Lutein, zeaxanthin, and goji berries extract	Cytoprotection	[62]
FB1 and OTA		24 h	Beetroot extract	Increased cell viability	[67]

AA: acrylamide, eIF2α: Eukaryotic Initiation Factor 2 alpha, GRP78: glucose regulated protein 78, LDH: lactate dehydrogenase, MAPKs: mitogen-activated protein kinases, P: Phospho, PERK: protein kinase, Pyk2: proline-rich tyrosine kinase 2 RNA (PKR)-like/Pancreatic ER Kinase, U-1240 MG: Human glioblastoma astrocytoma cells. Z-VAD-fmk: Z-Val-Ala-Asp (OMe)-FMK.

Regarding mycotoxins and possible strategies to mitigate the cytotoxicity of this compound, several studies have reported the cytoprotective capacity of some bioactive compounds, such as *Allium sativum* L. garlic against BEA and ZEA metabolites [60], coffee by-products against BEA [58], NAC against FB1 [65], lutein, zeaxanthin, and goji berries extract against BEA [62], and beetroot extract against FB1 and OTA [67] (Table 12).

10. Discussion

Overall, all AA studies were carried out in SH-SY5Y, except two that used U-1240 MG and PC12 cells, the range of concentrations went from 0.1 to 10 mM, and to highlight the techniques employed the most widely used was the MTT (13%), followed by the axonopathy assay (9%) and qPCR, Annexin V-FITC/PI apoptosis, GSH, and ROS assays (7% in all cases). Caspase-3 activity, Trypan blue, LDH assays, Calcein-AM, Neurite quantification, and phase-contrast microscopy were used in a proportion of 4% each. Finally, CCK8, SF, ELISA cytometry, DNA fragmentation detection, morphology assay, biochemical indicators, Western blot, immunocytochemistry, immunoprecipitation, immunofluorescence, and cytokine assessment were used in proportions of 2% for each one (Figure 2a).

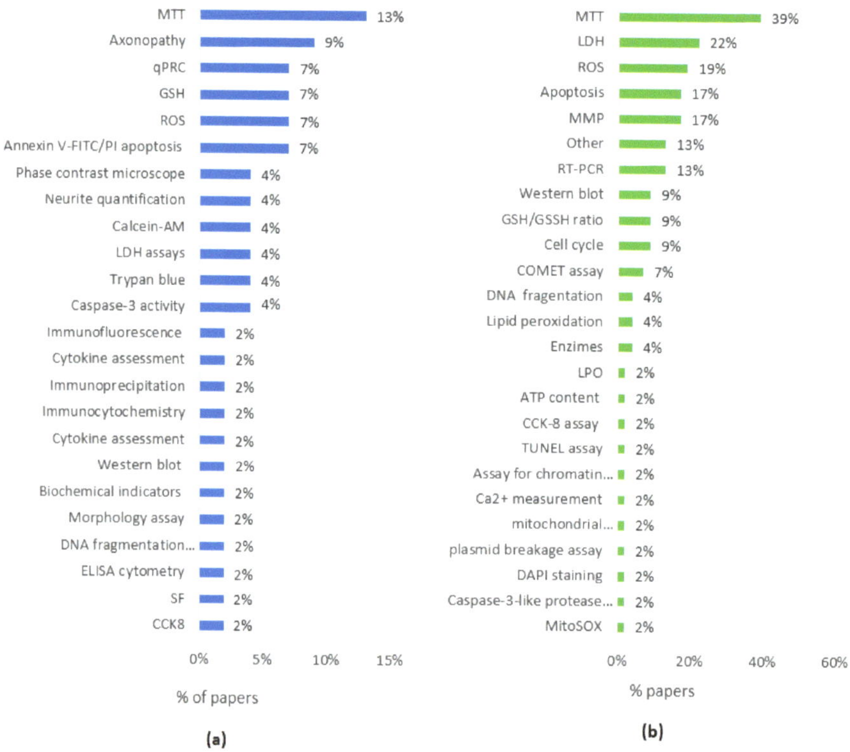

Figure 2. (**a**) Percentage of employed assays in SH-SY5Y cells treated with AA according to publications reviewed (n = 27). (**b**). Percentage of employed assays in SH-SY5Y cells treated with mycotoxins according to publications reviewed (n = 36).

Regarding mycotoxins, the main cells studied where SH-SY5Y, and the range of concentrations went from 0.009 to 200 μM. The most used techniques were MTT (39%), LDH (22%), ROS (19%), apoptosis, and MMP (17% in both cases), followed by RT-PCR (13%), Western blot, GSH/GSSH ratio, and cell cycle (9% in all cases). DNA fragmentation, lipid peroxidation, enzymes activity (4%), and the rest of the assays performed in SH-SY5Y cells were used in proportions of 2% for each one (Figure 2b).

As for the main MoA studied in SH-SY5Y cells treated with AA, cytotoxicity reached 32%, followed by axonopathy with a 20%, and the study of protein expression achieving 14% of the total of studies.

Concerning the mycotoxins, the main MoA studied in SH-SY5Y cells included cytotoxicity which reached 30%, mitochondrial membrane potential alterations at 16%, and oxidative stress at 15%.

On the other hand, several bioactive compounds were tested to modulate AA toxicity in this review; the use of curcumin, caffein, Z-VAD-fmk, and triphala was reported.

The most studied mycotoxin in SH-SY5Y cells was BEA, followed by ZEA and its metabolites, OTA and FB1.

11. Conclusions

Overall, we could conclude that the common toxic effects reported for both AA and several mycotoxins highlight an interesting field to investigate, not only because of their presence in the diet but also for the possibility of the merged effect on human health and its consequences at the neuronal level. Although there are several assays which report toxic effects, for these compounds the most used assays were MTT followed by axonopathy and qPCR, the less frequent assays were ELISA, SF, and CCK8, among others in the case of AA. For mycotoxins, MTT, LDH, ROS, apoptosis, MMP, and RT-PCR were the most employed techniques and MitoSOX, DAPI staining, Ca^{2+} measurement, and CCK-8 assay, among others, were less frequently used. The MoA reported were cytotoxicity, axonopathy, protein expression, apoptosis, increased cellular reactive oxygen species, and a reduction in intracellular GSH, impaired neurite outgrowth during differentiation in SH-SY5Y, and other neuronal cells, in a time- and concentration-dependent manner for AA and cytotoxicity, oxidative stress, gene expression, apoptosis, protein expression, cell cycle disruption, and genotoxicity in a in a time- and concentration-dependent manner for mycotoxins. Furthermore, similar to mycotoxins, AA modulated signaling pathways related to inflammation, oxidative stress, apoptosis, and differentiation through changes in the expression of proteins involved in different processes.

Although AA and mycotoxins neurotoxicity has been widely reported in the literature, several bioactive compounds naturally present in food products could reduce the toxic effects of AA and mycotoxins, according to the literature. On one hand, curcumin, caffein, Z-VAD-fmk, and triphala, tested in a similar percentage, have been shown to mitigate AA toxicity. On the other hand, *Allium sativum* L. garlic, coffee by-products, NAC, lutein, zeaxanthin, goji berries, and beetroot extract have shown to cytoprotect SH-SY5Y cells against mycotoxins like BEA, OTA, and FB1. Nevertheless, more in vivo and in vitro studies are needed to better investigate not only AA and mycotoxins' neuronal damage but also the possible beneficial role of bioactive compounds and how this mechanism of reduction in damage is happening.

12. Materials and Methods

For this review, a standard systematic literature search in PubMed, Scopus, and Web of Science was conducted by including the terms "acrylamide", "mycotoxins", "OTA", "DON", "BEA", "ZEA", "FB1", "ENN", "T-2", "neurotoxicity", "toxicity" "neuroblastoma", "in vitro", and "SH-SY5Y". The application of these search terms aimed to cover most of the literature regarding the study of the AA neurotoxicity research in the SH-SY5Y cell line. In order to discard unnecessary, incomplete, or irrelevant literature, the abstracts of the reports obtained were assessed. The search retrieved 97 articles, of which 59 were original, accessible, and AA-specific papers and thus were included in the analysis. The exclusion criteria for the remaining 38 articles were associated to the following: (i) written in a language different from English and/or not available online, (ii) representing a review, or (iii) not specific for AA and mycotoxins.

Author Contributions: Conceptualization, H.B., C.J. and A.J.-G.; methodology, H.B., C.J. and A.J.-G.; formal analysis, H.B., A.J.-G. and L.B.; investigation, H.B., A.J.-G. and L.B.; resources, H.B., C.J. and A.J.-G.; data curation, H.B., L.B. and A.J.-G.; writing—original draft preparation, H.B., A.J.-G. and L.B.; writing—review and editing, H.B., C.J. and A.J.-G.; visualization, H.B. and A.J.-G.; supervision, H.B. and A.J.-G.; project administration, H.B. and A.J.-G.; funding acquisition, H.B., C.J. and A.J.-G. All authors have read and agreed to the published version of the manuscript.

Funding: This research was funded by the Spanish Ministry of Science and Innovation grant number PID2020-115871RB-100-ALI, and Conselleria d´Educació, Universitats i Ocupació from Generalitat Valenciana CIAICO-2022/199 and project AICO/2021/037. L.B.T. would like to acknowledge the pre-PhD scholarship program from the Generalitat Valenciana (CIACIF/2021/203).

Institutional Review Board Statement: Not applicable.

Informed Consent Statement: Not applicable.

Data Availability Statement: Data is contained within the article.

Conflicts of Interest: The authors declare no conflict of interest.

References

1. Anese, M. Acrylamide in coffee and coffee substitutes. In *Acrylamide in Food: Analysis, Content and Potential Health Effects*; Academic Press Elsevier: Waltham, MA, USA, 2016. [CrossRef]
2. Guenther, H.; Anklam, E.; Wenzl, T.; Stadler, R.H. Acrylamide in Coffee: Review of Progress in Analysis, Formation and Level Reduction. *Food Addit. Contam.* **2007**, *24*, 60–70. [CrossRef] [PubMed]
3. Soares, C.M.D.; Alves, R.C.; Oliveira, M.B.P.P. Factors Affecting Acrylamide Levels in Coffee Beverages. *Coffee Health Dis. Prev.* **2015**, *32*, 217–224. [CrossRef]
4. Rannou, C.; Laroque, D.; Renault, E.; Prost, C.; Sérot, T. Mitigation Strategies of Acrylamide, Furans, Heterocyclic Amines and Browning during the Maillard Reaction in Foods. *Food Res. Int.* **2016**, *90*, 154–176. [CrossRef] [PubMed]
5. Taeymans, D.; Wood, J.; Ashby, P.; Blank, I.; Studer, A.; Stadler, R.H.; Gondé, P.; Eijck, P.; Lalljie, S.; Lingnert, H.; et al. A Review of Acrylamide: An Industry Perspective on Research, Analysis, Formation, and Control. *Crit. Rev. Food Sci. Nutr.* **2004**, *44*, 323–347. [CrossRef]
6. Tareke, E.; Rydberg, P.; Karlsson, P.; Eriksson, S.; Törnqvist, M. Analysis of Acrylamide, a Carcinogen Formed in Heated Foodstuffs. *J. Agric. Food Chem.* **2002**, *50*, 4998–5006. [CrossRef] [PubMed]
7. Xiang, J.; Liu, F.; Wang, B.; Chen, L.; Liu, W.; Tan, S. A Literature Review on Maillard Reaction Based on Milk Proteins and Carbohydrates in Food and Pharmaceutical Products: Advantages, Disadvantages, and Avoidance Strategies. *Foods* **2021**, *10*, 1998. [CrossRef] [PubMed]
8. LoPachin, R.M. Acrylamide Neurotoxicity: Neurological, Morhological and Molecular Endpoints in Animal Models. *Adv. Exp. Med. Biol.* **2005**, *561*, 21–37. [PubMed]
9. Tepe, Y.; Çebi, A. Acrylamide in Environmental Water: A Review on Sources, Exposure, and Public Health Risks. *Expo. Health* **2019**, *11*, 3–12. [CrossRef]
10. Michalak, J.; Czarnowska-Kujawska, M.; Klepacka, J.; Gujska, E. Zbieta Effect of Microwave Heating on the Acrylamide Formation in Foods. *Molecules* **2020**, *25*, 4140. [CrossRef]
11. Acrylamide (IARC Summary & Evaluation, Volume 60, 1994). Available online: https://inchem.org/documents/iarc/vol60/m60-11.html (accessed on 29 June 2023).
12. EUR-Lex—32017R2158—EN—EUR-Lex. Available online: https://eur-lex.europa.eu/legal-content/EN/TXT/?uri=CELEX:32017R2158 (accessed on 29 June 2023).
13. All News—ECHA. Available online: https://echa.europa.eu/-/one-hazardous-chemical-added-to-the-candidate-list (accessed on 29 June 2023).
14. Exon, J.H. A Review of the Toxicology of Acrylamide. *J. Toxicol. Environ. Health B Crit. Rev.* **2006**, *9*, 397–412. [CrossRef]
15. Friedman, M.; Mottram, D. *Chemistry and Safety of Acrylamide in Food*; Springer: New York, NY, USA, 2005; p. 561. [CrossRef]
16. Ghanayem, B.I.; McDaniel, L.P.; Churchwell, M.I.; Twaddle, N.C.; Snyder, R.; Fennell, T.R.; Doerge, D.R. Role of CYP2E1 in the Epoxidation of Acrylamide to Glycidamide and Formation of DNA and Hemoglobin Adducts. *Toxicol. Sci.* **2005**, *88*, 311–318. [CrossRef]
17. Nowak, A.; Zakłos-Szyda, M.; Zyzelewicz, D.; Koszucka, A.; Motyl, I. Acrylamide Decreases Cell Viability, and Provides Oxidative Stress, DNA Damage, and Apoptosis in Human Colon Adenocarcinoma Cell Line Caco-2. *Molecules* **2020**, *25*, 368. [CrossRef] [PubMed]
18. Eisenbrand, G. Revisiting the Evidence for Genotoxicity of Acrylamide (AA), Key to Risk Assessment of Dietary AA Exposure. *Arch. Toxicol.* **2020**, *94*, 2939–2950. [CrossRef] [PubMed]
19. Kopańska, M.; Łagowska, A.; Kuduk, B.; Banaś-Ząbczyk, A. Acrylamide Neurotoxicity as a Possible Factor Responsible for Inflammation in the Cholinergic Nervous System. *Int. J. Mol. Sci.* **2022**, *23*, 2030. [CrossRef] [PubMed]

20. LoPachin, R.M.; Gavin, T. Molecular Mechanism of Acrylamide Neurotoxicity: Lessons Learned from Organic Chemistry. *Environ. Health Perspect.* **2012**, *120*, 1650–1657. [CrossRef] [PubMed]
21. Tola, M.; Kebede, B. Occurrence, Importance and Control of Mycotoxins: A Review. *Cogent Food Agric.* **2016**, *2*, 1191103. [CrossRef]
22. Zain, M.E. Impact of Mycotoxins on Humans and Animals. *J. Saudi Chem. Soc.* **2011**, *15*, 129–144. [CrossRef]
23. El-Sayed, R.A.; Jebur, A.B.; Kang, W.; El-Demerdash, F.M. An Overview on the Major Mycotoxins in Food Products: Characteristics, Toxicity, and Analysis. *J. Future Foods* **2022**, *2*, 91–102. [CrossRef]
24. Wang, L.; Huang, Q.; Wu, J.; Wu, W.; Jiang, J.; Yan, H.; Huang, J.; Sun, Y.; Deng, Y. The metabolism and biotransformation of AFB1: Key enzymes and pathways. *Biochem. Pharmacol.* **2022**, *199*, 115005. [CrossRef]
25. Sun, Y.; Jiang, J.; Mu, P.; Lin, R.; Wen, J.; Deng, Y. Toxicokinetics and metabolism of deoxynivalenol in animals and humans. *Arch. Toxicol.* **2022**, *96*, 2639–2654. [CrossRef]
26. Abudayyak, M.; Karaman, E.F.; Ozden, S. Mechanisms Underlying Citrinin-Induced Toxicity via Oxidative Stress and Apoptosis-Mediated by Mitochondrial-Dependent Pathway in SH-SY5Y Cells. *Drug Chem. Toxicol.* **2022**, *46*, 944–954. [CrossRef]
27. Doi, K.; Uetsuka, K. Mechanisms of Mycotoxin-Induced Neurotoxicity through Oxidative Stress-Associated Pathways. *Int. J. Mol. Sci.* **2011**, *12*, 5213–5237. [CrossRef] [PubMed]
28. RASFF. The Rapid Alert System for Food and Feed- 2020 Annual Report. *European Commission 2020.* 2020. Available online: https://ec.europa.eu/food/system/files/2021-08/rasff_pub_annual-report_2020.pdf (accessed on 8 January 2024).
29. Xie, H.R.; Hu, L.S.; Li, G.Y. SH-SY5Y Human Neuroblastoma Cell Line: In Vitro Cell Model of Dopaminergic Neurons in Parkinson's Disease. *Chin. Med. J.* **2010**, *123*, 1086–1092. [CrossRef]
30. Kalinovskii, A.P.; Osmakov, D.I.; Koshelev, S.G.; Lubova, K.I.; Korolkova, Y.V.; Kozlov, S.A.; Andreev, Y.A. Retinoic Acid-Differentiated Neuroblastoma SH-SY5Y Is an Accessible In Vitro Model to Study Native Human Acid-Sensing Ion Channels 1a (ASIC1a). *Biology* **2022**, *11*, 167. [CrossRef] [PubMed]
31. Singh, J.; Kaur, G. Transcriptional Regulation of Polysialylated Neural Cell Adhesion Molecule Expression by NMDA Receptor Activation in Retinoic Acid-Differentiated SH-SY5Y Neuroblastoma Cultures. *Brain Res.* **2007**, *1154*, 8–21. [CrossRef]
32. IRIS Toxicological Review of Acrylamide (External Review Draft) | Risk Assessment Portal | US EPA. Available online: https://cfpub.epa.gov/ncea/risk/recordisplay.cfm?deid=187729 (accessed on 8 January 2024).
33. Rifai, L.; Saleh, F.A. A Review on Acrylamide in Food: Occurrence, Toxicity, and Mitigation Strategies. *Int. J. Toxicol.* **2020**, *39*, 93–102. [CrossRef]
34. Lineback, D.R.; Coughlin, J.R.; Stadler, R.H. Acrylamide in Foods: A Review of the Science and Future Considerations. *Annu. Rev. Food Sci. Technol.* **2012**, *3*, 15–35. [CrossRef]
35. Iqbal, S.Z. Mycotoxins in food, recent development in food analysis and future challenges; a review. *Curr. Opin. Food Sci.* **2021**, *42*, 237–247. [CrossRef]
36. Sebastià, A.; Pallarés, N.; Bridgeman, L.; Juan-García, A.; Castagnini, J.M.; Ferrer, E.; Barba, F.J.; Berrada, H. A Critical Review of Acrylamide Green Extraction and Determination in Food Matrices: Current Insights and Future Perspectives. *TrAC Trends Anal. Chem.* **2023**, *167*, 117267. [CrossRef]
37. Leite, M.; Freitas, A.; Barbosa, J.; Ramos, F. Comprehensive Assessment of Different Extraction Methodologies for Optimization and Validation of an Analytical Multi-Method for Determination of Emerging and Regulated Mycotoxins in Maize by UHPLC-MS/MS. *Food Chem. Adv.* **2023**, *2*, 100145. [CrossRef]
38. Bridgeman, L.; Juan, C.; Juan-García, A.; Berrada, H. Individual and Combined Effect of Acrylamide, Fumitremorgin C and Penitrem A on Human Neuroblastoma SH-SY5Y Cells. *Food Chem. Toxicol.* **2023**, *182*, 278–6915. [CrossRef]
39. Riss, T.; Niles, A.; Moravec, R.; Karassina, N.; Vidugiriene, J. Cytotoxicity Assays: In Vitro Methods to Measure Dead Cells. In *Assay Guidance Manual*; Eli Lilly & Company and the National Center for Advancing Translational Sciences: Bethesda, MD, USA, 2019.
40. Gupta, R.; Nanda, S.J. Cloud detection in satellite images with classical and deep neural network approach: A review. *Multimed. Tools Appl.* **2022**, *81*, 31847–31880. [CrossRef]
41. Yan, D.; Pan, X.; Yao, J.; Wang, D.; Wu, X.; Chen, X.; Shi, N.; Yan, H. MAPKs and NF-KB-Mediated Acrylamide-Induced Neuropathy in Rat Striatum and Human Neuroblastoma Cells SY5Y. *J. Cell Biochem.* **2019**, *120*, 3898–3910. [CrossRef] [PubMed]
42. Yan, D.; Wang, N.; Yao, J.; Wu, X.; Yuan, J.; Yan, H. Curcumin Attenuates the PERK-EIF2α Signaling to Relieve Acrylamide-Induced Neurotoxicity in SH SY5Y Neuroblastoma Cells. *Neurochem. Res.* **2022**, *47*, 1037–1048. [CrossRef] [PubMed]
43. Song, G.; Liu, Z.; Wang, L.; Shi, R.; Chu, C.; Xiang, M.; Tian, Q.; Liu, X. Protective Effects of Lipoic Acid against Acrylamide-Induced Neurotoxicity: Involvement of Mitochondrial Energy Metabolism and Autophagy. *Food Funct.* **2017**, *8*, 4657–4667. [CrossRef] [PubMed]
44. Chen, J.H.; Chou, C.C. Acrylamide Inhibits Cellular Differentiation of Human Neuroblastoma and Glioblastoma Cells. *Food Chem. Toxicol.* **2015**, *82*, 27–35. [CrossRef]
45. Attoff, K.; Kertika, D.; Lundqvist, J.; Oredsson, S.; Forsby, A. Acrylamide Affects Proliferation and Differentiation of the Neural Progenitor Cell Line C17.2 and the Neuroblastoma Cell Line SH-SY5Y. *Toxicol. Vitr.* **2016**, *35*, 100–111. [CrossRef] [PubMed]
46. Komoike, Y.; Matsuoka, M. Endoplasmic Reticulum Stress-Mediated Neuronal Apoptosis by Acrylamide Exposure. *Toxicol. Appl. Pharmacol.* **2016**, *310*, 68–77. [CrossRef] [PubMed]
47. Chen, J.H.; Wu, K.Y.; Chiu, I.M.; Tsou, T.C.; Chou, C.C. Acrylamide-Induced Astrogliotic and Apoptotic Responses in Human Astrocytoma Cells. *Toxicol. Vitr.* **2009**, *23*, 855–861. [CrossRef]

48. Okuno, T.; Matsuoka, M.; Sumizawa, T.; Igisu, H. Involvement of the Extracellular Signal-Regulated Protein Kinase Pathway in Phosphorylation of P53 Protein and Exerting Cytotoxicity in Human Neuroblastoma Cells (SH-SY5Y) Exposed to Acrylamide. *Arch. Toxicol.* **2006**, *80*, 146–153. [CrossRef]
49. Sumizawa, T.; Igisu, H. Apoptosis Induced by Acrylamide in SH-SY5Y Cells. *Arch. Toxicol.* **2007**, *81*, 279–282. [CrossRef]
50. Hartley, C.L.; Anderson, V.E.R.; Anderson, B.H.; Robertson, J. Acrylamide and 2,5-Hexanedione Induce Collapse of Neurofilaments in SH-SY5Y Human Neuroblastoma Cells to Form Perikaryal Inclusion Bodies. *Neuropathol. Appl. Neurobiol.* **1997**, *23*, 364–372. [CrossRef]
51. Frimat, J.P.; Sisnaiske, J.; Subbiah, S.; Menne, H.; Godoy, P.; Lampen, P.; Leist, M.; Franzke, J.; Hengstler, J.G.; Van Thriel, C.; et al. The Network Formation Assay: A Spatially Standardized Neurite Outgrowth Analytical Display for Neurotoxicity Screening. *Lab Chip* **2010**, *10*, 701–709. [CrossRef] [PubMed]
52. Nordin-Andersson, M.; Walum, E.; Kjellstrand, P.; Forsby, A. Acrylamide-Induced Effects on General and Neurospecific Cellular Functions during Exposure and Recovery. *Cell Biol. Toxicol.* **2003**, *19*, 43–51. [CrossRef] [PubMed]
53. Lourenssen, S.; Miller, K.G.; Blennerhassett, M.G. Discrete Responses of Myenteric Neurons to Structural and Functional Damage by Neurotoxins in Vitro. *Am. J. Physiol. Gastrointest Liver Physiol.* **2009**, *297*, G228–G239. [CrossRef] [PubMed]
54. Popova, D.; Karlsson, J.; Jacobsson, S.O.P. Comparison of Neurons Derived from Mouse P19, Rat PC12 and Human SH-SY5Y Cells in the Assessment of Chemical- and Toxin-Induced Neurotoxicity. *BMC Pharmacol. Toxicol.* **2017**, *18*, 42. [CrossRef] [PubMed]
55. Ban, M.; Shimoda, R.; Chen, J. Investigation of Nanoplastic Cytotoxicity Using SH-SY5Y Human Neuroblastoma Cells and Polystyrene Nanoparticles. *Toxicol. Vitr.* **2021**, *76*, 105225. [CrossRef] [PubMed]
56. Hallier-Vanuxeem, D.; Prieto, P.; Culot, M.; Diallo, H.; Landry, C.; Tähti, H.; Cecchelli, R. New Strategy for Alerting Central Nervous System Toxicity: Integration of Blood–Brain Barrier Toxicity and Permeability in Neurotoxicity Assessment. *Toxicol. Vitr.* **2009**, *23*, 447–453. [CrossRef] [PubMed]
57. Juan-García, A.; Caprioli, G.; Sagratini, G.; Mañes, J.; Juan, C. Coffee Silverskin and Spent Coffee Suitable as Neuroprotectors against Cell Death by Beauvericin and α-Zearalenol: Evaluating Strategies of Treatment. *Toxins* **2021**, *13*, 132. [CrossRef] [PubMed]
58. Juan, C.; de Simone, G.; Sagratini, G.; Caprioli, G.; Mañes, J.; Juan-García, A. Reducing the Effect of Beauvericin on Neuroblastoma SH-SY5Y Cell Line by Natural Products. *Toxicon* **2020**, *188*, 164–171. [CrossRef]
59. Agahi, F.; Font, G.; Juan, C.; Juan-García, A. Individual and Combined Effect of Zearalenone Derivates and Beauvericin Mycotoxins on SH-SY5Y Cells. *Toxins* **2020**, *12*, 212. [CrossRef] [PubMed]
60. Agahi, F.; Penalva-Olcina, R.; Font, G.; Juan-García, A.; Juan, C. Effects of Voghiera Garlic Extracts in Neuronal Human Cell Line against Zearalenone's Derivates and Beauvericin. *Food Chem. Toxicol.* **2022**, *162*, 112905. [CrossRef]
61. Pérez-Fuentes, N.; Alvariño, R.; Alfonso, A.; González-Jartín, J.; Gegunde, S.; Vieytes, M.R.; Botana, L.M. Single and Combined Effects of Regulated and Emerging Mycotoxins on Viability and Mitochondrial Function of SH-SY5Y Cells. *Food Chem. Toxicol.* **2021**, *154*, 112308. [CrossRef] [PubMed]
62. Montesano, D.; Juan-García, A.; Mañes, J.; Juan, C. Chemoprotective Effect of Carotenoids from Lycium Barbarum L. on SH-SY5Y Neuroblastoma Cells Treated with Beauvericin. *Food Chem. Toxicol.* **2020**, *141*, 111414. [CrossRef] [PubMed]
63. Kalagatur, N.K.; Abd_Allah, E.F.; Poda, S.; Kadirvelu, K.; Hashem, A.; Mudili, V.; Siddaiah, C. Quercetin Mitigates the Deoxynivalenol Mycotoxin Induced Apoptosis in SH-SY5Y Cells by Modulating the Oxidative Stress Mediators. *Saudi J. Biol. Sci.* **2021**, *28*, 465–477. [CrossRef]
64. Stockmann-Juvala, H.; Mikkola, J.; Naarala, J.; Loikkanen, J.; Elovaara, E.; Savolainen, K. Oxidative Stress Induced by Fumonisin B1 in Continuous Human and Rodent Neural Cell Cultures. *Free Radic. Res.* **2009**, *38*, 933–942. [CrossRef]
65. Paul, S.; Jakhar, R.; Bhardwaj, M.; Chauhan, A.K.; Kang, S.C. Fumonisin B1 Induces Poly (ADP-Ribose) (PAR) Polymer-Mediated Cell Death (Parthanatos) in Neuroblastoma. *Food Chem. Toxicol.* **2021**, *154*, 112326. [CrossRef] [PubMed]
66. Domijan, A.M.; Abramov, A.Y. Fumonisin B1 Inhibits Mitochondrial Respiration and Deregulates Calcium Homeostasis--Implication to Mechanism of Cell Toxicity. *Int. J. Biochem. Cell Biol.* **2011**, *43*, 897–904. [CrossRef]
67. Penalva-Olcina, R.; Juan, C.; Fernández-Franzón, M.; Juan-García, A. Effectiveness of Beetroot Extract in SH-SY5Y Neuronal Cell Protection against Fumonisin B1, Ochratoxin A and Its Combination. *Food Chem. Toxicol.* **2022**, *165*, 113164. [CrossRef]
68. Yoon, S.; Cong, W.T.; Bang, Y.; Lee, S.N.; Yoon, C.S.; Kwack, S.J.; Kang, T.S.; Lee, K.Y.; Choi, J.K.; Choi, H.J. Proteome Response to Ochratoxin A-Induced Apoptotic Cell Death in Mouse Hippocampal HT22 Cells. *Neurotoxicology* **2009**, *30*, 666–676. [CrossRef]
69. Pang, Y.; Zhang, L.; Liu, Q.; Peng, H.; He, J.; Jin, H.; Su, X.; Zhao, J.; Guo, J. NRF2/PGC-1α-Mediated Mitochondrial Biogenesis Contributes to T-2 Toxin-Induced Toxicity in Human Neuroblastoma SH-SY5Y Cells. *Toxicol. Appl. Pharmacol.* **2022**, *451*, 116167. [CrossRef]
70. Venkataramana, M.; Chandra Nayaka, S.; Anand, T.; Rajesh, R.; Aiyaz, M.; Divakara, S.T.; Murali, H.S.; Prakash, H.S.; Lakshmana Rao, P.V. Zearalenone Induced Toxicity in SHSY-5Y Cells: The Role of Oxidative Stress Evidenced by N-Acetyl Cysteine. *Food Chem. Toxicol.* **2014**, *65*, 335–342. [CrossRef]
71. Frangiamone, M.; Alonso-Garrido, M.; Font, G.; Cimbalo, A.; Manyes, L. Pumpkin Extract and Fermented Whey Individually and in Combination Alleviated AFB1- and OTA-Induced Alterations on Neuronal Differentiation in Vitro. *Food Chem. Toxicol.* **2022**, *164*, 113011. [CrossRef]
72. Yu, C.P.; Pan, Y.L.; Wang, X.L.; Xin, R.; Li, H.Q.; Lei, Y.T.; Zhao, F.F.; Zhang, D.; Zhou, X.R.; Ma, W.W.; et al. Stimulating the Expression of Sphingosine Kinase 1 (SphK1) Is Beneficial to Reduce Acrylamide-Induced Nerve Cell Damage. *Ecotoxicol. Environ. Saf.* **2022**, *237*, 113511. [CrossRef]

73. Ning, W.; Li, S.; Tsering, J.; Ma, Y.; Li, H.; Ma, Y.; Ogbuehi, A.C.; Pan, H.; Li, H.; Hu, S.; et al. Protective Effect of Triphala against Oxidative Stress-Induced Neurotoxicity. *Biomed. Res. Int.* **2021**, *2021*, 6674988. [CrossRef]
74. Agahi, F.; Juan, C.; Font, G.; Juan-García, A. Neurotoxicity of Zearalenone's Metabolites and Beauvericin Mycotoxins via Apoptosis and Cell Cycle Disruption. *Toxicology* **2021**, *456*, 152784. [CrossRef]
75. Agahi, F.; Álvarez-Ortega, N.; Font, G.; Juan-García, A.; Juan, C. Oxidative Stress, Glutathione, and Gene Expression as Key Indicators in SH-SY5Y Cells Exposed to Zearalenone Metabolites and Beauvericin. *Toxicol. Lett.* **2020**, *334*, 44–52. [CrossRef]
76. Agahi, F.; Juan-García, A.; Font, G.; Juan, C. Study of Enzymatic Activity in Human Neuroblastoma Cells SH-SY5Y Exposed to Zearalenone's Derivates and Beauvericin. *Food Chem. Toxicol.* **2021**, *152*, 112227. [CrossRef] [PubMed]
77. Forsby, A. Neurite Degeneration in Human Neuronal SH-SY5Y Cells as an Indicator of Axonopathy. *Neuromethods* **2011**, *56*, 255–268.
78. Nakagawa-Yagi, Y.; Choi, D.K.; Ogane, N.; Shimada, S.I.; Seya, M.; Momoi, T.; Ito, T.; Sakaki, Y. Discovery of a Novel Compound: Insight into Mechanisms for Acrylamide-Induced Axonopathy and Colchicine-Induced Apoptotic Neuronal Cell Death. *Brain Res.* **2001**, *909*, 8–19. [CrossRef] [PubMed]
79. Nordin-Andersson, M.; Forsby, A.; Heldring, N.; DeJongh, J.; Kjellstrand, P.; Walum, E. Neurite Degeneration in Differentiated Human Neuroblastoma Cells. *Toxicol. Vitr.* **1998**, *12*, 557–560. [CrossRef] [PubMed]
80. Attoff, K.; Johansson, Y.; Cediel-Ulloa, A.; Lundqvist, J.; Gupta, R.; Caiment, F.; Gliga, A.; Forsby, A. Acrylamide Alters CREB and Retinoic Acid Signalling Pathways during Differentiation of the Human Neuroblastoma SH-SY5Y Cell Line. *Sci. Rep.* **2020**, *10*, 16714. [CrossRef] [PubMed]
81. Hinojosa, M.G.; Johansson, Y.; Cediel-Ulloa, A.; Ivanova, E.; Gabring, N.; Gliga, A.; Forsby, A. Evaluation of MRNA Markers in Differentiating Human SH-SY5Y Cells for Estimation of Developmental Neurotoxicity. *Neurotoxicology* **2023**, *97*, 65–77. [CrossRef] [PubMed]
82. Sharma, R.; Gettings, S.M.; Hazell, G.; Bourbia, N. In Vitro Study of Ochratoxin A in the Expression of Genes Associated with Neuron Survival and Viability. *Toxicology* **2023**, *483*, 153376. [CrossRef]
83. Zhang, X.; Boesch-Saadatmandi, C.; Lou, Y.; Wolffram, S.; Huebbe, P.; Rimbach, G. Ochratoxin A induces apoptosis in neuronal cells. *Genes Nutr.* **2009**, *4*, 41. [CrossRef]
84. Sumizawa, T.; Igisu, H. Release of Heat Shock Proteins from Human Neuroblastoma Cells Exposed to Acrylamide. *J. Toxicol. Sci.* **2008**, *33*, 117–122. [CrossRef]
85. Stockmann-Juvala, H.; Naarala, J.; Loikkanen, J.; Vähäkangas, K.; Savolainen, K. Fumonisin B1-Induced Apoptosis in Neuroblastoma, Glioblastoma and Hypothalamic Cell Lines. *Toxicology* **2006**, *225*, 234–241. [CrossRef] [PubMed]

toxins MDPI

Systematic Review

Machine Learning Applied to the Detection of Mycotoxin in Food: A Systematic Review

Alan Inglis [1,*], Andrew C. Parnell[1], Natarajan Subramani [2] and Fiona M. Doohan [2]

[1] Hamilton Institute, Eolas Building, Maynooth University, W23 F2H6 Maynooth, Kildare, Ireland; andrew.parnell@mu.ie
[2] School of Biology and Environmental Science, University College Dublin, D04 C1P1 Dublin, Ireland; natarajan.subramani@ucd.ie (N.S.); fiona.doohan@ucd.ie (F.M.D.)
* Correspondence: alan.inglis@mu.ie

Abstract: Mycotoxins, toxic secondary metabolites produced by certain fungi, pose significant threats to global food safety and public health. These compounds can contaminate a variety of crops, leading to economic losses and health risks to both humans and animals. Traditional lab analysis methods for mycotoxin detection can be time-consuming and may not always be suitable for large-scale screenings. However, in recent years, machine learning (ML) methods have gained popularity for use in the detection of mycotoxins and in the food safety industry in general due to their accurate and timely predictions. We provide a systematic review on some of the recent ML applications for detecting/predicting the presence of mycotoxin on a variety of food ingredients, highlighting their advantages, challenges, and potential for future advancements. We address the need for reproducibility and transparency in ML research through open access to data and code. An observation from our findings is the frequent lack of detailed reporting on hyperparameters in many studies and a lack of open source code, which raises concerns about the reproducibility and optimisation of the ML models used. The findings reveal that while the majority of studies predominantly utilised neural networks for mycotoxin detection, there was a notable diversity in the types of neural network architectures employed, with convolutional neural networks being the most popular.

Keywords: machine learning; predictive model; mycotoxin; food safety; systematic review

Key Contribution: Recent developments in machine learning present promising approaches to improve the precision and efficiency of detecting mycotoxins. This review comprehensively gathers and examines the latest research at the juncture of machine learning and mycotoxin detection in food items. It offers a detailed assessment of the methods used, accomplishments, and potential future developments.

1. Introduction

Mycotoxins are a group of naturally occurring toxic chemical compounds produced by certain species of moulds (fungi) during growth on various crops and foodstuffs, including cereals, nuts, spices, and dairy products [1]. The ingestion of certain mycotoxins has been linked to a range of harmful health impacts on both humans and animals, from short-term poisoning to long-term consequences such as liver cancer and, in some cases, death [2–4]. Mycotoxins are secondary metabolites (that is, compounds produced by an organism that are not essential for its primary life processes) and are often produced during the pre-harvest, harvest, and storage phases under favourable conditions of humidity and temperature [3,5]. The most prevalent mycotoxins include aflatoxins, tricothecenes, fumonisins, zearalenones, ochratoxins, and patulin, and are produced by certain plant-pathogenic species of *Aspergillus*, *Fusarium*, and *Penicillium* [6]. Mycotoxin contamination in crop products has been found to vary significantly across different geographical locations and is influenced by annual weather conditions [7,8]. However, since 2012, there has been a noted increase in the occurrence of mycotoxins in Europe, with the impacts of climate change being most likely a contributing factor [9,10]. An estimated 60–80% of the world's crop supply is contaminated by mycotoxins, and an estimated 20% of those crops surpass the legally mandated food safety thresholds set by the European Union (EU) [11].

With the world's food supply chain being highly interconnected, the presence of mycotoxins not only endangers human health but also has an impact on the stability of agricultural markets and trade [3,12]. The economic impact of mycotoxin contamination is substantial, with a global estimate in the billions of euros for detection, regulation enforcement, and mitigation efforts to manage mycotoxin presence in food and feeds annually [13]. It is estimated that, between 2010 and 2019, approximately 75 million tonnes of wheat in Europe, which constitutes 5% of the wheat intended for human consumption, surpassed the maximum threshold for DON contamination. This excess led to the reclassification of this contaminated wheat grain as 'animal feed', resulting in an economic loss of around EUR 3 billion [14]. Additionally, [15] shows that, between 2010 and 2020, aflatoxins were responsible for the demotion of 4.2% of wheat intended for food, which potentially represented an additional economic loss of EUR 2.5 billion. As a result, the detection and management of mycotoxins in crops and food products is crucial for ensuring food safety and safeguarding consumer health worldwide as well as contributing to economic stability.

According to [16], the standard methodology for mycotoxin detection comprises three main steps: sampling, sample preparation, and analytical determination. Chromatographic techniques, such as liquid chromatography mass spectrometry (LC–MS), high-performance liquid chromatography (HPLC), and gas chromatography mass spectrometry (GC–MS), along with immunoassay-based methods like enzyme-linked immunosorbent assays (ELISAs), are widely recognised as the most prevalent analytical approaches for the detection of mycotoxins [17,18]. The mycotoxin level in a bulk load is determined by measuring a sample taken from the food source. From this, the concentration of mycotoxins in the entire load is assumed to be the same as the concentration of the sample. However, these techniques often require extensive sample preparation, sophisticated equipment, and highly trained personnel, leading to significant costs and time delays in the analytical process. Furthermore, the varied and intricate nature of different foods requires customised detection methods, which can add complexity to the screening process [19,20].

While traditional detection methods such as LC–MS, HPLC, GC–MS, and ELISA generate reliable data, they often result in large, complex datasets that require extensive interpretation and analysis. Machine learning (ML) approaches for both the detection and prediction of the presence of mycotoxins have seen a rise in recent years as an alternative to traditional detection methods (see Figure 1). At its core, ML employs statistical methods to create algorithms that allow computers to learn from data and make decisions based on identified patterns and inferences, without being explicitly programmed for each specific task. ML methods offer a sophisticated approach to deciphering the complex patterns hidden within the data and are adept at processing and analysing large datasets and extracting meaningful patterns that are not immediately apparent. By leveraging ML algorithms, researchers can gain deeper insights into the data and offer a significant advantage, when compared with traditional lab analysis, in terms of efficiency, cost, and scalability, as well as maintaining or improving the accuracy of mycotoxin detection [21].

ML methods can be, broadly, broken into three categories, that is, supervised learning (SL), unsupervised learning (UL), and reinforcement learning (RL). In SL, an algorithm is trained using a dataset that includes both inputs and the corresponding outputs. The model learns to associate the inputs with the outputs. After training, the model can apply this learned relationship to predict the outputs for new, unseen inputs [22]. In UL, an algorithm is presented with only the input data and identifies patterns and structures in the data based only on the inputs. After training, it can classify new inputs based on the patterns it has found. In RL, an algorithm learns to make decisions by performing actions to achieve a goal.

It processes feedback through rewards or penalties associated with its actions, using this information to develop a decision-making framework that aims to maximise rewards [23].

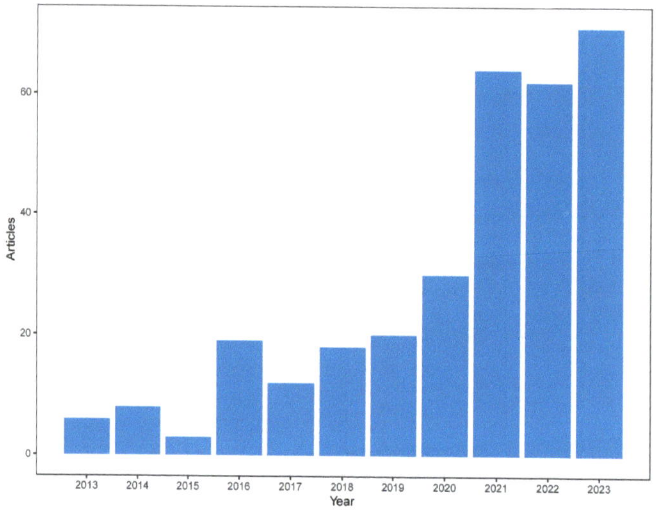

Figure 1. Number of publications between 2013 and 2023 found by our systematic search criteria in Scopus.

Within these categories, many different types of ML models exist and are used based on the specificity of the problem. The most popular of these models, as found by this research, are discussed in detail below. Although ML applications in food safety and mycotoxin detection are widespread, there appears to be a lack of comprehensive reviews that cover the broad spectrum of ML methodologies specifically tailored to mycotoxin analysis, as most studies tend to concentrate on individual techniques. For example, Ref. [24] uses neural networks (NNs) for the prediction of contamination from the mycotoxin fumonisin in corn. Additionally, NNs have been used to forecast the accumulation of the trichothecene mycotoxin deoxynivalenol (DON) in barley seeds [25] and to predict fungal growth [26]. For a comprehensive review of the use of NNs in food science, see Ref. [27]; for a review of ML methods in general in the field of food safety, see Ref. [28]; and in agriculture, see Ref. [21].

ML techniques can alleviate some of the current burdens of mycotoxin detection by providing an efficient and low-cost solution [29]. Additionally, with the impact of climate change, the need for these models to provide reliable predictions at the farm level is increasingly crucial, especially in terms of food safety and health. In this work, we present a comprehensive systematic review of some of the more popular ML techniques used in the detection and prediction of mycotoxin on a range of foods and crops. Our review also identifies critical areas in the current body of work that warrant attention. A notable concern is the often insufficient discussion on the selection and tuning of hyperparameters in ML models, which is crucial for understanding and replicating study results. This lack of details creates issues with the reproducibility of the reviewed methods and also hinders the advancement and application of these techniques.

The organisation of our article is as follows: In Section 2, we provide details regarding our literature search methodology. This includes a description of the search criteria and keywords and discussing the prevalence of each ML method. In Section 3, we provide a short introduction to the ML process and describe some of the common terms. In Section 4, we give a brief introduction to the main ML algorithms used (and their hyperparameters) and discuss the outcomes of the articles reviewed based on the type of machine learning model used. Finally, in Section 5, we provide some concluding remarks.

2. Literature Search Methodology

The literature search for this review was primarily conducted using Scopus (https://www.scopus.com, URL accessed on 10 November 2023), a widely recognised academic search engine that indexes scholarly articles across various disciplines. To ensure the relevance of the research, the search was restricted to articles published within the last 10 years (since November 2023). This time frame was chosen to capture the most recent advances and trends in the application of machine learning to mycotoxin detection in crops. The search engine was used to identify key studies, reviews, and seminal works pertinent to the topic at hand.

Search Criteria and Overview

A comprehensive search was conducted on the Scopus database and focused on publications between the years 2013 and 2023. The search was conducted using the primary keyword *"mycotoxin"* in combination with these machine learning-related terms: *"artificial intelligence"*, *"bagging"*, *"Bayesian network"*, *"boosting"*, *"decision tree"*, *"deep learning"*, *"ensemble"*, *"gradient boost"*, *"k-means"*, *"k-nearest neighbour"*, *"knn"*, *"machine learning"*, *"neural network"*, *"principal component analysis"*, *"random forest"*, *"supervised learning"*, *"support vector machine"*, *"SVM"*, and *"unsupervised learning"*. The search terms were motivated by a similar search used in a review of machine learning for the monitoring and prediction of food safety by [28]. This strategy was employed to ensure a wide coverage of potential articles at the intersection of mycotoxin detection and machine learning methodologies.

This search yielded 313 documents on Scopus. Figure 1 shows the results obtained from Scopus over the years 2013 to 2023. There is a general increasing trend across the years, with a marked rise after the year 2021.

To limit the search further, only peer-reviewed articles in English in the fields of agricultural and biological sciences, environmental science, computer science, and mathematics were chosen. This reduced the search size to 91. After examining the abstracts of all the 91 articles, 30 were selected for their relevance and included in this study. A flow diagram demonstrating our selection process can be found in Appendix A. From these articles, the predominant ML technique used was neural networks (NNs), followed by random forests (RFs) and gradient boosting (GB), and then support vector machines (SVMs), decision trees (DTs), and Bayesian networks (BNs). Figure 2 shows the frequency of each ML algorithm used in the literature.

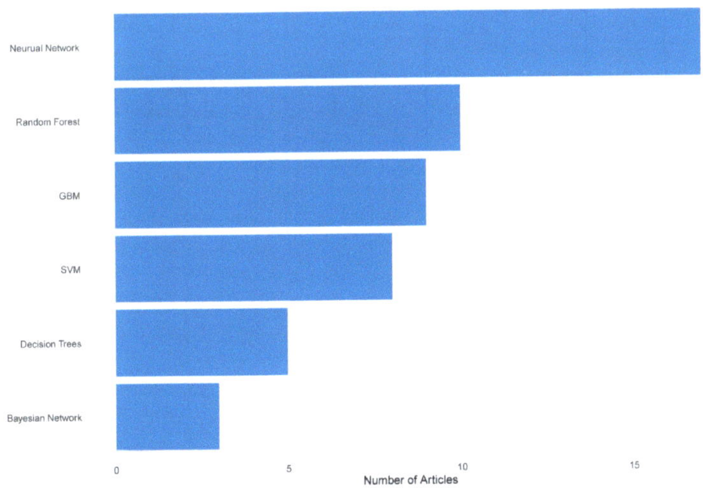

Figure 2. Most popular machine learning methods reviewed in this work.

3. A Brief Introduction to Machine Learning

In this section, we provide a general overview of the ML process. This foreknowledge is useful when discussing the ML approaches reviewed later in this document, though those already with experience in this topic may skip this section. To begin, we describe the typical process of creating an ML model.

3.1. Typical Machine Learning Process

Figure 3 shows a typical ML process for unsupervised and supervised learning methods.

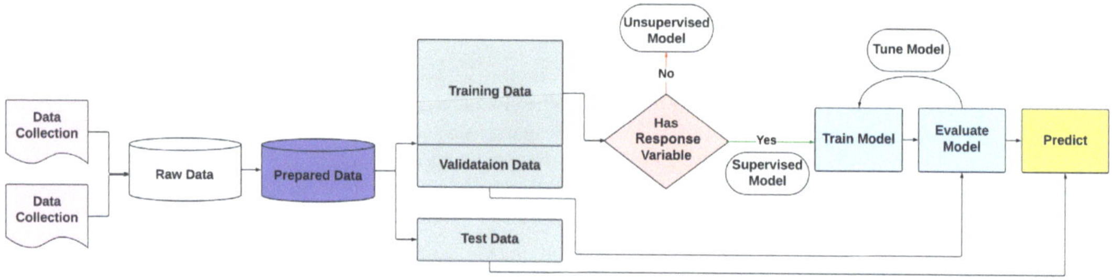

Figure 3. Typical machine learning process.

We can break up the process outlined in Figure 3 into five distinct steps. These are as follows:

1. Data Collection: The process starts with the collection of raw data, which can be from many sources or sites.
2. Data Preparation: These raw data are then prepared for analysis. This process typically involves cleaning and formatting the data.
3. Data Splitting: After preparation, the data can be split into three parts. These are training data, validation data, and test data (discussed more below).
4. Model Selection: Depending on the type of data, either an unsupervised or supervised learning model (or models) is chosen.
5. Model Training, Evaluation, and Prediction: This process involves training the model with training data, optimising the hyperparameters of the model using the validation data, and then evaluating its overall performance using test data.

3.2. Training, Validation, and Test Data

In ML, validation and test data are crucial for developing and evaluating models. Validation data are a separate subset of the original data, not used in training the model (see Figure 3). It helps in fine-tuning the model's parameters (known as hyperparameters), which are pre-set configurations of the model. This fine-tuning of hyperparameters during the validation process is essential to optimise the model's performance. One common technique used during this process is regularisation. Regularisation involves adding a penalty to the model's complexity, which helps prevent overfitting by ensuring that the model generalises well to new, unseen data rather than just memorising the training data. Validation also assists in selecting the best version of the model by providing feedback on its performance. This step is essential to prevent overfitting, ensuring that the model learns to generalise from the data and makes accurate predictions on new, unseen data. Test data are used after the model has been trained and validated (see Figure 3). It is another distinct subset of the dataset, not used in either training or validation. The test data are used to evaluate the final model's performance, providing an unbiased assessment of how well the model is likely to perform in real-world scenarios.

In all the referenced studies we cover below, model performance is quantified by evaluating the model performance on the test dataset, unless otherwise stated. Sometimes authors also report the training or validation dataset performance, but for the reasons outlined above, these should be discarded as a measure of model performance. The common performance metrics used in these studies include the following:

- R^2: This statistic measures the proportion of the variance in the dependent variable that can be explained by the independent variables in the model. An R^2 value closer to 1 indicates that the model accounts for a significant amount of the variance in the dependent variable.
- MSE and RMSE: Mean Square Error (MSE) is the average of the squares of the errors, which are the differences between predicted and actual values. Lower MSE values indicate a better fit of the model to the data. Root Mean Square Error (RMSE) is the square root of MSE. It has the same units as the quantity being estimated (for regression problems) and provides a measure of the differences between a model's predicted values and the actual observed values. Like MSE, a lower RMSE is better.
- Accuracy: This metric is commonly used for classification tasks and represents the ratio of correctly predicted observations to the total observations. High accuracy indicates that the model can correctly classify instances with high reliability.
- AUC: Area Under the Receiver Operating Characteristic Curve (AUC) is used in binary classification to measure a model's ability to distinguish between classes. An AUC of 1 represents perfect classifier performance, while an AUC of 0.5 denotes a model with no discriminative power.

4. Application of Machine Learning to Mycotoxin Data

In this section, we first include a brief discussion on common data types in mycotoxin detection. We then discuss the most common ML algorithms (from Figure 2) and review their application to mycotoxin data. Each subsection is dedicated to a single ML method in which we describe the basic algorithm, how it makes predictions/detections, some advantages and disadvantages of the algorithm, and finally a review of the literature using these methods. In cases where the reviewed studies employ multiple machine learning models, we categorise each paper based on the highest-performing model used in that particular work.

4.1. Types of Data Used in Mycotoxin Detection

In the context of ML applications for mycotoxin detection, the literature highlights the use of various data types, including weather parameters (temperature, rainfall, and relative humidity), crop phenology, agronomic data, and spectral imaging. Additionally, spatiotemporal data, which include information collected over time and across different spatial locations, play a vital role in understanding and predicting mycotoxin contamination by incorporating key environmental variables and temporal dynamics. Each type of data offers unique characteristics and applications. Understanding the context and conditions under which these data are collected is essential for interpreting the results and evaluating the effectiveness of different ML models.

4.1.1. Weather Data

Weather variables, including temperature, relative humidity, precipitation, and carbon dioxide levels, play a significant role in mycotoxigenic fungal growth and subsequent mycotoxin formation on agricultural commodities [30–32]. ML models can leverage historical and real-time weather data to predict the likelihood of mycotoxin contamination. For example, continuous monitoring of these variables in the field can help create more dynamic and responsive models. Incorporating these factors allows for a more comprehensive understanding of the conditions that favour mycotoxin contamination and can improve the predictive power of ML models.

Ref. [33] proposed a Convolutional Neural Network model based on CO2 respiration rate and the visual appearance of mold formation for classifying mycotoxin contamination in wheat grains stored in sealed containers, which achieved an accuracy of 83.3%. Ref. [34] constructed a predictive model that incorporated multiple data sources, such as historical records of aflatoxin and fumonisin in corn, daily weather conditions, satellite imagery, dynamic geospatial soil characteristics, and land usage information. Using both a gradient boosting machine and a neural network, the study demonstrated that the NN models exhibited high class-specific accuracy for predicting mycotoxin levels over a 1-year period, with accuracies of 73% for aflatoxin and 85% for fumonisin, demonstrating their efficacy in forecasting annual mycotoxin levels.

4.1.2. Agronomic Data

The impact of agronomic factors on mycotoxin occurrence has been extensively studied in various research. These factors include previous crop details, the use of fungicides, cropping patterns, and cultivar selection, all of which have been found to significantly affect mycotoxin levels [35–37]. In a study by Ref. [38], data on cropping system factors were used as input variables to predict aflatoxins and fumonisins in corn. Additionally, soil properties, when combined with meteorological data and historical aflatoxin content, have been used in gradient boosting machine models to distinguish aflatoxin-contaminated corn [39].

4.1.3. Crop Phenology and Cultivar-Specific Data

Another important aspect of spatiotemporal data is the inclusion of specific cultivars. Different crop varieties can exhibit varying levels of susceptibility to fungal colonisation and mycotoxin contamination [37,40]. Including data on specific cultivars in ML models can help tailor predictions and interventions to the particular characteristics of each crop variety. Certain wheat varieties may be more resistant to Fusarium head blight, while others might be more prone to infection. By incorporating cultivar-specific data, ML models can provide more accurate risk assessments and suggest more effective mitigation strategies [41]. This approach enhances the precision of mycotoxin contamination forecasts and supports targeted agricultural practices, such as selecting the most resistant varieties for planting in high-risk areas. Additionally, integrating crop phenology data, such as growth stages and development timelines, can improve the temporal accuracy of predictions [42].

4.1.4. Spectral Data

Spectral data are one of the most common types used in mycotoxin detection, valued for their non-invasive nature. This data type involves capturing the reflectance or absorbance of light at various wavelengths from the material being analysed. Spectral data can be further categorised into multispectral and hyperspectral data, each offering different levels of detail and information.

Multispectral Imaging: This imaging technique captures data at a few specific wavelength bands, making it effective for distinguishing between different materials based on their spectral signatures. Unlike hyperspectral imaging, which captures continuous spectral information across a wide range of wavelengths, multispectral imaging focuses on discrete bands, making data collection and processing less complex while still providing valuable information for specific applications. For instance, multispectral images can be captured in controlled greenhouse environments, where conditions such as temperature, humidity, and lighting are regulated to optimise data quality. This controlled setting allows for consistent and repeatable measurements, crucial for precise analysis. An example of this application is a study [43] that used hyperspectral data to detect Fusarium head blight in wheat under greenhouse conditions, demonstrating the potential of spectral imaging in plant pathology. Moreover, multispectral imaging can be integrated with advanced computational techniques for enhanced analysis. In another study, Ref. [44] used ML combined with multispectral imaging and image processing techniques to detect aflatoxin contamination in figs.

Hyperspectral Imaging: Hyperspectral imaging is a technique that captures data across a continuous spectrum of wavelengths, providing significantly more detailed information compared with multispectral imaging. This method is particularly valuable for the precise identification of toxigenic fungal contaminants and mycotoxins [45]. Hyperspectral images can be acquired using various platforms, including ground-based systems and unmanned aerial vehicles (UAVs). UAV hyperspectral imagery showed to effectively monitor Fusarium head blight in wheat fields, highlighting its potential for large-scale agricultural monitoring [46]. In another study, Ref. [47] used a visible and near-infrared hyperspectral imaging system operating in the range of 400–900 nm under ultraviolet excitation. They successfully differentiated spectral characteristics between corn kernels inoculated with aflatoxigenic A. flavus strains and naturally infected kernels from the same field. Furthermore, Ref. [48] explored the combination of fluorescence and reflectance visible and near-infrared hyperspectral images for detecting aflatoxin contamination in inoculated corn kernels in the field.

Ground vs. Intact Material: The context in which spectral data are collected can also vary. In some cases, imaging occurs on ground material, where samples are collected and analysed in a laboratory setting. This approach allows for controlled conditions and high-resolution data. In other instances, imaging is performed on intact material, such as whole peanut grains [49], to assess contamination directly in the field or during processing.

4.1.5. Limitations in Image Analysis

While image analysis using spectral data is a powerful tool for detecting mycotoxins, there are notable limitations and challenges. One significant factor is that visual features of an image, such as plant damage or fungal presence, may not always directly correlate with the presence of specific mycotoxins [50]. This is particularly relevant when different species of fungi, capable of producing various mycotoxins, are involved [51]. For example, certain fungi can cause visible damage or contamination on crops, which may be detected by ML models. However, these visual features might not indicate the presence of the specific mycotoxin of interest [50]. As a result, models focusing on plant damage or fungal contamination might not accurately reflect the levels of regulated mycotoxins. This discrepancy underscores the importance of integrating spectral imaging features that are more closely associated with the specific mycotoxins being regulated. Addressing this challenge requires combining image analysis with other data types, such as chemical analysis or molecular techniques, to improve the specificity and accuracy of mycotoxin detection. By doing so, ML models can better distinguish between general fungal contamination and the presence of specific harmful mycotoxins.

4.2. Neural Networks

Neural networks (NNs), first introduced by [52], are a class of machine learning algorithms modelled loosely after the human brain [53]. They are designed to identify patterns and make predictions by learning from data and can be used for supervised or unsupervised problems. NNs are made up of interconnected nodes and edges, where the nodes represent the *neurons* and the edges are the links between the neurons. The nodes are organised into layers, where the first layer is called the input layer, the last layer is the output layer, and all intermediate layers are called hidden layers. Typically, in an NN, the data are fed to the input layer; then one or more hidden layers perform computations and learn from the data, and finally, predictions (or classifications) are provided by the output layer. A simple diagram of an NN can be seen in Figure 4.

Every neuron in a hidden layer applies a weighted sum of the inputs to transform the data. This is followed by a function, referred to as an *activation* function [53]. The network fine-tunes the weights associated with each neuron by employing optimisation algorithms throughout the training phase. There are numerous hyperparameters associated with NNs. Some of the main hyperparameters include (i) the learning rate, which determines how much the weights are changed at each iteration; (ii) the number of epochs, which refers to how many times the entire training dataset is passed forward and backward through the neural network; (iii) the batch size, which controls the number of training examples used in one iteration; and (iv) activation functions like ReLU (Rectified Linear Unit), sigmoid, and tanh, which determine the output value of a node given an input or a set of inputs. After training, the NN is capable of generating predictions for new, unseen data by passing the input across the layers to produce an output.

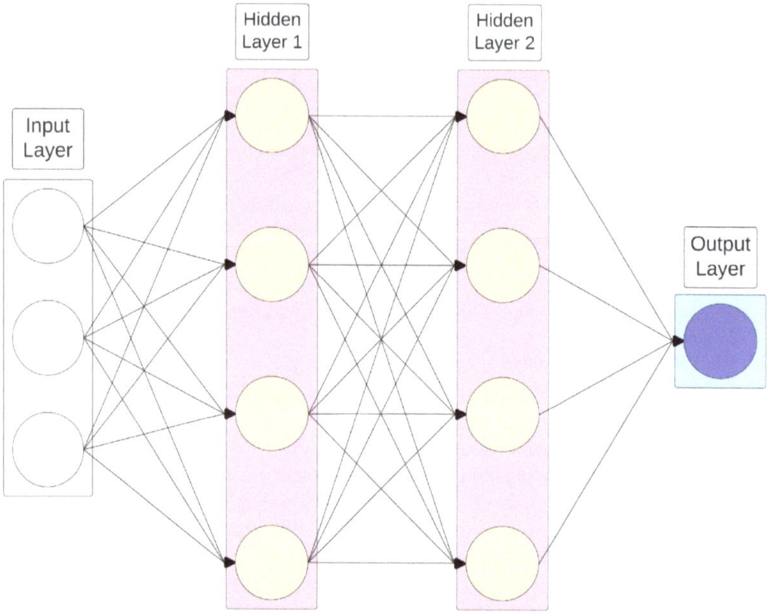

Figure 4. Basic neural network structure, showing an input layer, two hidden layers, and an output layer, where each circle represents a neuron, and these neurons are interconnected by lines symbolising neural connections. The input layer receives the initial data, which are then processed through successive hidden layers using weights and activation functions, refining the information before it reaches the output layer.

Like all machine learning models, NNs come with their own set of advantages and disadvantages. For example, NNs excel at identifying and modelling non-linear interactions present in data, which are common in biological processes. They are also flexible and can handle a wide range of data types, such as numerical and categorical, text, and image data. Despite their advantages, neural networks also have limitations. One of the major limitations is interpretability. NNs are considered *black-box* algorithms, meaning that it is difficult to understand why specific predictions are being made [54]. Second, like many of the other ML approaches we cover, they are not probabilistic models, making it hard to accurately quantify the uncertainty in the predictions. Overfitting can also be an issue for NNs. Without appropriate regularisation, NNs can become too complex, capturing the noise in the training data instead of generalising to the underlying pattern [55]. Finally, training large NNs requires a significant amount of computing power. The computational cost of NNs will increase with the complexity of the model [56]. In the following subsections, we review the use of NNs on different types of mycotoxin data.

4.2.1. NNs Applied to Spatiotemporal Data

NNs have been widely applied to spatiotemporal data, despite them not forming part of the traditional suite of spatiotemporal analytics techniques. In the field of mycotoxin study, NNs have been used for a variety of tasks and data types. For example, Ref. [38] used data from several sites in Northern Italy over the years 2005 to 2018. Their goal was to predict the presence of mycotoxins (specifically, aflatoxin and fumonisins) using NNs in corn. In their work, they trained two NNs to predict if the contamination levels were above legal thresholds at the time of harvest. Both models performed well, achieving an accuracy of greater than 75% on the test data. However, they recommend, for future research, that improvements can be made to the modelling by taking into account the co-occurrence of aflatoxin and fumonisins in corn and their complex interaction, which may be due to the effects of climate change.

Ref. [57] applied NNs to analyse the concentration of mycotoxins in winter wheat grain. They examined 23 winter wheat genotypes with different Fusarium resistances from three different sites in Poland during the years 2011 to 2013. They developed three NN models; however, only two of these are concerned with the detection of mycotoxins, that is, the DONANN model, which is used to detect DON, and the NIVANN model, which examines the nivalenol content. The DONANN and NIVANN models were designed using an automatic network designer using Statistica v7.1 software [58], and were evaluated among a set of 10,000 generated networks. The performance of these models was assessed on several statistical metrics, but the primary focus was on the correlation coefficient (which, in this case, would be the correlation between the predicted values from the model and the actual observed values) and the mean absolute error (MAE), which is the absolute differences between the predicted values and the actual values. For the best-performing DONANN model, a low MAE of 0.37 was reported; however, the correlation coefficient was exceptionally high at 0.99, indicating an almost perfect linear relationship between the predicted and actual values. The best-performing NIVANN model, while exhibiting a slightly lower correlation coefficient of 0.81 and an MAE of 0.02, still performed within acceptable ranges. The architecture of the created models was designed as a multi-layer perceptron (MLP) type of NN, with two hidden layers. Despite reporting training, validation, and test errors, the authors did not specify the dataset on which the correlation and MAE metrics were based.

In a novel application of NNs, Ref. [59] used a transformer-based deep learning method, called *GPTransformer*. A transformer-based deep learning algorithm refers to a type of NN architecture that relies on a mechanism called *attention* to boost the performance of the model [60]. In their work, the authors proposed a transformer-based genomic prediction model for predicting Fusarium head blight disease levels and associated DON concentration in barley data collected in three locations in Canada over the years 2014 to 2015. One of their goals was to compare the accuracy of the GPTransformer model to existing genomic prediction methods such as decision tree algorithms (DT), linear regression (LReg), and traditional statistical algorithms like best linear unbiased prediction (BLUP). The authors used the Pearson correlation coefficient (PCC) as a measure of performance, which calculates the linear relation between the true output and the predicted output. They showed that the GPTransformer model (and all of the used ML models) did not significantly outperform the statistical method of BLUP in terms of predictive accuracy. However, GPTransformer did perform better than both the DT and LReg methods. The authors note that the ML methods used are able to capture non-additive genetic elements, and as such, the predictions provided might include some of these interactions in their estimations.

4.2.2. NNs Applied to Spectral Data

Hyperspectral (or just spectral) data refer to the capture and processing of information from across the electromagnetic spectrum [61]. Refs. [43,62,63] applied NN classification algorithms to pixels of hyperspectral image data to examine wheat for Fusarium head blight infection. Each author used a convolutional NN (CNN), which captures spatial patterns or motifs by identifying and calculating weights from the images according to how often the motif appears.

In Ref. [43], the authors investigated four distinct methods for converting hyperspectral imaging data. They then evaluated the performance of eight different CNN models in classifying pixels as either healthy or infected with Fusarium head blight. The effectivenesses of these models were compared based on their classification accuracy. They found that a particular type of CNN called *DarkNet 19* [64] performed the best, with an accuracy of close to 100% across all data conversion methods, on both the validation and test data. For Ref. [63], tests showed that the CNN model is effective in detecting images that contain the blight and achieved an R^2 value of 0.80, and the mean average accuracy for the testing dataset was 92%. In Ref. [62], the authors compared the accuracies of the different NNs to determine which is the best at identifying diseased regions of the wheat kernel. They showed that a two-dimensional convolutional bidirectional gated recurrent unit NN performed the best, with an accuracy of 84.6% on the validation dataset and an F1 score and accuracy of 0.75 and 74.3%, respectively, on the test data.

Ref. [49] used a combination of hyperspectral data and NNs to detect aflatoxin in peanuts. They showed the CNN's efficacy in classifying infected peanuts and achieved a test set accuracy of 95%. They later expanded their work and used a one-dimensional CNN (1D-CNN) to classify aflatoxin infection in corn and peanuts. This time, they achieved accuracies of 96.4% for peanuts and 92.1% for corn [65].

In a research conducted by [66], infrared (IR) spectroscopy and ML algorithms were used to detect fungal contamination in corn. In their study, 183 naturally infected samples (contaminated with different *Fusarium* DON species and at different concentrations) were obtained from the seed production Linz of Austria (SBL) and from the Cereal Research Centre of Hungary (CRC). The authors assessed several classification ML models, including multi-layer perceptron (MLP) neural networks, random forests, support vector machines, and adaptive boosting, for their accuracy in correctly classifying contaminated from non-contaminated samples. Their results showed that the MLP approach correctly classified 94% of the non-contaminated samples and 91% of the contaminated samples. The authors note that while this approach yields promising results, these findings are specific to a contamination threshold of 1250 mg/kg, which is the EU regulatory limit, and that subsequent research will aim to evaluate the performance of the classification methods across various contamination levels.

4.2.3. NNs with an Electronic Nose

An electronic nose (e-nose) is a device intended to detect chemical compounds in gasses. E-noses have been extensively used in the detection of aflatoxins [67,68], fumonisins [69], and DON [70] in corn. However, Ref. [71] used an e-nose supported by NNs for the detection of aflatoxin and fumonisins in corn. In their work, they compared three different approaches, that is, NN, logistic regression (LR), and discriminant analysis (DA), to examine the e-nose's ability to discriminate between samples contaminated with concentrations either exceeding or falling below legal thresholds on data spanning 5 years. They showed that all methodologies achieve an accuracy of above 70%, with the NN performing the best with an accuracy of 78% for aflatoxin detection and 77% for fumonisin detection. They went on to suggest that the e-nose, when supported by an NN, can provide a fast screening tool for classifying samples.

4.2.4. NN Summary

Neural Networks have been widely adopted as the ML algorithm of choice for analysing mycotoxin data, especially in the field of hyperspectral imaging. However, as of yet, there seems to be a gap between research applications and the wider use in industry. The application of NNs in hyperspectral data for mycotoxin detection (and food safety in general) is a relatively new process, and the implementation of an NN approach to hyperspectral data in industrial quality control faces various challenges, mainly due to hardware limitations, such as the cost of operating imaging equipment [72]. However, in research, NNs for use in hyperspectral imaging have seen an increase in popularity with many of the reviewed works being widely cited, for example, Refs. [62,63].

4.3. Random Forests

A random forest (RF) [73] is an ensemble learning method used for classification and regression. The RF algorithm creates a *forest* of decision trees, where each tree in the forest is built from a sample drawn with replacement (that is, a bootstrap sample) from the training set and selects splits from a random subset of features.

While Section 4.6.1 provides a comprehensive examination of decision trees, this section offers a concise introduction to familiarise readers with the basic concepts and terminologies associated with decision trees. Figure 5 shows an example of a single decision tree. In constructing each decision tree, the root node is the starting point, and it represents the entire dataset, which gets split based on a feature that provides the best separation according to a certain criterion [like Gini impurity [74]]. The decision nodes are the points where the data are split further. Each decision node represents a decision rule on a specific feature. The process continues recursively until a stopping criterion is met, such as reaching the tree's maximum depth, attaining a minimum sample count in a leaf, or achieving adequate purity within the leaf nodes. The leaf/terminal nodes represent the final output of the decision process. Each branch/sub-tree represents a possible outcome of the decision made at the decision node, leading to further sub-trees or leaf nodes.

For RF classification tasks, each tree in the forest votes for a class, and the class receiving the majority of votes becomes the model's prediction. For regression tasks, the forest takes the average of the outputs by individual trees. Figure 6 shows a summary of the RF algorithm.

One of the main advantages of using RFs is their versatility. They are capable of performing both regression and classification tasks, as well as handling large datasets. Additionally, they require very little tuning and can perform well without much hyper-parameter optimisation. Some of the main hyperparameters associated with RF include the following: (i) Number of trees: this is the number of trees in the forest. Generally, more trees increase performance but also increase the computational cost. (ii) Maximum depth of trees: the maximum depth of each tree. Deeper trees can model more complex patterns but might lead to overfitting. (iii) Minimum samples split: the smallest number of samples needed to split an internal node. Setting higher values helps prevent the model from learning overly specific patterns, which can lead to overfitting. As with NNs, RFs are a black-box algorithm, and so interpretability can be an issue. Each decision tree upon which the RF is built can be easy to interpret, but since RFs consist of a large number of decision trees averaged together, the decision process by which a prediction is made can be somewhat opaque.

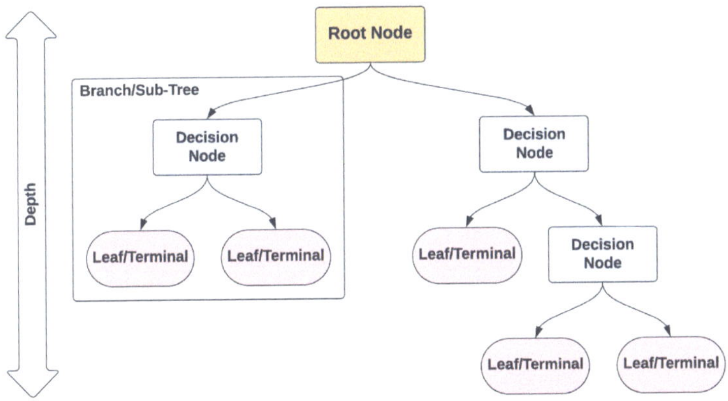

Figure 5. Decision tree process demonstrating the structure of a decision tree, including the root node, branching to decision nodes, and culminating in leaf/terminal nodes. The depth of the tree is indicated, showing the levels of decision making from the root to the leaves.

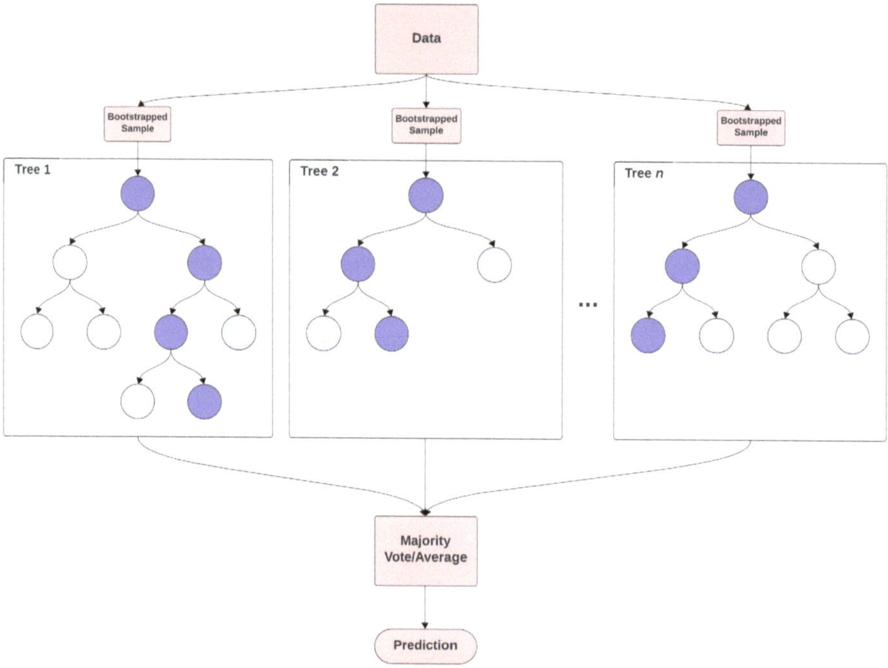

Figure 6. The random forest algorithm constructs an ensemble of decision trees, with each tree built from a unique bootstrapped sample of the original dataset. Nodes are colored light blue to represent the regular decision nodes of the trees. Distinct paths through each tree are shown, highlighted by the darker blue nodes, and represent a sequence of decisions made from the root to a leaf node based on the input features. The final prediction of the random forest is determined by aggregating the predictions of all trees, using majority voting for classification tasks or mean prediction for regression tasks.

4.3.1. RFs for Spectral Data

As with NNs, RFs have been applied to hyperspectral data. For example, Ref. [75] used a RF classification model to classify corn silage for high or low mycotoxin contamination using near-infrared spectroscopy (NIR). In their study, 155 samples were collected from several sites in the Po Valley (Italy) and from Sardinia over the years 2017 to 2019. Their aim was to develop qualitative models capable of distinguishing corn silage based on either the total concentrations or the total counts of various groups of mycotoxins (in this case, *Fusarium* and *Penicillium* toxins). To evaluate various classification strategies, different distinct threshold levels were established for each mycotoxin contamination. These thresholds were used to categorise each sample as having either a high or low contamination level in relation to these specified values. To predict the contamination level, an RF classification model was fitted, using the wavelength of light as the predictors, and achieved an out-of-sample accuracy of above 90% for the classification of both Fusarium and penicillium toxins.

In a 2023 study, Ref. [76] utilised NIR spectroscopy for detecting DON in oat samples from Spain and Sweden collected over the years 2021–2022. The authors applied two different transformation techniques to the spectral data and examined which allowed for greater classification of the data using four different ML algorithms (k-nearest neighbours, naïve Bayes, NN, and RF). Both preprocessing transformation methods achieved similar results for all ML methods, with RFs performing the best with an accuracy of 77.8% and an area under the curve (AUC) of around 0.77. However, they noted that other similar studies have been conducted that achieved a higher classification accuracy, such as [77].

In a similar study, Ref. [78] constructed a biosensor array for identifying mycotoxins in peanuts and corn, produced by Aspergillus flavus, using six ML models, including partial least square determination analysis (sPLS-DA), linear support vector machine (svmLinear), radial support vector machine (svmRadial), RF, NN, and high-dimensional discriminant analysis (HDDA). The authors used the classification models for three separate purposes: to distinguish healthy from infected samples, to distinguish the pre-mould status in infected samples, and to distinguish between infected peanuts or corn samples. To distinguish the pre-mould status, the aim was to create a three-class model to predict either the control or 1 or 2 days after inoculation. Their approach achieved a reported 100% accuracy in distinguishing healthy from infected samples and RF accuracies of 95% and 98% in identifying pre-mould status in peanuts and corn, respectively. However, such high levels of accuracy warrant further investigation, as such high accuracy rates can often be indicative of issues in the experimental design, such as the creation of non-representative test sets or overfitting, especially if the test sets are not properly randomised.

4.3.2. RFs for Mycotoxin Treatment

ML models in mycotoxin treatment can be used to predict mycotoxin contamination risk and optimise mitigation strategies. This application can boost accuracy in prediction and effectiveness in deploying targeted anti-fungal treatments. In a study conducted by [79], the authors employed machine learning techniques to predict the growth of *Fusarium culmorum* and *Fusarium proliferatum*, as well as their production of mycotoxins, in environments where ethylene vinyl alcohol copolymer films are used. These films contain pure components of essential oils, which are used to inhibit the growth of the fungi and their mycotoxin production. In their work, they studied fungal growth on corn in vitro and modelled the fungal growth and toxin production under different environmental scenarios and with different treatments applied. The ML models used were NNs, RF, extreme gradient boosted trees (XGB), and multiple linear regression (MLR). The performance of the ML methods was assessed using the root mean square error (RMSE). It was found that RF performed the best in predicting the growth rates of *Fusarium culmorum* and *Fusarium proliferatum* and mycotoxin production, having consistently the lowest RMSE value.

Ref. [80] evaluated the anti-fungal properties of specific lactic acid bacteria strains against *Fusarium* species found in cereals. To achieve this, various machine learning algorithms, including NN, RF, XGB, and MLR, were employed to predict the extent of fungal growth inhibition resulting from the application of the tested lactic acid bacteria strains. As with the previous study, the RMSE was the metric used to assess the performance of the model, in conjunction with the R^2 value. In this work, both RF and XGB showed comparable performances, reporting similar RMSE (0.0604 and 0.0581, respectively) and R^2 values (0.992 and 0.992, respectively) on the test data, in predicting the percentage of growth inhibition.

Several other studies exist on the topic of using ML models (and specifically RF) to predict mycotoxin growth in the presence of treatments. In the interest of brevity and space, we name them here but do not provide additional details of the studies. In each of these studies, the authors used multiple ML models, with a general consensus that RF models performed the best at their given tasks. See Refs. [81–83] for more details.

4.3.3. Random Forest Summary

RFs have emerged as a robust and versatile tool in the field of mycotoxin detection and treatment and have gained popularity due to their ease of use, computational speed, and predictive performance. These studies collectively underline the significant potential of RF in enhancing food safety measures, although it is crucial to acknowledge the necessity for rigorous validation and testing to ensure the reliability of these models.

4.4. Gradient Boosting

Gradient boosting (GB) [84] builds on the concept of boosting, where weak learners are converted into strong ones through an iterative process. The GB framework builds boosted regression models by sequentially training a weak classifier (such as a linear regression or simple decision tree) successively on the data using the residuals from previous model fits (as shown in Figure 7). This process ensures that each new weak classifier addresses the inaccuracies of its predecessors, thereby enhancing the prediction accuracy. The final model aggregates the outputs from all these weak classifiers to form a robust, 'strong' classifier through an ensemble approach. The term *gradient* in gradient boosting refers to the method's use of gradient descent, a numerical optimisation algorithm, to minimise the loss or the difference between the actual and predicted values.

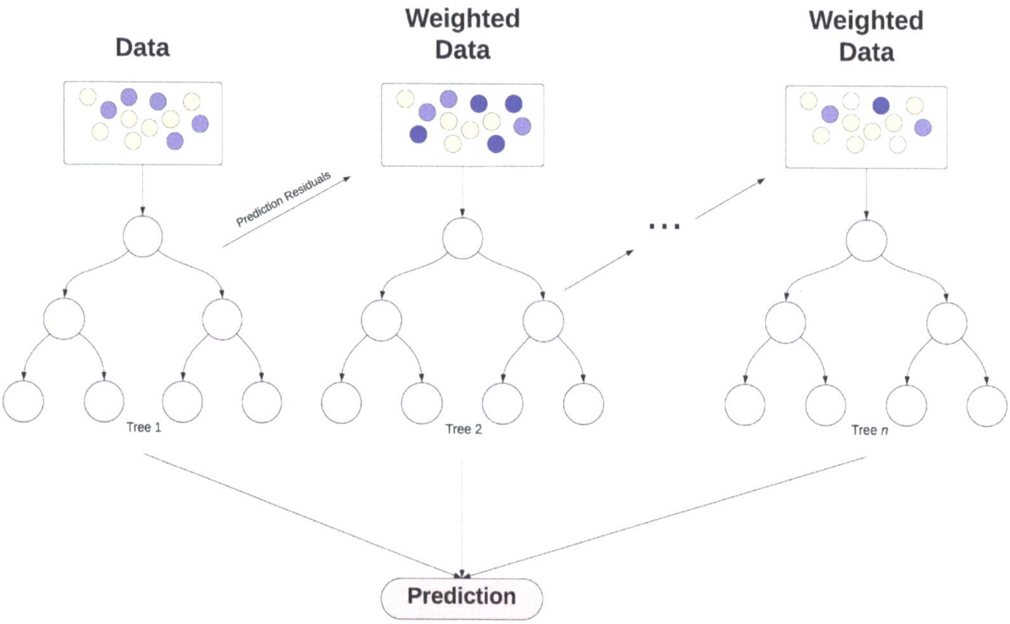

Figure 7. Gradient boosting process. Here, the weak learners are trees that are trained sequentially on weighted data with iteratively adjusted weights based on previous prediction errors. The light yellow circles represent data points with lower residuals (errors), the light blue circles represent data points with moderate residuals, and the dark blue circles represent data points with higher residuals from previous model predictions. The pink circles within the trees indicate the decision nodes of each weak learner. The final prediction is made by aggregating the outputs from all weak learners.

In gradient boosting, when the weak learners are decision trees, each tree is grown in a greedy manner, but unlike random forests, trees are grown sequentially. After the first tree is built and predictions are made, the errors (residuals) from those predictions are used to build the next tree. The subsequent tree aims to predict the residuals from the previous tree. This process is continued, with each new tree correcting the residuals of the ensemble of all previous trees. The final prediction is made by summing the predictions from all trees, which can be thought of as a weighted vote where trees that reduce the error the most have more influence.

An advantage of GB models is their strong predictive capability and adaptability, especially in dealing with complex non-linear relationships between independent variables and the dependent variable. They adapt to various prediction problems by supporting different loss functions, making them suitable for both regression and classification tasks.

However, these models have their challenges. Without careful tuning and regularisation, there is a risk of overfitting, a problem exacerbated by noisy data [85]. Additionally, their sequential boosting process is computationally intensive and time-consuming compared with methods like random forests that build trees in parallel. This complexity can be a significant drawback in scenarios where computational resources or time are limited. Some of the main hyperparameters associated with GB are as follows: (i) Number of weak learners: this defines the number of boosting stages or learners to be created. More learners can lead to a more powerful model, but also increase the risk of overfitting and raise computational cost. (ii) Learning rate: this parameter scales the contribution of each learner. A smaller learning rate requires more weak learners but can yield a more generalised model. In the case of the weak learner being trees, (iii) the maximum depth of trees determines the maximum depth of each individual tree. Deeper trees can model more complex patterns but can also lead to overfitting. An extension of a GBM model is called eXtreme Gradient Boosting (XGB) [86], with the key difference between the two being performance. In general, XGB models are faster and have better optimisation. Additionally, XGB models have the ability to deal with missing values.

4.4.1. GB for Spatiotemporal Data

In a study by [87], the authors designed a program for aflatoxin monitoring in feed products (peanuts and soy beans), while considering both the performance of the model and the cost of monitoring. In the study, they applied four different ML algorithms (namely, GB, LR, SVM, and DT) to historical data concerning monitoring for the presence of aflatoxins in feed products. The data were collected from several sites around the world, including China, Brazil, and Argentina, over the years 2005 to 2018. The ML algorithms were compared to predict which feed batches are high risk and which should be considered for further aflatoxin analysis. In their work, they found that all the ML models performed well and used several error metrics to assess their models. They obtained an accuracy of over 90% for all models and an AUC and recall of over 0.8 and 0.6, respectively. However, the XGB model performed better than all other models, and the authors proposed a reduction to the monitoring cost of up to 96% for the years 2016 to 2018.

In Ref. [88], the authors proposed to use un-targeted metabolomics and ML techniques to mine biomarkers of the species *Aspergillus* on peanut data collected from several sites in China over the years 2013 to 2018. They initially used an RF model to determine *Aspergillus* species with 97.8% accuracy. They then went on to use XGB to create a decision rule to help regulators in evaluating risk prioritisation with a claimed accuracy of 87.2%. However, the authors noted that they built the XGB model using only a single tree and used this tree to create an operable decision workflow for risk assessment. Although using a single tree can reduce complexity, it also increases the likelihood of less robust predictions. Part of the strength of XGB (and GBM) models is that they iteratively correct the mistakes of previous trees, a process that is lost if only a single tree is used.

Ref. [39] conducted a study with the objective of evaluating the performance of GBM models to predict the presence of aflatoxins in corn at two risk thresholds, that is, 20 ppb and 5 ppb. These cut-off values were chosen based on the U.S. Food and Drug Administration's (FDA) action level for corn (20 ppb) [89], whereas the lower cut off is based on the European standard of 5 ppb [90]. Additionally, the authors performed feature engineering, which is the process of transforming raw data into meaningful and informative features with the intention of enhancing the performance of ML algorithms [91]. The data used were historical climate, soil, and aflatoxin data, collected in several sites in Iowa in the years 2010, 2011, 2012, and 2021. As the data had many missing values, the authors used an imputation method; however, they noted that data from the months of January, February, and December had to be excluded from the model as there were too many missing values to accurately impute the data. The authors reported that the GBM model performed well, achieving high accuracy rates of 96.8% for the 20 ppb threshold and 90.3% for the 5 ppb threshold. The study highlighted the significant influence of the vegetative index (which is a quantitative measure that uses satellite imagery to assess the amount and health of plant life in a specific area) in August on aflatoxins risk for both thresholds, indicating the critical environmental and ecological impact of drought conditions during this month. Additionally, predictors related to soil properties (such as hydraulic conductivity, pH, and bulk density) were found to potentially affect aflatoxin contamination levels before harvest.

4.4.2. GB for Spectral Data

Ref. [92] conducted a study on aflatoxin and fumonisin contamination in a single kernel corn. They argued that bulk sampling of the corn may not produce accurate results, and thus focus solely on single kernels. In their study, they performed measurements to show the skewness of the data and calculated weighted sums of toxin contamination. Additionally, they aimed to improve single kernel classification performance through the use of different ML applications. Their methodology was to take corn kernels that were already contaminated and scan them using the NIR technique. The samples were then ground and measured for both toxins using the ELISA method (discussed in Section 1). In their work, they used five different ML models to classify both mycotoxins. They are GBM, RF, least absolute shrinkage and selection operator (LASSO), elastic-net regularised generalised linear models (GLMNETs), and support vector machines (SVMs). They additionally applied ML algorithms for classifying each individual mycotoxin. For aflatoxin, they used bagged AdaBoost, linear discriminant analysis (LDA), and penalised logistic regression (PLR). For fumonisin classification, GBM and penalised discriminant analysis (PDA) were used. For aflatoxin, they found that GBM was the best-performing model, with an accuracy of 83%, on both the training and the test data. For fumonisin, the PDA model performed the best with an accuracy of 86% on the test data. However, the authors noted that, for future studies, opportunities for better classification exist, including increasing the proportion of samples so the algorithm can learn the characteristics of contaminated corn kernels better.

4.4.3. Gradient Boosting Summary

The application of GBM models across various datasets, from spatiotemporal to spectral data, demonstrate their versatility and potential in predicting mycotoxin contamination in agricultural products. While GBM models generally exhibit high accuracy, there are criticisms concerning the robustness of these models when applied with limited trees, as in the case of [88], or when handling datasets with substantial missing values, as noted by [39]. The high accuracy rates reported should be examined for potential overfitting or lack of generalisation to broader datasets. The approach of ref. [92] to single kernel analysis opens avenues for improved precision in toxin detection, but also indicates the need for larger sample sizes to enhance model learning.

4.5. Support Vector Machines

Support vector machines (SVMs) [93] are a set of supervised learning methods used for classification, regression, and outlier detection. To make predictions, SVMs identify the optimal hyperplane that maximises the margin between the two classes (where the margin is defined as the distance between the nearest data points of each class and the dividing hyperplane). The data points that are closest to the hyperplane and that influence its position and orientation are known as support vectors, as they *support* or define the hyperplane. Figure 8 illustrates an SVM in action. One of the key advantages of SVMs is their versatility as they can be used on a variety of data types, and are particularly useful for image recognition [94]. Additionally, they are memory efficient since they only use a subset of training points, called support vectors, in the decision function. However, SVMs require careful tuning of the hyperparameters and an appropriate kernel choice. A kernel is a function used to transform data into a higher-dimensional space. By projecting the data into a higher dimension, a kernel makes it possible to find a hyperplane that can effectively separate the classes. Some of the common kernels include [95]:

1. Linear: No non-linear transformation, suitable for linearly separable data.
2. Polynomial: Suitable for non-linearly separable data, involves higher degree terms of the features.
3. Radial basis function: Good for non-linear data, uses a Gaussian distribution.
4. Sigmoid: Similar to the sigmoid function in logistic regression.

Additional hyperparameters include the following: (i) Gamma: This is needed for all kernels except linear. It determines the extent of the influence that a single training example has. Low values indicate a wide reach, and high values indicate a close reach. A high gamma value can cause the model to overfit. (ii) Degree: This is only relevant for a polynomial kernel. It defines the degree of the polynomial used in the kernel. A higher degree can model more complex relationships but increases the risk of overfitting. (iii) Coef0: This is a parameter for polynomial and sigmoid kernels that adjusts the independent term in the kernel function. It is often called the kernel bias.

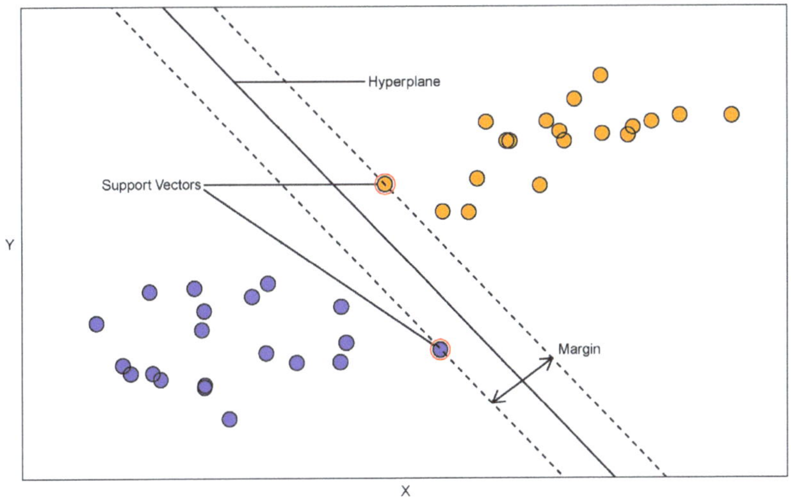

Figure 8. Support vector machine process. The diagram illustrates the SVM's method of finding the optimal hyperplane that maximises the margin between two classes, depicted by the blue and orange points. The support vectors, which are the data points closest to the decision boundary, define the margin.

4.5.1. SVMs for Spectral Data

In the review of the literature concerning the use of SVMs in mycotoxin detection, it was found that they were overwhelmingly used for image recognition and, as such, primarily used spectral data. For example, ref. [45] used several ML models (SVM, NN, and LR) for the classification of Fusarium head blight in wheat, using spectral data. The data were collected in the years 2020 to 2021 at a single site in Belgium, with the experiment using eight varieties of wheat. They found that the SVM model outperformed both the NN and LR method in classifying contaminated wheat in every variety, with a classification accuracy of 96.5% on the test data (with NN and LR achieving accuracies of 82.9% and 82.5%, respectively).

In a similar study, Ref. [96] used three different imaging methods alongside ML classification models to test ground corn samples for the presence of aflatoxin and fumonisin, both as individual contaminants and in combination. Two classification models were used, partial least squares-discriminant analysis (PLS-DA) and SVM, using specific threshold values for each mycotoxin. The naturally contaminated corn samples were obtained from the Office of Texas State Chemist, which in turn collected the samples from different feed companies located around Texas. They found that the SVM performed better than the PLS-DA with classification accuracies of 89.1%, 71.7%, and 95.7% for each imaging technique. The imaging method with the highest accuracy was the short-wave infrared (SWIR) method.

In a study concerning the detection of *Aspergillus parasiticus* in corn kernels using NIR hyperspectral imaging, conducted by [97], the authors used SVMs to compare the performances of multiple different preprocessing and imaging techniques. For their study, corn kernels were harvested from Hefei City, Anhui Province, China, in 2015. Each day (for a period 7 days), 36 sterilised corn kernels were inoculated with *Aspergillus parasiticus* and were grouped into four groups depending on the day of inoculation. From this, an SVM was used to determine which groups were infected using different preprocessing techniques. Additionally, this study examined the orientation of the kernel in the image to determine if this property had an effect on predictive performance. They found that the best preprocessing method was a combination of the standard normal variate (SNV) and moving average smoothing (MAS) methods, with an accuracy of 91.67% for detecting contaminated kernels using the validation data. They also found that the performance of the classified models was influenced by orientation; however, the models built using data from a mix of kernels with their germs facing both up and down still achieved an accuracy of 84.38% on the validation data.

4.5.2. Support Vector Machine Summary

In the reviewed work, SVMs demonstrated considerable accuracy in mycotoxin detection through spectral data analysis. However, as with other ML methods reviewed, the consistently high classification accuracy reported raises questions about potential overfitting and the representativeness of the datasets used. Moreover, factors such as kernel orientation (which refers to the way in which the kernel function transforms the input data into a higher-dimensional space to find an optimal boundary between classes) significantly influenced SVM performance, indicating that model robustness may be context dependent. The choice of kernel and its parameters, like orientation, scale, and type, is critical in shaping the decision surface and, thus, the SVM's ability to generalise from training to unseen data.

4.6. Other ML Methods

In this section, we cover the remaining ML methods. These include decision trees and Bayesian networks and have been grouped together as they make up a minority of the reviewed work. As such, they are not separated by the type of data used, and all data types are discussed together.

4.6.1. Decision Trees

Decision tree (DT) learning is a type of non-parametric supervised learning algorithm used for both classification and regression tasks [74,98]. A DT is a flowchart-like structure, resembling a tree structure with branches representing decision paths and leaves (or terminal nodes) representing predicted outcomes (see Figure 5 in Section 4.3). A DT splits the data into subsets based on the value of input features. Splits are chosen to maximise the separation of the classes based on measures like Gini impurity or information gain [74]. This process continues recursively until a stopping criterion is met, resulting in a tree where each path represents a decision pathway that leads to a predicted outcome. The advantages of decision trees include their simplicity, interpretability, and ability to handle both numerical and categorical data. However, DTs have a tendency to overfit, especially when a tree is particularly deep [74]. This can be mitigated by pruning the tree or setting a maximum depth of the tree via the use of hyperparameters. As this method is a tree-based approach, there is an overlap with RF and GB in terms of hyperparameters. Some of these include maximum depth, minimum samples split, and minimum samples leaf (i.e., the minimum number of samples needed to be at a leaf node. Setting this parameter can ensure that each leaf node represents a reasonable number of samples, which can smooth the model, particularly for regression tasks, and prevent overfitting).

The use of DTs in the field of mycotoxin detection is quite varied. For example, in a study conducted by [99], in which they assessed the use of an electronic nose to identify DON contamination of wheat samples, an extension of decision trees called Classification and Regression Trees (CART) [74] was used to classify samples based on four thresholds of DON contamination (1750, 1250, 750, and 500 µg/kg). For this study, 214 wheat samples were collected from Northern Italy during the years 2014–2015 and 2017–2018. For the threshold values of ≥1250 µg/kg, the accuracy of sample classification was the highest, ranging between 88% and 92%. The lower thresholds of ≤750 µg/kg were found to be the least accurate, with an accuracy of <83%. The authors proposed that the reduced sensitivity of the instrument at lower DON concentrations might explain this drop in accuracy.

Ref. [99] examined the classification of DON mycotoxin-contaminated corn and peanuts at regulatory limits using spectral data. The spectral data were analysed using a bootstrap-aggregated (bagged) DT approach, focusing on the protein and carbohydrate absorption bands of the spectrum. The corn samples were obtained by Saatbau Linz (Linz, Austria) and the Cereal Research Centre (Szeged, Hungary). For the peanuts, 92 different infected samples were purchased from public markets in Tanzania, Mozambique, and Burkina Faso. The authors demonstrated that the DT method could classify corn samples at the 1750 and 500 µg/kg thresholds for DON with accuracies of 79% and 85%, respectively. Additionally, it was able to classify peanut samples for aflatoxin at 8 µg/kg with a 77% accuracy.

In a study related to identifying and predicting risks related to the presence of fumonisins in breakfast cereal products, Ref. [100] developed a model specifically designed to predict the risk of fumonisin contamination, with a particular emphasis on a mixture of ingredients. In their research, fifty-eight distinct breakfast products were purchased from local grocery stores in Florence, Italy, during 2019. The selection criteria for purchasing breakfast products included (i) products with packaging sizes ranging from 200 to 500 g, including both plastic and non-plastic materials; (ii) items sourced from retail shops; and (iii) products primarily made of wheat, corn, dry fruits, rice, and oats. Principal component analysis (PCA) and k-means clustering were employed to explore the connection between cereal ingredients, their composition and packaging, and the concentration of fumonisins. The findings suggested that the fumonisin concentration might be linked to complex non-linear interactions among various factor variables. To explore this potential and identify the factors most closely linked with high concentrations, DTs were employed. Two decision trees (DTs) were developed, with the first indicating a relationship between high concentrations of fumonisins and cereal products rich in corn, particularly when combined with high levels of sodium or rice. The second tree highlighted a link between corn and either high sodium or high-fat concentrations. In both models, the presence of plastic packaging appeared to mitigate the concentration of fumonisins to a certain degree.

4.6.2. Bayesian Network

Bayesian networks (BN) are a type of probabilistic graphical model that uses Bayesian statistics to represent and infer the conditional dependencies between different variables in a dataset [101]. The networks are structured as a directed acyclic graph (DAG), with feature nodes representing variables and edges indicating probabilistic relationships between them. Predictions in BNs are made through a process called probabilistic inference, which involves calculating the likelihood of certain outcomes based on known information and the network's structure. In contrast with linear regression models, BN models excel at analysing variable dependencies, handling non-linear interactions, and incorporating diverse types of data [102]. The strengths of BN include the handling of uncertainty, the integration of prior knowledge with observed data (thereby enhancing the model's predictive capabilities), and interpretability. However, some disadvantages of using BNs exist. As the number of variables increases, the complexity of the network and the computational resources required for inference can grow exponentially.

In a study aimed for predicting DON contamination in wheat, ref. [103] compared three different modelling approaches. These are a mixed effect LR method, a mechanistic model (which simulates the mechanisms of plant and fungus development stages and their interactions) adapted to the current data, and a BN. These were all used to predict DON contamination. The data used were collected in the Netherlands over the years 2001 to 2013. The results of the experiments showed that all three models performed well, with the LR method performing the best, achieving an accuracy of 88% for detecting DON contamination. However, the authors noted that this model is greatly reliant on both the specific location and the available data, and it requires that all input data be present. The mechanistic model achieved an accuracy of 80%, while the BN achieved an 86% accuracy. However, the authors noted that the BN is easier to implement when the data are incomplete, when compared with the other methods.

Ref. [104] constructed transcriptional regulatory networks (TRNs) using a BN algorithm called the *module network* algorithm. TRNs are complex systems in biology that describe the relationships and interactions between various proteins and genes involved in the process of *transcription* [105], where transcription is the process by which the information encoded in a section of the DNA is transcribed to produce a complementary RNA strand. The goal of their work was to understand how specific gene groups (modules) in the fungus *Fusarium graminearum* regulate biological processes. The authors reported that their network inference is of high credibility, with 81.8% of the evaluable modules classified as high or moderate confidence based on their validation against a variety of evidence sources. This suggests a robust alignment of the inferred network with the existing understanding of the biological processes within *Fusarium graminearum*.

4.6.3. Summary of Other ML Methods

Decision trees have shown varying degrees of effectiveness in detecting mycotoxins, as evidenced by diverse research outcomes. The use of CART to classify contaminated wheat samples achieved higher accuracy at certain thresholds but showed diminished performance at lower contamination levels. A bagged DT approach showed moderate success, suggesting that while DTs are capable classifiers, their accuracy can vary significantly based on the mycotoxin levels and sample types. The application of these methods includes potential issues with model sensitivity, particularly at lower toxin concentrations, and a reliance on the quality of the data. These factors underscore the need for a careful calibration and validation of DTs in diverse settings for reliable mycotoxin detection.

BNs have shown effectiveness in mycotoxin detection, as demonstrated in various studies, but with some limitations. Ref. [103] compared BNs with other models for predicting DON contamination in wheat, achieving a respectable 86% accuracy. However, they highlighted BNs' advantage in handling incomplete data, a significant benefit over other methods like logistic regression. The reviewed applications show BNs' flexibility and efficiency, though their performance can be contingent on data completeness and specific biological contexts, which may limit their broader applicability.

4.7. *Summary and Comparison of Case Studies*

To provide a comprehensive overview of the specific case studies discussed, here, we include a summary table in Table 1 that highlights the key findings by describing the data types, ML models used, application contexts, and reported accuracies. In cases where more than one ML model is used, the highest-performing model is reported in the accuracy column.

Examining Table 1, we can see that the most frequently used ML model in the reviewed studies is the neural network, with convolutional neural networks also being highly prevalent. The most common data type used is spatiotemporal data, followed by hyperspectral data. The research covers a range of crops, including corn, wheat, barley, peanuts, and oats, with a primary focus on detecting contaminants such as aflatoxin, fumonisins, and Fusarium head blight. However, the most commonly studied crop is corn. Many studies achieved high accuracy rates, often above 90%, showcasing the potential of ML models to enhance mycotoxin detection in agriculture. However, it is important to consider that these high accuracies may be influenced by the controlled environments of individual laboratories, which can lead to overfitting and potentially less reliable performance in real-world applications (see Section 5 for a discussion on this).

Table 1. Summary of reviewed mycotoxin detection studies. In cases where more than one ML model is used, the highest-performing model is reported.

Study	Data Type	ML Model	Application Context	Accuracy
Camardo et al., 2021 [38]	Spatiotemporal	NN	Corn in Northern Italy (2005–2018)	>75%
Niedbala et al., 2020 [57]	Spatiotemporal	NN	Winter wheat in Poland (2011–2013)	DONANN: 99%, NIVANN: 81%
Jubair et al., 2021 [59]	Spatiotemporal	GPTransformer	Barley in Canada (2014–2015)	Not significantly better than BLUP
Rangarajan et al., 2022 [43]	Hyperspectral	CNN (DarkNet 19)	Wheat for Fusarium head blight	100%
Qiu et al., 2019 [63]	Hyperspectral	CNN	Wheat for Fusarium head blight	92%
Jin et al., 2018 [62]	Hyperspectral	CNN (2D conv. bidirectional GRU)	Wheat for Fusarium head blight	84.6%
Han et al., 2019 [49]	Hyperspectral	CNN	Peanuts for aflatoxin	95%
Han et al., 2019 [49]	Hyperspectral	CNN	Corn and peanuts for aflatoxin	96% corn, 92% peanuts
Gao et al., 2021 [65]	Hyperspectral	1D-CNN	Peanuts and corn for aflatoxin	Peanuts: 96.4%, corn: 92.1%
Öner et al., 2019 [66]	Infrared spectroscopy	MLP NN, RF, SVM, adaptive boosting	Corn for fungal contamination	MLP: 91%
Leggieri et al., 2021 [71]	E-nose	NN, LR, DA	Corn for aflatoxin and fumonisins	NN: 78% (aflatoxin), 77% (fumonisins)
Ghilardelli et al., 2022 [75]	NIR	RF	Corn silage	>90%
Teixidó et al., 2023 [76]	NIR	RF	Oat for DON	77.8%
Ma et al., 2023 [78]	Biosensor	sPLS-DA, svmLinear, svmRadial, RF, NN, HDDA	Peanuts and corn	RF: 95–98%
Tarazona et al., 2021 [79]	Spatiotemporal	NN, RF, XGB, MLR	Corn	RF: Best performance
Mateo et al., 2023 [80]	Spatiotemporal	NN, RF, XGB, MLR	Cereals for lactic acid bacteria	RF, XGB: Similar performance
Chávez et al., 2022 [92]	NIR	GBM, RF, LASSO, GLMNET, SVM	Single kernel corn	GBM: 83%
Liu et al., 2018 [103]	Spatiotemporal	BN, LR, mechanistic	Wheat for DON	LR: 88%, BN: 86%
Guo et al., 2020 [104]	TRNs	BN	Fusarium graminearum	High confidence modules: 81.8%
Kim et al., 2024 [33]	Weather	CNN	Wheat grains stored in sealed containers	83.3%
Castano et al., 2023 [34]	Weather	NN	Corn in the US	Aflatoxin: 73%, fumonisin: 85%
Branstad et al., 2023 [39]	Spatiotemporal	GBM	Corn in Iowa	20 ppb: 96.8%, 5 ppb: 90.3%
Wang et al., 2022 [87]	Spatiotemporal	XGB	Peanuts and soybeans in China, Brazil, Argentina	>90%
Xie et al., 2022 [88]	Spatiotemporal	XGB	Peanuts in China	98%
Xie et al., 2022 [88]	Spatiotemporal	XGB with decision rule	Peanuts in China	87.2%
Zhao et al., 2017 [97]	Hyperspectral	SVM	Corn in China	91.67%
Kim et al., 2023 [96]	Hyperspectral	SVM	Ground corn samples from Texas	95.7%
Almoujahed et al., 2022 [45]	Spectral	SVM	Fusarium head blight in wheat in Belgium, 2020–2021	97%
Kos et al., 2016 [99]	Spatiotemporal	CART	Wheat in Italy	88–92%
Purchase et al., 2023 [100]	Spatiotemporal	DT	Breakfast cereals in Italy	High fumonisin risk in high sodium or high-fat cereals

5. Conclusions

Our research focuses on highlighting and evaluating different ML models for monitoring and predicting the presence of mycotoxins in common crops. We conducted an extensive literature review of over 30 studies performed within the years 2013 to 2023. The number of publications in each field has grown significantly over the 10 years reviewed; however, the application of ML in the area of monitoring and predicting mycotoxins is still in its infancy, and despite the promise of ML methods in mycotoxin detection, their adoption in industry has been cautious. This is likely due to the high operational costs associated with advanced techniques like hyperspectral imaging, as opposed to the use of ML methods themselves. The prevalence of such data-intensive methods raises questions about the feasibility of widespread implementation, particularly in resource-constrained settings.

We found that the most common data type was spectral or image data, and as such, the most common ML method used was NNs, as they can be readily applied to image data. RFs were the second most popular ML method and have gained traction due to their robustness and ease of implementation. Additionally, most of the studies reviewed used classification ML techniques to distinguish contaminated from healthy crops. The high predictive accuracy reported in the reviewed studies suggests that these methods represent a promising approach for mycotoxin detection and enhancing food safety in general. However, a point to note is that the reported high accuracy of the ML model's predictions, often exceeding 90%, may not fully account for the homogeneity of training and test sets within individual laboratories. This homogeneity can result in overfitting, where models appear highly accurate in a controlled setting but may not perform as well under the variable conditions of real-world applications.

Although this work focused on the application of the most popular ML methods, numerous other ML and statistical techniques have been applied to mycotoxin detection data. For example, in a study by [106], classification models such as partial least squares-discriminant analysis (PLS-DA) and principal component-linear discriminant analysis (PC-LDA) were employed to distinguish between wheat samples with high and low contamination. Additionally, statistical techniques like PCA are often used as a dimension reduction method. Refs. [107–109] used PCA when dealing with high-dimensional image data.

A critical bottleneck in the development of ML applications for food safety is the lack of detailed hyperparameter descriptions, which further complicates the landscape, as these parameters are crucial for the replication and validation of ML models. Without clear reporting on hyperparameter tuning, the ability to reproduce results and validate findings becomes challenging, hindering the progression towards robust and reliable ML applications in food safety. The majority of the reviewed studies do not provide open access to code, and many have limited access to data, further impeding the reproducibility of the described methods.

Despite these challenges, the future prospects of ML in food safety are promising. As the field matures, there is a need for standardisation in reporting practices and for developing models that can reliably perform across diverse laboratory conditions and datasets. Extensive research could be conducted that directly compares different ML models under a standardised set of hyperparameters, providing clearer insights into the most effective techniques in specific contexts related to mycotoxin detection.

As the field is growing, there are numerous avenues for future work. One such avenue is model interpretability. Given the critical nature of food safety, future research could also focus on improving the interpretability of ML models. Techniques like SHAP (SHapley Additive exPlanations) [110] and LIME (Local Interpretable Model-Agnostic Explanations) [111] can be used to make the models' decisions more transparent and trustworthy. Furthermore, addressing the current bottlenecks, such as the high operational costs and the need for data standardisation, will be crucial. Future research should explore cost-effective techniques and advocate for open-access datasets and standardised reporting practices to enhance reproducibility and application in diverse settings.

Author Contributions: Conceptualization, A.I. and A.C.P.; investigation, A.I.; data curation, A.I.; writing—original draft preparation, A.I.; writing—review and editing, A.C.P., F.M.D. and N.S.; visualization, A.I.; supervision, A.C.P. All authors have read and agreed to the published version of the manuscript.

Funding: This work was conducted as part of the Mycotox-I project, which is kindly supported by the Department of Agriculture, Food, and the Marine (DAFM) and the Department of Agriculture, Environment, and Rural Affairs (DAERA), grant number 2021R460. Andrew Parnell's work was supported by the SFI Centre for Research Training in Foundations of Data Science 18/CRT/6049 and the SFI Research Centre award 12/RC/2289_P2. For the purpose of open access, the author has applied a CC BY public copyright licence to any author-accepted manuscript version arising from this submission.

Institutional Review Board Statement: Not applicable.

Informed Consent Statement: Not applicable.

Data Availability Statement: No new data were created or analysed in this study. Data sharing is not applicable to this article.

Conflicts of Interest: The authors declare no conflicts of interest.

Abbreviations

The following abbreviations are used in this manuscript:

NN	Neural Network
CNN	Convolutional Neural Network
RF	Random Forest
GBM	Gradient Boosted Machine
XGB	eXtreme Gradient Boosted Machine
DT	Decision Trees
CART	Classification and Regression Trees
SVM	Support Vector Machine
BM	Bayesian Models
BN	Bayesian Network
LDA	Linear Discriminant Analysis
PDA	Penalised Discriminant Analysis
LReg	Linear Regression
LR	Logistic Regression
MLR	Multiple Linear Regression
LASSO	Least Absolute Shrinkage and Selection Operator
GLMNET	Elastic-Net Regularized Generalised Linear Models
PLS-DA	Partial Least Squares-Discriminant Analysis
sPLS-DA	Sparse Partial Least Squares-Discriminant Analysis
PCA	Principal Component Analysis
MLP	Multi-Layer Perceptron
BLUP	Best Linear Unbiased Prediction
PCC	Pearson Correlation Coefficient
RMSE	Root Mean Square Error
R^2	Coefficient of Determination
AUC	Area Under the Curve
NIR	Near-Infrared Spectroscopy
DON	Deoxynivalenol

Appendix A

The quality of review has been assessed according to PRISMA guidelines [112]. This review has not been registered in a public registry. Figure A1 shows a flow chart demonstrating the selection process used in this work.

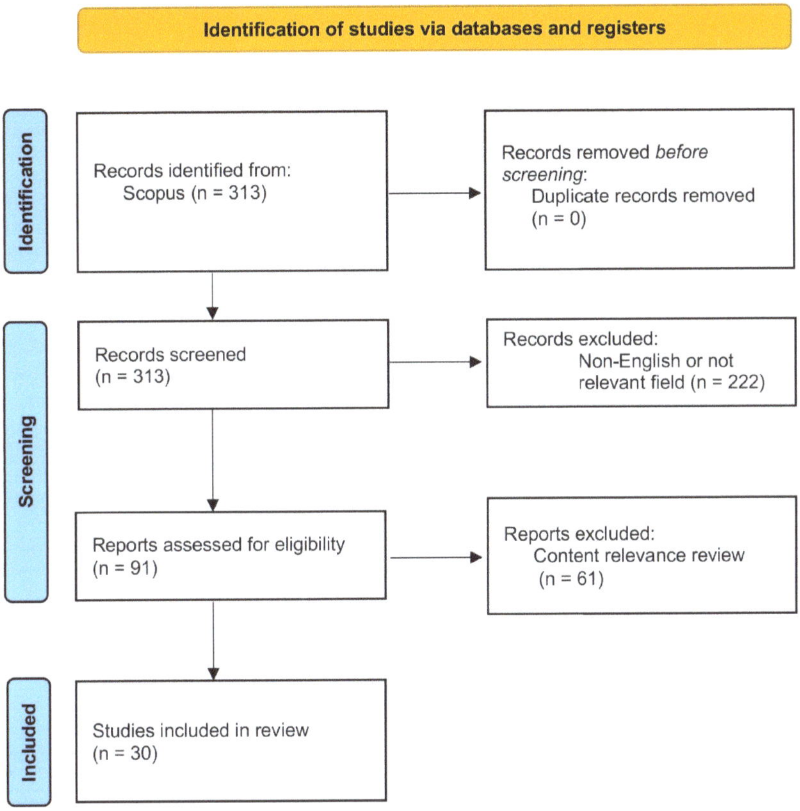

Figure A1. PRISMA flowchart of literature search strategy.

References

1. The World Health Organization (WHO). Food Safety. The World Health Organization. 2023. Available online: https://www.who.int/news-room/fact-sheets/detail/mycotoxins (accessed on 5 November 2023).
2. Mavrommatis, A.; Giamouri, E.; Tavrizelou, S.; Zacharioudaki, M.; Danezis, G.; Simitzis, P.E.; Zoidis, E.; Tsiplakou, E.; Pappas, A.C.; Georgiou, C.A.; et al. Impact of mycotoxins on animals' oxidative status. *Antioxidants* **2021**, *10*, 214. [CrossRef] [PubMed]
3. Marroquín-Cardona, A.; Johnson, N.; Phillips, T.; Hayes, A. Mycotoxins in a changing global environment—A review. *Food Chem. Toxicol.* **2014**, *69*, 220–230. [CrossRef]
4. Liu, Y.; Wu, F. Global burden of aflatoxin-induced hepatocellular carcinoma: A risk assessment. *Environ. Health Perspect.* **2010**, *118*, 818–824. [CrossRef]
5. Van der Fels-Klerx, H.; Liu, C.; Focker, M.; Montero-Castro, I.; Rossi, V.; Manstretta, V.; Magan, N.; Krska, R. Decision support system for integrated management of mycotoxins in feed and food supply chains. *World Mycotoxin J.* **2022**, *15*, 119–133. [CrossRef]
6. Tola, M.; Kebede, B. Occurrence, importance and control of mycotoxins: A review. *Cogent Food Agric.* **2016**, *2*, 1191103. [CrossRef]
7. Logrieco, A.; Battilani, P.; Leggieri, M.C.; Jiang, Y.; Haesaert, G.; Lanubile, A.; Mahuku, G.; Mesterházy, A.; Ortega-Beltran, A.; Pasti, M.; et al. Perspectives on global mycotoxin issues and management from the MycoKey Maize Working Group. *Plant Dis.* **2021**, *105*, 525–537. [CrossRef]
8. Leggieri, M.C.; Lanubile, A.; Dall'Asta, C.; Pietri, A.; Battilani, P. The impact of seasonal weather variation on mycotoxins: Maize crop in 2014 in northern Italy as a case study. *World Mycotoxin J.* **2020**, *13*, 25–36. [CrossRef]
9. Zingales, V.; Taroncher, M.; Martino, P.A.; Ruiz, M.J.; Caloni, F. Climate change and effects on molds and mycotoxins. *Toxins* **2022**, *14*, 445. [CrossRef] [PubMed]
10. Medina, A.; Akbar, A.; Baazeem, A.; Rodriguez, A.; Magan, N. Climate change, food security and mycotoxins: Do we know enough? *Fungal Biol. Rev.* **2017**, *31*, 143–154. [CrossRef]
11. Eskola, M.; Kos, G.; Elliott, C.T.; Hajšlová, J.; Mayar, S.; Krska, R. Worldwide contamination of food-crops with mycotoxins: Validity of the widely cited 'FAO estimate' of 25%. *Crit. Rev. Food Sci. Nutr.* **2020**, *60*, 2773–2789. [CrossRef]

12. Alshannaq, A.; Yu, J.H. Occurrence, toxicity, and analysis of major mycotoxins in food. *Int. J. Environ. Res. Public Health* **2017**, *14*, 632. [CrossRef] [PubMed]
13. Wu, F. Global impacts of aflatoxin in maize: Trade and human health. *World Mycotoxin J.* **2015**, *8*, 137–142. [CrossRef]
14. Johns, L.E.; Bebber, D.P.; Gurr, S.J.; Brown, N.A. Emerging health threat and cost of Fusarium mycotoxins in European wheat. *Nat. Food* **2022**, *3*, 1014–1019. [CrossRef]
15. Latham, R.L.; Boyle, J.T.; Barbano, A.; Loveman, W.G.; Brown, N.A. Diverse mycotoxin threats to safe food and feed cereals. *Essays Biochem.* **2023**, *67*, 797–809. [PubMed]
16. Whitaker, T.B. Standardisation of mycotoxin sampling procedures: An urgent necessity. *Food Control* **2003**, *14*, 233–237. [CrossRef]
17. Anfossi, L.; Giovannoli, C.; Baggiani, C. Mycotoxin detection. *Curr. Opin. Biotechnol.* **2016**, *37*, 120–126. [CrossRef] [PubMed]
18. Maragos, C.M. Emerging technologies for mycotoxin detection. *J. Toxicol. Toxin Rev.* **2004**, *23*, 317–344. [CrossRef]
19. Soares, R.R.; Ricelli, A.; Fanelli, C.; Caputo, D.; de Cesare, G.; Chu, V.; Aires-Barros, M.R.; Conde, J.P. Advances, challenges and opportunities for point-of-need screening of mycotoxins in foods and feeds. *Analyst* **2018**, *143*, 1015–1035. [CrossRef] [PubMed]
20. Renaud, J.B.; Miller, J.D.; Sumarah, M.W. Mycotoxin testing paradigm: Challenges and opportunities for the future. *J. AOAC Int.* **2019**, *102*, 1681–1688. [CrossRef] [PubMed]
21. Liakos, K.G.; Busato, P.; Moshou, D.; Pearson, S.; Bochtis, D. Machine learning in agriculture: A review. *Sensors* **2018**, *18*, 2674. [CrossRef]
22. Baştanlar, Y.; Özuysal, M. Introduction to machine learning. In *miRNomics: MicroRNA Biology and Computational Analysis*; Humana Press: Totowa, NJ, USA, 2014; pp. 105–128.
23. Alpaydin, E. *Introduction to Machine Learning*; MIT Press: Cambridge, MA, USA, 2020.
24. Torelli, E.; Firrao, G.; Bianchi, G.; Saccardo, F.; Locci, R. The influence of local factors on the prediction of fumonisin contamination in maize. *J. Sci. Food Agric.* **2012**, *92*, 1808–1814. [CrossRef] [PubMed]
25. Mateo, F.; Gadea, R.; Mateo, E.M.; Jiménez, M. Multilayer perceptron neural networks and radial-basis function networks as tools to forecast accumulation of deoxynivalenol in barley seeds contaminated with *Fusarium culmorum*. *Food Control* **2011**, *22*, 88–95. [CrossRef]
26. Panagou, E.Z.; Kodogiannis, V.S. Application of neural networks as a non-linear modelling technique in food mycology. *Expert Syst. Appl.* **2009**, *36*, 121–131. [CrossRef]
27. Zhou, L.; Zhang, C.; Liu, F.; Qiu, Z.; He, Y. Application of deep learning in food: A review. *Compr. Rev. Food Sci. Food Saf.* **2019**, *18*, 1793–1811. [CrossRef] [PubMed]
28. Wang, X.; Bouzembrak, Y.; Lansink, A.O.; van der Fels-Klerx, H. Application of machine learning to the monitoring and prediction of food safety: A review. *Compr. Rev. Food Sci. Food Saf.* **2022**, *21*, 416–434. [CrossRef] [PubMed]
29. Bernardes, R.C.; De Medeiros, A.; da Silva, L.; Cantoni, L.; Martins, G.F.; Mastrangelo, T.; Novikov, A.; Mastrangelo, C.B. Deep-learning approach for fusarium head blight detection in wheat seeds using low-cost imaging technology. *Agriculture* **2022**, *12*, 1801. [CrossRef]
30. Magan, N.; Medina, A. Integrating gene expression, ecology and mycotoxin production by Fusarium and Aspergillus species in relation to interacting environmental factors. *World Mycotoxin J.* **2016**, *9*, 673–684. [CrossRef]
31. Verheecke-Vaessen, C.; Diez-Gutierrez, L.; Renaud, J.; Sumarah, M.; Medina, A.; Magan, N. Interacting climate change environmental factors effects on Fusarium langsethiae growth, expression of Tri genes and T-2/HT-2 mycotoxin production on oat-based media and in stored oats. *Fungal Biol.* **2019**, *123*, 618–624. [CrossRef]
32. Natarajan, S.; Balachandar, D.; Paranidharan, V. Inhibitory effects of epiphytic *Kluyveromyces marxianus* from Indian senna (*Cassia angustifolia* Vahl.) on growth and aflatoxin production of *Aspergillus flavus*. *Int. J. Food Microbiol.* **2023**, *406*, 110368. [CrossRef]
33. Kim, Y.; Kang, S.; Ajani, O.S.; Mallipeddi, R.; Ha, Y. Predicting early mycotoxin contamination in stored wheat using machine learning. *J. Stored Prod. Res.* **2024**, *106*, 102294. [CrossRef]
34. Castano-Duque, L.; Winzeler, E.; Blackstock, J.M.; Liu, C.; Vergopolan, N.; Focker, M.; Barnett, K.; Owens, P.R.; van der Fels-Klerx, H.; Vaughan, M.M.; et al. Dynamic geospatial modeling of mycotoxin contamination of corn in Illinois: unveiling critical factors and predictive insights with machine learning. *Front. Microbiol.* **2023**, *14*, 1283127. [CrossRef] [PubMed]
35. Orlando, B.; Barrier-Guillot, B.; Gourdain, E.; Maumene, C. Identification of agronomic factors that influence the levels of T-2 and HT-2 toxins in barley grown in France. *World Mycotoxin J.* **2010**, *3*, 169–174. [CrossRef]
36. Edwards, S.G. Influence of agricultural practices on Fusarium infection of cereals and subsequent contamination of grain by trichothecene mycotoxins. *Toxicol. Lett.* **2004**, *153*, 29–35. [CrossRef] [PubMed]
37. Edwards, S.G.; Jennings, P. Impact of agronomic factors on Fusarium mycotoxins in harvested wheat. *Food Addit. Contam. Part A* **2018**, *35*, 2443–2454. [CrossRef] [PubMed]
38. Camardo Leggieri, M.; Mazzoni, M.; Battilani, P. Machine learning for predicting mycotoxin occurrence in maize. *Front. Microbiol.* **2021**, *12*, 661132. [CrossRef] [PubMed]
39. Branstad-Spates, E.H.; Castano-Duque, L.; Mosher, G.A.; Hurburgh, C.R., Jr.; Owens, P.; Winzeler, E.; Rajasekaran, K.; Bowers, E.L. Gradient boosting machine learning model to predict aflatoxins in Iowa corn. *Front. Microbiol.* **2023**, *14*, 1248772. [CrossRef] [PubMed]
40. Wegulo, S. Factors influencing deoxynivalenol accumulation in small grain cereals. *Toxins* **2012**, *4*, 1157–1180. [CrossRef] [PubMed]

41. Dhakal, K.; Sivaramakrishnan, U.; Zhang, X.; Belay, K.; Oakes, J.; Wei, X.; Li, S. Machine learning analysis of hyperspectral images of damaged wheat kernels. *Sensors* **2023**, *23*, 3523. [CrossRef] [PubMed]
42. Wang, X.; Liu, C.; van der Fels-Klerx, H. Regional prediction of multi-mycotoxin contamination of wheat in Europe using machine learning. *Food Res. Int.* **2022**, *159*, 111588. [CrossRef]
43. Rangarajan, A.K.; Whetton, R.L.; Mouazen, A.M. Detection of fusarium head blight in wheat using hyperspectral data and deep learning. *Expert Syst. Appl.* **2022**, *208*, 118240. [CrossRef]
44. Kalkan, H.; Güneş, A.; Durmuş, E.; Kuşçu, A. Non-invasive detection of aflatoxin-contaminated figs using fluorescence and multispectral imaging. *Food Addit. Contam. Part A* **2014**, *31*, 1414–1421. [CrossRef] [PubMed]
45. Almoujahed, M.B.; Rangarajan, A.K.; Whetton, R.L.; Vincke, D.; Eylenbosch, D.; Vermeulen, P.; Mouazen, A.M. Detection of fusarium head blight in wheat under field conditions using a hyperspectral camera and machine learning. *Comput. Electron. Agric.* **2022**, *203*, 107456. [CrossRef]
46. Liu, L.; Dong, Y.; Huang, W.; Du, X.; Ma, H. Monitoring wheat fusarium head blight using unmanned aerial vehicle hyperspectral imagery. *Remote Sens.* **2020**, *12*, 3811. [CrossRef]
47. Hruska, Z.; Yao, H.; Kincaid, R.; Brown, R.; Cleveland, T.; Bhatnagar, D. Fluorescence excitation–emission features of aflatoxin and related secondary metabolites and their application for rapid detection of mycotoxins. *Food Bioprocess Technol.* **2014**, *7*, 1195–1201. [CrossRef]
48. Zhu, F.; Yao, H.; Hruska, Z.; Kincaid, R.; Brown, R.L.; Bhatnagar, D.; Cleveland, T.E. Integration of fluorescence and reflectance visible near-infrared (VNIR) hyperspectral images for detection of aflatoxins in corn kernels. *Trans. ASABE* **2016**, *59*, 785–794.
49. Han, Z.; Gao, J. Pixel-level aflatoxin detecting based on deep learning and hyperspectral imaging. *Comput. Electron. Agric.* **2019**, *164*, 104888. [CrossRef]
50. Del Fiore, A.; Reverberi, M.; Ricelli, A.; Pinzari, F.; Serranti, S.; Fabbri, A.A.; Bonifazi, G.; Fanelli, C. Early detection of toxigenic fungi on maize by hyperspectral imaging analysis. *Int. J. Food Microbiol.* **2010**, *144*, 64–71. [CrossRef] [PubMed]
51. Serranti, S.; Cesare, D.; Bonifazi, G. The development of a hyperspectral imaging method for the detection of Fusarium-damaged, yellow berry and vitreous Italian durum wheat kernels. *Biosyst. Eng.* **2013**, *115*, 20–30. [CrossRef]
52. McCulloch, W.S.; Pitts, W. A logical calculus of the ideas immanent in nervous activity. *Bull. Math. Biophys.* **1943**, *5*, 115–133. [CrossRef]
53. Gurney, K. *An Introduction to Neural Networks*; CRC Press: Boca Raton, FL, USA, 2018.
54. Montavon, G.; Samek, W.; Müller, K.R. Methods for interpreting and understanding deep neural networks. *Digit. Signal Process.* **2018**, *73*, 1–15. [CrossRef]
55. Srivastava, N.; Hinton, G.; Krizhevsky, A.; Sutskever, I.; Salakhutdinov, R. Dropout: a simple way to prevent neural networks from overfitting. *J. Mach. Learn. Res.* **2014**, *15*, 1929–1958.
56. Canziani, A.; Paszke, A.; Culurciello, E. An analysis of deep neural network models for practical applications. *arXiv* **2016**, arXiv:1605.07678.
57. Niedbała, G.; Kurasiak-Popowska, D.; Stuper-Szablewska, K.; Nawracała, J. Application of artificial neural networks to analyze the concentration of ferulic acid, deoxynivalenol, and nivalenol in winter wheat grain. *Agriculture* **2020**, *10*, 127. [CrossRef]
58. StatSoft, Inc. STATISTICA (Data Analysis Software System), Version 7.1. 2020. Available online: http://www.statsoft.com (accessed on 28 April 2024).
59. Jubair, S.; Tucker, J.R.; Henderson, N.; Hiebert, C.W.; Badea, A.; Domaratzki, M.; Fernando, W. GPTransformer: A transformer-based deep learning method for predicting Fusarium related traits in barley. *Front. Plant Sci.* **2021**, *12*, 761402. [CrossRef] [PubMed]
60. Vaswani, A.; Shazeer, N.; Parmar, N.; Uszkoreit, J.; Jones, L.; Gomez, A.N.; Kaiser, Ł.; Polosukhin, I. Attention is all you need. In *Advances in Neural Information Processing Systems*; Curran Associates Inc.: Red Hook, NY, USA, 2017; Volume 30.
61. Grahn, H.; Geladi, P. *Techniques and Applications of Hyperspectral Image Analysis*; John Wiley & Sons: Hoboken, NJ, USA, 2007.
62. Jin, X.; Jie, L.; Wang, S.; Qi, H.J.; Li, S.W. Classifying wheat hyperspectral pixels of healthy heads and Fusarium head blight disease using a deep neural network in the wild field. *Remote Sens.* **2018**, *10*, 395. [CrossRef]
63. Qiu, R.; Yang, C.; Moghimi, A.; Zhang, M.; Steffenson, B.J.; Hirsch, C.D. Detection of fusarium head blight in wheat using a deep neural network and color imaging. *Remote Sens.* **2019**, *11*, 2658. [CrossRef]
64. Redmon, J.; Farhadi, A. YOLO9000: Better, faster, stronger. In Proceedings of the IEEE Conference on Computer Vision and Pattern Recognition, Honolulu, HI, USA, 21–26 July 2017; pp. 7263–7271.
65. Gao, J.; Zhao, L.; Li, J.; Deng, L.; Ni, J.; Han, Z. Aflatoxin rapid detection based on hyperspectral with 1D-convolution neural network in the pixel level. *Food Chem.* **2021**, *360*, 129968. [CrossRef]
66. Oener, T.; Thiam, P.; Kos, G.; Krska, R.; Schwenker, F.; Mizaikoff, B. Machine learning algorithms for the automated classification of contaminated maize at regulatory limits via infrared attenuated total reflection spectroscopy. *World Mycotoxin J.* **2019**, *12*, 113–122. [CrossRef]
67. Ottoboni, M.; Pinotti, L.; Tretola, M.; Giromini, C.; Fusi, E.; Rebucci, R.; Grillo, M.; Tassoni, L.; Foresta, S.; Gastaldello, S.; et al. Combining E-nose and lateral flow immunoassays (LFIAs) for rapid occurrence/co-occurrence aflatoxin and fumonisin detection in maize. *Toxins* **2018**, *10*, 416. [CrossRef]
68. Campagnoli, A.; Cheli, F.; Savoini, G.; Crotti, A.; Pastori, A.; Dell'Orto, V. Application of an electronic nose to detection of aflatoxins in corn. *Vet. Res. Commun.* **2009**, *33*, 273–275. [CrossRef]

69. Gobbi, E.; Falasconi, M.; Torelli, E.; Sberveglieri, G. Electronic nose predicts high and low fumonisin contamination in maize cultures. *Food Res. Int.* **2011**, *44*, 992–999. [CrossRef]
70. Lippolis, V.; Pascale, M.; Cervellieri, S.; Damascelli, A.; Visconti, A. Screening of deoxynivalenol contamination in durum wheat by MOS-based electronic nose and identification of the relevant pattern of volatile compounds. *Food Control* **2014**, *37*, 263–271. [CrossRef]
71. Leggieri, M.C.; Mazzoni, M.; Fodil, S.; Moschini, M.; Bertuzzi, T.; Prandini, A.; Battilani, P. An electronic nose supported by an artificial neural network for the rapid detection of aflatoxin B1 and fumonisins in maize. *Food Control* **2021**, *123*, 107722. [CrossRef]
72. Saha, D.; Manickavasagan, A. Machine learning techniques for analysis of hyperspectral images to determine quality of food products: A review. *Curr. Res. Food Sci.* **2021**, *4*, 28–44. [CrossRef] [PubMed]
73. Breiman, L. Random forests. *Mach. Learn.* **2001**, *45*, 5–32. [CrossRef]
74. Breiman, L.; Friedman, J.; Olshen, R.; Stone, C. *Classification and Regression Trees*; Wadsworth: Hartford, CT, USA, 1984.
75. Ghilardelli, F.; Barbato, M.; Gallo, A. A preliminary study to classify corn silage for high or low mycotoxin contamination by using near infrared spectroscopy. *Toxins* **2022**, *14*, 323. [CrossRef] [PubMed]
76. Teixido-Orries, I.; Molino, F.; Femenias, A.; Ramos, A.J.; Marín, S. Quantification and classification of deoxynivalenol-contaminated oat samples by near-infrared hyperspectral imaging. *Food Chem.* **2023**, *417*, 135924. [CrossRef]
77. Femenias, A.; Gatius, F.; Ramos, A.J.; Sanchis, V.; Marín, S. Near-infrared hyperspectral imaging for deoxynivalenol and ergosterol estimation in wheat samples. *Food Chem.* **2021**, *341*, 128206. [CrossRef]
78. Ma, J.; Guan, Y.; Xing, F.; Eltzov, E.; Wang, Y.; Li, X.; Tai, B. Accurate and non-destructive monitoring of mold contamination in foodstuffs based on whole-cell biosensor array coupling with machine-learning prediction models. *J. Hazard. Mater.* **2023**, *449*, 131030. [CrossRef]
79. Tarazona, A.; Mateo, E.M.; Gómez, J.V.; Gavara, R.; Jiménez, M.; Mateo, F. Machine learning approach for predicting *Fusarium culmorum* and *F. proliferatum* growth and mycotoxin production in treatments with ethylene-vinyl alcohol copolymer films containing pure components of essential oils. *Int. J. Food Microbiol.* **2021**, *338*, 109012. [CrossRef]
80. Mateo, E.M.; Tarazona, A.; Aznar, R.; Mateo, F. Exploring the impact of lactic acid bacteria on the biocontrol of toxigenic *Fusarium* spp. and their main mycotoxins. *Int. J. Food Microbiol.* **2023**, *387*, 110054. [CrossRef] [PubMed]
81. Mateo, E.M.; Gómez, J.V.; Tarazona, A.; García-Esparza, M.Á.; Mateo, F. Comparative analysis of machine learning methods to predict growth of *F. sporotrichioides* and production of T-2 and HT-2 toxins in treatments with ethylene-vinyl alcohol films containing pure components of essential oils. *Toxins* **2021**, *13*, 545. [CrossRef] [PubMed]
82. Tarazona, A.; Mateo, E.M.; Gómez, J.V.; Romera, D.; Mateo, F. Potential use of machine learning methods in assessment of *Fusarium culmorum* and *Fusarium proliferatum* growth and mycotoxin production in treatments with antifungal agents. *Fungal Biol.* **2021**, *125*, 123–133. [CrossRef] [PubMed]
83. Srinivasan, R.; Lalitha, T.; Brintha, N.; Sterlin Minish, T.; Al Obaid, S.; Alharbi, S.A.; Sundaram, S.; Mahilraj, J. Predicting the Growth of *F. proliferatum* and *F. culmorum* and the Growth of Mycotoxin Using Machine Learning Approach. *BioMed Res. Int.* **2022**, *2022*, 9592365. [CrossRef] [PubMed]
84. Friedman, J.H. Greedy function approximation: A gradient boosting machine. *Ann. Stat.* **2001**, *29*, 1189–1232. [CrossRef]
85. Natekin, A.; Knoll, A. Gradient boosting machines, a tutorial. *Front. Neurorobot.* **2013**, *7*, 21. [CrossRef] [PubMed]
86. Chen, T.; Guestrin, C. Xgboost: A scalable tree boosting system. In Proceedings of the 22nd ACM SIGKDD International Conference on Knowledge Discovery and Data Mining, San Francisco, CA, USA, 13–17 August 2016; pp. 785–794.
87. Wang, X.; Bouzembrak, Y.; Oude Lansink, A.; Van der Fels-Klerx, H. Designing a monitoring program for aflatoxin B1 in feed products using machine learning. *NPJ Sci. Food* **2022**, *6*, 40. [CrossRef] [PubMed]
88. Xie, H.; Wang, X.; van der Hooft, J.J.; Medema, M.H.; Chen, Z.Y.; Yue, X.; Zhang, Q.; Li, P. Fungi population metabolomics and molecular network study reveal novel biomarkers for early detection of aflatoxigenic Aspergillus species. *J. Hazard. Mater.* **2022**, *424*, 127173. [CrossRef] [PubMed]
89. FDA. Guidance for Industry: Action Levels for Poisonous or Deleterious Substances in Human Food and Animal Feed. 2020. Available online: https://www.fda.gov/regulatory-information/search-fda-guidance-documents/guidance-industry-action-levels-poisonous-or-deleterious-substances-human-food-and-animal-feed#afla (accessed on 5 November 2023).
90. EFSA. *Aflatoxins (Sum of B1, B2, G1, G2) in Cereals and Cereal-Derived Food Products*; EFSA: Parma, Italy, 2013.
91. Zheng, A.; Casari, A. *Feature Engineering for Machine Learning: Principles and Techniques for Data Scientists*; O'Reilly Media, Inc.: Newton, MA, USA, 2018.
92. Chavez, R.A.; Cheng, X.; Herrman, T.J.; Stasiewicz, M.J. Single kernel aflatoxin and fumonisin contamination distribution and spectral classification in commercial corn. *Food Control* **2022**, *131*, 108393. [CrossRef]
93. Vapnik, V. *The Nature of Statistical Learning Theory*; Springer Science & Business Media: Berlin/Heidelberg, Germany, 1999.
94. Burges, C.J. A tutorial on support vector machines for pattern recognition. *Data Min. Knowl. Discov.* **1998**, *2*, 121–167. [CrossRef]
95. Cristianini, N.; Shawe-Taylor, J. *An Introduction to Support Vector Machines and Other Kernel-Based Learning Methods*; Cambridge University Press: Cambridge, UK, 2000.
96. Kim, Y.K.; Baek, I.; Lee, K.M.; Kim, G.; Kim, S.; Kim, S.Y.; Chan, D.; Herrman, T.J.; Kim, N.; Kim, M.S. Rapid Detection of Single-and Co-Contaminant Aflatoxins and Fumonisins in Ground Maize Using Hyperspectral Imaging Techniques. *Toxins* **2023**, *15*, 472. [CrossRef] [PubMed]

97. Zhao, X.; Wang, W.; Chu, X.; Li, C.; Kimuli, D. Early detection of *Aspergillus parasiticus* infection in maize kernels using near-infrared hyperspectral imaging and multivariate data analysis. *Appl. Sci.* **2017**, *7*, 90. [CrossRef]
98. Quinlan, J.R. Induction of decision trees. *Mach. Learn.* **1986**, *1*, 81–106. [CrossRef]
99. Kos, G.; Sieger, M.; McMullin, D.; Zahradnik, C.; Sulyok, M.; Öner, T.; Mizaikoff, B.; Krska, R. A novel chemometric classification for FTIR spectra of mycotoxin-contaminated maize and peanuts at regulatory limits. *Food Addit. Contam. Part A* **2016**, *33*, 1596–1607. [CrossRef] [PubMed]
100. Purchase, J.; Donato, R.; Sacco, C.; Pettini, L.; Rookmin, A.D.; Melani, S.; Artese, A.; Purchase, D.; Marvasi, M. The association of food ingredients in breakfast cereal products and fumonisins production: Risks identification and predictions. *Mycotoxin Res.* **2023**, *39*, 165–175. [CrossRef] [PubMed]
101. Jensen, F.V.; Nielsen, T.D. *Bayesian Networks and Decision Graphs*; Springer: Berlin/Heidelberg, Germany, 2007; Volume 2.
102. Buriticá, J.A.; Tesfamariam, S. Consequence-based framework for electric power providers using Bayesian belief network. *Int. J. Electr. Power Energy Syst.* **2015**, *64*, 233–241. [CrossRef]
103. Liu, C.; Manstretta, V.; Rossi, V.; Van der Fels-Klerx, H. Comparison of three modelling approaches for predicting deoxynivalenol contamination in winter wheat. *Toxins* **2018**, *10*, 267. [CrossRef] [PubMed]
104. Guo, L.; Ji, M.; Ye, K. Dynamic network inference and association computation discover gene modules regulating virulence, mycotoxin and sexual reproduction in *Fusarium graminearum*. *BMC Genom.* **2020**, *21*, 179. [CrossRef] [PubMed]
105. Babu, M.M.; Luscombe, N.M.; Aravind, L.; Gerstein, M.; Teichmann, S.A. Structure and evolution of transcriptional regulatory networks. *Curr. Opin. Struct. Biol.* **2004**, *14*, 283–291. [CrossRef]
106. De Girolamo, A.; von Holst, C.; Cortese, M.; Cervellieri, S.; Pascale, M.; Longobardi, F.; Catucci, L.; Porricelli, A.C.R.; Lippolis, V. Rapid screening of ochratoxin A in wheat by infrared spectroscopy. *Food Chem.* **2019**, *282*, 95–100. [CrossRef]
107. Shen, F.; Zhao, T.; Jiang, X.; Liu, X.; Fang, Y.; Liu, Q.; Hu, Q.; Liu, X. On-line detection of toxigenic fungal infection in wheat by visible/near infrared spectroscopy. *LWT* **2019**, *109*, 216–224. [CrossRef]
108. Jha, S.N.; Jaiswal, P.; Kaur, J.; Ramya, H. Rapid detection and quantification of aflatoxin B1 in milk using fourier transform infrared spectroscopy. *J. Inst. Eng. Ser. A* **2021**, *102*, 259–265. [CrossRef]
109. Milićević, D.; Petronijević, R.; Petrović, Z.; Djinović-Stojanović, J.; Jovanović, J.; Baltić, T.; Janković, S. Impact of climate change on aflatoxin M1 contamination of raw milk with special focus on climate conditions in Serbia. *J. Sci. Food Agric.* **2019**, *99*, 5202–5210. [CrossRef]
110. Shapley, L.S. *A Value for n-Person Games*; Princeton University Press: Princeton, NJ, USA, 1953.
111. Ribeiro, M.T.; Singh, S.; Guestrin, C. "Why should I trust you?" Explaining the predictions of any classifier. In Proceedings of the 22nd ACM SIGKDD International Conference on Knowledge Discovery and Data Mining, San Francisco, CA, USA, 13–17 August 2016; pp. 1135–1144.
112. Page, M.J.; McKenzie, J.E.; Bossuyt, P.M.; Boutron, I.; Hoffmann, T.C.; Mulrow, C.D.; Shamseer, L.; Tetzlaff, J.M.; Akl, E.A.; Brennan, S.E. The PRISMA 2020 statement: an updated guideline for reporting systematic reviews *Br. Med. J. Publ. Group* **2021**, *372*, n71. [CrossRef] [PubMed]

toxins

MDPI

Article

First Report on Mycotoxin Contamination of Hops (*Humulus lupulus* L.)

Ivana Dodlek Šarkanj [1], Nada Vahčić [2], Ksenija Markov [2], Josip Haramija [3], Natalija Uršulin-Trstenjak [1], Krunoslav Hajdek [4], Michael Sulyok [5], Rudolf Krska [5,6] and Bojan Šarkanj [1,*]

1 Department of Food Technology, University North, Trg dr. Žarka Dolinara 1, HR-48000 Koprivnica, Croatia; idsarkanj@unin.hr (I.D.Š.); natalija.ursulin-trstenjak@unin.hr (N.U.-T.)
2 Faculty of Food Technology and Biotechnology, University of Zagreb, Pierottijeva 6, HR-10000 Zagreb, Croatia; nvahcic@pbf.hr (N.V.); ksenija.markov@pbf.hr (K.M.)
3 Koprivnica Branch, State Inspectorate, Florijanski trg 18, HR-48000, Koprivnica, Croatia; josip.haramija@dirh.hr
4 Department of Packaging, Recycling and Environmental Protection, University North, Trg dr. Žarka Dolinara 1, HR-48000 Koprivnica, Croatia; khajdek@unin.hr
5 Institute of Bioanalytics and Agro-Metabolomics, Department of Agrobiotechnology (IFA-Tulln), University of Natural Resources and Life Sciences, Vienna, Konrad Lorenz Str. 20, AT-3430 Tulln, Austria; michael.sulyok@boku.ac.at (M.S.); rudolf.krska@boku.ac.at (R.K.)
6 Institute for Global Food Security, School of Biological Sciences, Queen's University Belfast, University Road, Belfast BT7 1NN, UK
* Correspondence: bsarkanj@unin.hr

Abstract: The presence of mycotoxins and other toxic metabolites in hops (*Humulus lupulus* L.) was assessed for the first time. In total, 62 hop samples were sampled in craft breweries, and analyzed by a multi-toxin LS-MS/MS method. The study collected samples from craft breweries in all of the Croatian counties and statistically compared the results. Based on previous reports on *Alternaria* spp. and *Fusarium* spp. contamination of hops, the study confirmed the contamination of hops with these toxins. *Alternaria* toxins, particularly tenuazonic acid, were found in all tested samples, while *Fusarium* toxins, including deoxynivalenol, were present in 98% of samples. However, no *Aspergillus* or *Penicillium* metabolites were detected, indicating proper storage conditions. In addition to the *Alternaria* and *Fusarium* toxins, abscisic acid, a drought stress indicator in hops, was also detected, as well as several unspecific metabolites. The findings suggest the need for monitoring, risk assessment, and potential regulation of *Alternaria* and *Fusarium* toxins in hops to ensure the safety of hop usage in the brewing and pharmaceutical industries. Also, four local wild varieties were tested, with similar results to the commercial varieties for toxin contamination, but the statistically significant regional differences in toxin occurrence highlight the importance and need for targeted monitoring.

Keywords: hops; *Humulus lupulus* L.; mycotoxins; *Alternaria* spp.; *Fusarium* spp.; LC-MS/MS

Key Contribution: First confirmation of the widespread contamination of hops with *Alternaria* and *Fusarium* toxins: a neglected risk in the brewing and pharmaceutical industry.

1. Introduction

Hops (*Humulus lupulus* L.) are an essential ingredient in brewing, imparting a bitter flavor and floral aroma to beer. The first written record of the utilization of hops in brewing dates back to 736 in a monastery document from the Hallertau region in Bavaria, Germany [1]. Since then, hops have been well documented for their role in flavoring, preserving, and stabilizing beer. The female inflorescences of the hop plant, known as cones or strobili, are particularly valued for their production of secondary metabolites, such as terpenes, sesquiterpenes, and prenylated phenolic compounds, which contribute to the beer's bitterness and aromatic qualities [2]. These compounds also exhibit antiseptic properties, enhancing the beer's shelf life and safety [3]. However, the cultivation and storage of hops is not without microbial challenges. One significant concern is the contamination of hops with (mycotoxigenic) fungi [4], which can produce mycotoxins that pose health risks to consumers. Mycotoxins are toxic secondary metabolites produced by certain species of fungi [5], and their presence in food and beverages is a serious safety issue, particularly in a changing climate [6]. The biological control of these fungi and their toxins is a critical area of research, with studies focusing on pre-harvest approaches to mitigate the risk of contamination [7], and also ensuring stability and prevention of infection during storage [8]. Hops have been shown to be contaminated by all the main genera of mycotoxin-producing fungi [4]: *Alternaria* spp., *Aspergillus* spp., *Fusarium* spp., and *Penicillium* spp. [9]. There are several reports of *Alternaria alternata* infections of hops [4,10–12], a known producer of *Alternaria* toxins, recently regulated in the EU (EC 553/2022 [13]). The infection symptoms include necrotic lesions on the tips of bracts and bracteoles of developing cones [14]. The disease is called *Alternaria* cone disorder (ACD) and is widespread in hop yards and other agricultural systems worldwide. There is still no reported occurrence of any of the *Alternaria* toxins in hops, although their presence has been implied due to frequent infection reports. *Aspergillus* spp. is one of the fungi genera that has still not been isolated in naturally infected hops [4]; moreover, hop extracts have shown great antifungal properties against *Aspergillus* spp. [15,16]. *Fusarium* spp. is one of the most often reported mycotoxin-producing species in hops. It is known to cause hop fusariosis (HF) [17], *Fusarium* canker (FC) wilting, cankers in the crown, foliar necrosis and death of infected plants [18], and hop wilt (HW) [19]. Several *Fusarium* species have been identified as the causal agents of HW, including *F. oxysporum*, *F. culmorum*, *F. solani*, *F. proliferatum*, and *F. acuminatum* [19]. *F. oxysporum* and *F. culmorum* were most frequently isolated, and both are known mycotoxin producers [20,21]. The *Fusarium* fungi first colonize the underground plant parts (roots, crown, and rootstocks) and the basal part of the stem, from where they disperse and attack the neighboring vascular tissues. The interrupted delivery of water and nutrients to the terminal plant parts causes chlorosis, necrosis, and wilting, first of the apical leaves, and then, of the lower leaves. Due to the relatively high occurrence of *Fusarium* infection, there have also been qualitative and quantitative PCR assays developed for the detection of *Fusarium* spp. in hops [22]. Although a high incidence of *Fusarium* fungi is well documented, no data on the occurrence of *Fusarium* mycotoxins have been published. The growth of *Penicillium* spp. has rarely been reported on hops. It has been shown that hop extracts inhibit the growth of several *Penicillium* spp. [15,16,23] but, on the other hand, *Penicillium* spp. was isolated as a probable source of diastatic enzymes driving "hop creep" in dry-hopped beer [24,25]. After identification of the presence of *Penicillium* spp., there were no data on the presence of their mycotoxins in hops. Interestingly, hop essential oils and extracts possess antifungal properties, which can be leveraged to combat the growth of mycotoxigenic fungi [26]. Isoxanthohumol, an isoprene flavonoid found in hops, has demonstrated significant antifungal activity against phytopathogenic fungi such as *Botrytis cinerea*. This compound disrupts the metabolic processes of the fungi, affecting their carbohydrate metabolism and hindering ATP generation by inhibiting respiration. Additionally, isoxanthohumol causes membrane lipid peroxidation, accelerating the death of fungal cells. These findings suggest that hops not only contribute to the sensory qualities of beer but also have the potential to enhance its safety by reducing the risk of fungal contamination [27]. In addition, extracts of *H. lupulus* inhibited the mycelial growth of *F. culmorum*, a pathogenic fungus that causes root rot in wheat [28]. The oils and extracts of *H. lupulus* also showed activity against the fungus *Trichophyton mentagrophytes* var. *interdigitale* [27]. There is a lack of testing for mycotoxin occurrence, although there is enough evidence that mycotoxin-producing fungi can contaminate the hops. The purpose of this study was to investigate the possibility of contamination of the hops by mycotoxins, due to sufficient evidence of the presence of mycotoxin-producing fungi in the published literature. The results constitute additional research in the field and can contribute to the development of risk assessments for hop products, and possible changes in hop production, processing, and storage to cope with a mycotoxin contamination threat. Also, by using a wide range of state-of-the-art multi-toxin methods, additional information on the presence of plant, bacterial, and unspecific toxins are tested, for further risk assessment and increased food safety.

2. Results and Discussion

The results of this study confirmed the presence of multiple mycotoxins associated with the fungal genera previously detected in hop samples [4,10–12,14,17–19,21]. The results are divided by the main producers of the toxins/metabolites. A summary of all the results is given in Figure 1 and they are also divided by the producers. All of the collected samples were used in craft breweries in Croatia and four local wild varieties were collected in the market. The data were also used for creating a heat map (Figure 2) and comparing the data between counties by using Kruskal–Wallis ANOVA. All the results showed that there was a statistically significant ($p < 0.05$) difference between the distributions of the data in different counties.

	Alternaria spp.								*Fusarium* spp.																plant metabolite	unspecific					
name of hops	Altenusin [µg/kg]	Alternariol [µg/kg]*	Alternariolmethylether [µg/kg]	Altersetin [µg/kg]	Infectopyrone [µg/kg]	Tentoxin [µg/kg]*	Tenuazonic acid [µg/kg]*	sum Alternaria [µg/kg]	Beauvericin [µg/kg]**	Bikaverin [µg/kg]	Butenolid [µg/kg]	Culmorin [µg/kg]	Deoxynivalenol [µg/kg]***	Enniatin B [µg/kg]**	Enniatin B1 [µg/kg]**	Enniatin B2 [µg/kg]**	Enniatin B3 [µg/kg]**	Fumonisin B1 [µg/kg]***	HT-2 toxin [µg/kg]***	Moniliformin [µg/kg]**	Nivalenol [µg/kg]**	Siccanol [µg/kg]	T-2 toxin [µg/kg]***	sum Fusarium [µg/kg]	Abscisic acid [µg/kg]	3-Nitropropionic acid [µg/kg]	Brevianamid F [µg/kg]	Citreorosein [µg/kg]	cyclo(L-Pro-L-Val) [µg/kg]	Tryptophol [µg/kg]	sum unspecific [µg/kg]
Amarino 2021	<LOD	<LOD	<LOD	9.41	<LOD	<LOD	11.4	20.8	<LOD	<LOD	4.35	<LOD	<LOD	0.049	0.041	<LOD	<LOD	<LOD	<LOD	<LOD	<LOD	<LOD	<LOD	4.44	3286	1.27	0.16	<LOD	1.14	14.0	16.6
Amber Ale	<LOD	<LOD	0.36	20.85	<LOD	<LOD	511	532	<LOD	<LOD	<LOD	<LOD	<LOD	0.13	0.055	<LOD	<LOD	14.3	<LOD	<LOD	<LOD	186	<LOD	201	2787	<LOD	0.83	<LOD	1.67	26.2	28.7
Bobek	<LOD	<LOD	0.37	9.12	<LOD	<LOD	1164	1174	<LOD	<LOD	<LOD	27.1	26.8	1.91	0.647	<LOD	<LOD	15.7	<LOD	<LOD	<LOD	249	<LOD	321	3930	<LOD	0.50	0.20	<LOD	43.5	44.2
Cascade	<LOD	<LOD	0.02	14.05	<LOD	<LOD	3.74	17.8	<LOD	<LOD	<LOD	<LOD	<LOD	<LOD	<LOD	<LOD	<LOD	14.9	<LOD	<LOD	15.7	<LOD	<LOD	30.7	3158	<LOD	0.32	<LOD	0.18	6.58	7.08
Cascade	<LOD	<LOD	0.12	8.72	<LOD	<LOD	9.72	18.6	<LOD	<LOD	4.74	<LOD	57.8	0.030	<LOD	<LOD	<LOD	<LOD	<LOD	<LOD	<LOD	<LOD	<LOD	62.6	3466	<LOD	0.11	<LOD	0.81	13.5	14.4
Cascade	<LOD	<LOD	0.04	6.54	<LOD	<LOD	68.3	74.9	<LOD	<LOD	<LOD	12.1	<LOD	<LOD	<LOD	<LOD	<LOD	15.0	<LOD	<LOD	<LOD	<LOD	<LOD	27.1	1439	1.19	0.58	<LOD	<LOD	13.9	15.7
Cascade (Aba)	<LOD	<LOD	0.03	10.58	<LOD	<LOD	9.49	20.1	<LOD	<LOD	8.51	<LOD	<LOD	<LOD	<LOD	<LOD	<LOD	<LOD	<LOD	<LOD	<LOD	<LOD	<LOD	8.51	2019	<LOD	<LOD	<LOD	<LOD	5.30	5.30
Cascade 2020	<LOD	0.0338	0.21	5.98	<LOD	<LOD	5.63	11.9	<LOD	<LOD	<LOD	<LOD	<LOD	<LOD	<LOD	<LOD	<LOD	<LOD	<LOD	<LOD	<LOD	<LOD	<LOD	<LOD	3864	<LOD	<LOD	0.12	<LOD	0.48	0.60
Celeia	<LOD	<LOD	0.02	11.19	<LOD	<LOD	783	795	<LOD	<LOD	<LOD	1053	<LOD	0.36	0.161	<LOD	<LOD	<LOD	<LOD	<LOD	<LOD	442	<LOD	1495	3053	<LOD	0.23	<LOD	<LOD	18.7	18.9
Centennial	<LOD	<LOD	0.13	10.64	<LOD	<LOD	14.1	24.8	<LOD	<LOD	2.03	68.8	<LOD	<LOD	<LOD	<LOD	<LOD	16.0	<LOD	<LOD	<LOD	<LOD	<LOD	86.8	3843	<LOD	0.38	<LOD	0.92	20.2	21.5
Centennial	<LOD	<LOD	0.43	12.24	<LOD	<LOD	10.2	22.9	<LOD	<LOD	<LOD	4.31	<LOD	<LOD	<LOD	<LOD	<LOD	17.2	<LOD	<LOD	<LOD	<LOD	<LOD	21.5	3416	<LOD	<LOD	<LOD	2.74	63.8	66.5
Centennial (Aba)	<LOD	<LOD	<LOD	18.18	<LOD	<LOD	34.6	52.7	<LOD	<LOD	<LOD	<LOD	<LOD	<LOD	<LOD	<LOD	<LOD	<LOD	<LOD	17.4	<LOD	<LOD	<LOD	17.4	3587	<LOD	<LOD	<LOD	1.86	9.13	11.0
Centennial 2020	<LOD	<LOD	0.34	7.62	<LOD	<LOD	31.6	39.6	<LOD	<LOD	7.82	23.6	<LOD	0.23	0.003	<LOD	<LOD	<LOD	<LOD	10.7	<LOD	<LOD	<LOD	42.3	2861	<LOD	<LOD	<LOD	1.54	32.9	34.4
Challenger 2021	<LOD	0.2163	0.53	12.95	309.3	<LOD	376	699	<LOD	<LOD	8.12	46.9	<LOD	0.51	0.310	<LOD	<LOD	<LOD	<LOD	<LOD	<LOD	112	<LOD	168	3578	0.91	0.87	0.67	4.13	37.7	44.2
Chinook	<LOD	<LOD	0.63	2.81	<LOD	<LOD	6.92	10.4	<LOD	<LOD	<LOD	197	2.79	<LOD	<LOD	<LOD	<LOD	15.7	<LOD	<LOD	<LOD	<LOD	<LOD	215	3224	<LOD	<LOD	<LOD	<LOD	23.6	23.6
Chinook (Aba)	<LOD	<LOD	0.06	9.87	<LOD	<LOD	13.4	23.3	<LOD	<LOD	<LOD	153	<LOD	<LOD	<LOD	<LOD	<LOD	15.1	<LOD	<LOD	<LOD	40.9	<LOD	209	1990	<LOD	0.30	<LOD	2.04	16.4	18.7
Citra	<LOD	<LOD	0.49	9.90	<LOD	<LOD	11.6	22.0	<LOD	<LOD	<LOD	<LOD	<LOD	<LOD	<LOD	<LOD	<LOD	19.2	<LOD	<LOD	<LOD	25.5	<LOD	44.7	2549	<LOD	0.25	<LOD	<LOD	87.8	88.1
Citra 2021	<LOD	<LOD	0.36	17.68	<LOD	<LOD	4.49	22.5	<LOD	<LOD	5.25	<LOD	21.3	0.022	<LOD	<LOD	<LOD	<LOD	<LOD	<LOD	<LOD	<LOD	<LOD	26.6	3634	<LOD	0.06	0.20	<LOD	58.3	58.6
Columbus	<LOD	<LOD	<LOD	24.86	<LOD	<LOD	23.0	47.9	<LOD	<LOD	6.26	93.2	<LOD	0.058	<LOD	<LOD	<LOD	<LOD	<LOD	<LOD	<LOD	<LOD	<LOD	99.5	3579	0.32	0.94	<LOD	9.35	4.26	14.9
Eldorado	<LOD	<LOD	0.10	15.36	<LOD	<LOD	31.3	46.8	<LOD	<LOD	<LOD	90.9	<LOD	0.008	0.080	<LOD	<LOD	16.1	<LOD	<LOD	<LOD	<LOD	<LOD	107	4341	<LOD	0.14	<LOD	2.72	31.2	34.0
Fuggle Bio	15.84	<LOD	0.73	14.85	173.9	<LOD	56.5	262	<LOD	<LOD	<LOD	30.9	15.1	0.92	0.379	<LOD	<LOD	15.9	<LOD	<LOD	<LOD	89.1	<LOD	152	2202	0.51	1.05	<LOD	3.26	62.1	67.0
Gerweiss Magnum	<LOD	<LOD	0.74	6.72	<LOD	<LOD	473	480	0.06	<LOD	29.2	12.5	<LOD	<LOD	0.597	0.208	<LOD	14.3	13.1	30.1	<LOD	162	0.58	265	2816	<LOD	0.68	<LOD	1.41	34.0	36.1
Golding	<LOD	<LOD	<LOD	19.55	<LOD	<LOD	864	883	0.29	<LOD	<LOD	27.3	<LOD	0.32	0.239	<LOD	<LOD	15.8	<LOD	<LOD	<LOD	515	<LOD	559	2982	<LOD	0.29	<LOD	0.15	2.34	2.77
H.M.180	<LOD	<LOD	<LOD	13.60	<LOD	<LOD	34.5	48.1	0.15	<LOD	4.93	<LOD	17.7	1.68	0.698	<LOD	<LOD	<LOD	<LOD	<LOD	<LOD	92.4	<LOD	118	3187	<LOD	<LOD	<LOD	<LOD	4.92	4.92
Hallertauer Magnum	<LOD	<LOD	0.71	9.12	<LOD	<LOD	351	361	0.88	<LOD	26.2	13.6	18.1	2.54	0.750	<LOD	<LOD	16.2	<LOD	<LOD	<LOD	197	2.37	278	2790	<LOD	0.35	0.05	2.07	27.5	30.0
Hopsi, Saaz	<LOD	<LOD	0.18	13.74	<LOD	<LOD	42.9	56.8	<LOD	<LOD	10.8	<LOD	<LOD	0.11	0.095	<LOD	<LOD	<LOD	<LOD	<LOD	4.39	64.7	<LOD	80.1	3947	<LOD	0.18	<LOD	0.64	21.6	22.5
Local Wild	<LOD	<LOD	0.20	17.39	<LOD	<LOD	91.8	109	<LOD	<LOD	<LOD	<LOD	<LOD	0.19	0.026	<LOD	<LOD	16.4	<LOD	<LOD	<LOD	84.4	<LOD	101	2499	<LOD	0.46	0.04	0.39	16.5	17.4
Local wild	<LOD	<LOD	0.46	13.27	<LOD	<LOD	21.2	35.0	0.13	<LOD	7.70	<LOD	<LOD	0.13	<LOD	<LOD	<LOD	<LOD	<LOD	<LOD	<LOD	27.1	<LOD	35.0	3094	<LOD	0.91	<LOD	3.42	27.4	31.7
Local wild	<LOD	<LOD	<LOD	26.62	<LOD	<LOD	145	172	<LOD	<LOD	<LOD	154	<LOD	0.043	<LOD	<LOD	<LOD	14.8	<LOD	<LOD	33.9	123	<LOD	326	3515	<LOD	1.11	<LOD	2.86	30.4	34.4
Local wild 2019	<LOD	<LOD	0.04	23.42	<LOD	<LOD	22.0	45.4	<LOD	<LOD	<LOD	<LOD	33.2	2.00	0.314	<LOD	<LOD	<LOD	<LOD	<LOD	55.1	<LOD	<LOD	90.6	4013	<LOD	0.29	<LOD	<LOD	21.7	22.0
Magnum	<LOD	<LOD	0.79	6.55	<LOD	<LOD	711	718	0.62	<LOD	26.0	12.2	<LOD	2.50	0.970	<LOD	<LOD	15.8	<LOD	30.4	<LOD	198	<LOD	287	2582	<LOD	0.47	<LOD	2.05	29.4	31.9
Magnum	<LOD	<LOD	0.64	8.77	<LOD	<LOD	292	301	0.20	<LOD	39.1	<LOD	<LOD	1.36	0.452	<LOD	<LOD	<LOD	20.3	<LOD	<LOD	135	0.20	196	2923	<LOD	0.37	<LOD	2.16	34.6	37.1
Magnum	<LOD	<LOD	0.85	5.62	<LOD	<LOD	349	356	<LOD	<LOD	39.4	19.1	<LOD	2.95	0.791	0.199	0.002	14.9	<LOD	<LOD	<LOD	187	<LOD	264	3016	<LOD	0.67	<LOD	1.83	43.9	46.4
Magnum (Aba)	<LOD	<LOD	0.82	5.30	<LOD	<LOD	377	383	<LOD	<LOD	31.2	29.9	<LOD	1.07	0.403	<LOD	<LOD	16.1	24.3	29.1	<LOD	151	0.72	284	3096	<LOD	0.60	<LOD	2.37	34.9	37.9
Magnum 2018	<LOD	<LOD	0.35	25.41	<LOD	<LOD	349	374	0.06	<LOD	9.22	44.8	<LOD	1.34	0.588	<LOD	<LOD	<LOD	<LOD	<LOD	<LOD	172	<LOD	228	3226	<LOD	0.86	<LOD	6.44	21.0	28.3
Mandarina Bavaria	<LOD	<LOD	0.68	2.48	<LOD	<LOD	47.6	50.8	<LOD	<LOD	<LOD	<LOD	<LOD	0.004	<LOD	<LOD	<LOD	16.2	<LOD	<LOD	<LOD	76.5	<LOD	92.7	2355	<LOD	0.23	<LOD	0.09	42.4	42.7

Figure 1. *Cont.*

name of hops	*Alternaria* spp.								*Fusarium* spp.					
	Altenusin [µg/kg]	Alternariol [µg/kg]*	Alternariolmethylether [µg/k	Altersetin [µg/kg]	Infectopyrone [µg/kg]	Tentoxin [µg/kg]*	Tenuazonic acid [µg/kg]*	sum Alternaria [µg/kg]	Beauvericin [µg/kg]**	Bikaverin [µg/kg]	Butenolid [µg/kg]	Culmorin [µg/kg]	Deoxynivalenol [µg/kg]***	Enniatin B [µg/kg]**
Mandarina Bravaria (Ale)	<LOD	<LOD	<LOD	2.51	<LOD	<LOD	41.7	44.2	<LOD	<LOD	<LOD	<LOD	<LOD	0.15
Mandarina Bravaria 2021	<LOD	<LOD	0.19	3.69	<LOD	<LOD	35.3	39.2	<LOD	<LOD	<LOD	88.6	<LOD	<LOD
Mosaic	<LOD	<LOD	0.86	11.00	<LOD	<LOD	4.77	16.6	<LOD	<LOD	<LOD	0.81	<LOD	0.002
Mosaic 2019	<LOD	<LOD	0.64	12.44	<LOD	<LOD	6.37	19.4	<LOD	<LOD	<LOD	8.64	<LOD	<LOD
Mosaic 2020	23.04	<LOD	0.84	10.80	<LOD	20.0	4.86	59.6	<LOD	<LOD	<LOD	10.1	<LOD	<LOD
Mosaic 2021	<LOD	<LOD	0.27	13.63	<LOD	<LOD	7.34	21.2	<LOD	<LOD	<LOD	304	<LOD	<LOD
Nugget	<LOD	<LOD	0.75	13.71	<LOD	<LOD	25.6	40.0	<LOD	<LOD	<LOD	10.4	<LOD	<LOD
Premium	<LOD	<LOD	0.12	17.01	<LOD	<LOD	207	224	<LOD	<LOD	<LOD	<LOD	<LOD	0.022
Saaz	<LOD	<LOD	0.24	12.64	<LOD	<LOD	47.0	59.9	<LOD	<LOD	<LOD	<LOD	<LOD	0.094
Saaz 2021	<LOD	<LOD	<LOD	17.09	95.1	<LOD	47.3	160	<LOD	<LOD	<LOD	406	<LOD	0.20
Sabro 2020	34.37	<LOD	<LOD	25.07	<LOD	<LOD	3.13	62.6	<LOD	<LOD	<LOD	582	<LOD	<LOD
Simcoe 2020	<LOD	0.0015	0.20	15.61	<LOD	<LOD	10.1	25.9	<LOD	<LOD	9.17	<LOD	<LOD	<LOD
Sladek	<LOD	<LOD	0.23	24.69	<LOD	<LOD	164	189	<LOD	<LOD	<LOD	<LOD	<LOD	0.082
Staročeško Magnum	<LOD	<LOD	1.02	6.22	<LOD	<LOD	310	317	<LOD	<LOD	51.8	12.2	<LOD	1.11
Styrian Aurora	<LOD	<LOD	0.32	16.51	<LOD	<LOD	55.8	72.7	<LOD	0.38	<LOD	<LOD	<LOD	0.16
Styrian Aurora 2021	<LOD	<LOD	0.18	20.08	<LOD	<LOD	61.4	81.7	<LOD	<LOD	<LOD	<LOD	<LOD	0.26
Styrian Cardinal (Ale)	<LOD	<LOD	0.33	13.80	<LOD	<LOD	158	172	<LOD	<LOD	<LOD	30.0	<LOD	0.035
Styrian Fox 2021	<LOD	<LOD	0.24	8.57	<LOD	<LOD	72.5	81.3	<LOD	<LOD	<LOD	6.63	<LOD	0.13
Styrian Goldine	<LOD	<LOD	0.06	12.47	<LOD	<LOD	678	691	0.07	<LOD	<LOD	<LOD	<LOD	0.39
Styrian Golding 2021	<LOD	<LOD	0.08	15.30	<LOD	20.1	688	724	0.46	0.37	<LOD	<LOD	<LOD	0.32
Styrian Golding Celeia	<LOD	<LOD	<LOD	22.66	<LOD	<LOD	693	715	<LOD	<LOD	<LOD	<LOD	<LOD	0.56
Styrian Wolf 2021	<LOD	<LOD	<LOD	<LOD	<LOD	<LOD	5.28	5.28	<LOD	<LOD	<LOD	<LOD	<LOD	0.013
Yakima Cascade	<LOD	<LOD	0.15	6.87	<LOD	<LOD	7.23	14.2	<LOD	<LOD	<LOD	16.3	<LOD	<LOD
Yakima Chinook	<LOD	<LOD	0.75	5.76	<LOD	<LOD	8.68	15.2	<LOD	<LOD	4.02	46.9	<LOD	<LOD
Yakima Citra	<LOD	<LOD	0.29	5.86	<LOD	<LOD	8.54	14.7	<LOD	<LOD	<LOD	<LOD	9.99	<LOD
Yakima Columbus	<LOD	<LOD	0.07	10.28	<LOD	<LOD	13.1	23.5	<LOD	<LOD	<LOD	53.6	768	0.015
average	24.41	0.0839	0.3772	12.71	192.8	20.09	172.8	196.8	0.292	0.373	15.99	108.5	97.09	0.585
sd	9.342	0.1159	0.284	6.212	108.3	0.079	264.3	270.8	0.278	0.003	14.62	208.6	236.2	0.807
min	15.84	0.0015	0.0166	2.475	95.14	20.03	3.13	5.285	0.059	0.371	2.026	0.805	2.792	0.002
max	34.37	0.2163	1.0152	26.62	309.3	20.14	1164	1174	0.881	0.375	51.78	1053	768	2.954
median	23.04	0.0338	0.3221	12.44	173.9	20.09	38.52	54.78	0.176	0.373	8.506	29.94	19.74	0.185
IQR	9.266	0.1074	0.5009	7.941	107.1	0.056	260.3	268.1	0.338	0.002	20.99	78.02	15.81	0.868
above LOD	3	3	51	61	3	2	62	62	10	2	21	34	10	41
prevalence	5%	5%	82%	98%	5%	3%	100%	100%	16%	3%	34%	55%	16%	66%

name of hops	*Fusarium* spp.										plant metabolite	unspecific					
	Enniatin B1 [µg/kg]**	Enniatin B2 [µg/kg]**	Enniatin B3 [µg/kg]**	Fumonisin B1 [µg/kg]***	HT-2 toxin [µg/kg]****	Moniliformin [µg/kg]**	Nivalenol [µg/kg]**	Siccanol [µg/kg]	T-2 toxin [µg/kg]****	sum Fusarium [µg/kg]	Abscisic acid [µg/kg]	3-Nitropropionic acid [µg/kg]	Brevianamid F [µg/kg]	Citreorosein [µg/kg]	cyclo(L-Pro-L-Val) [µg/kg]	Tryptophol [µg/kg]	sum unspecific [µg/kg]µ
Mandarina Bravaria (Ale)	<LOD	<LOD	<LOD	15.0	<LOD	<LOD	<LOD	<LOD	<LOD	15.1	2742	<LOD	0.35	<LOD	0.51	2.90	3.76
Mandarina Bravaria 2021	<LOD	<LOD	<LOD	<LOD	<LOD	<LOD	<LOD	<LOD	<LOD	88.6	2355	<LOD	0.42	<LOD	<LOD	17.6	18.0
Mosaic	<LOD	<LOD	<LOD	<LOD	<LOD	<LOD	<LOD	47.1	<LOD	47.9	3669	<LOD	<LOD	<LOD	<LOD	32.6	32.6
Mosaic 2019	<LOD	<LOD	<LOD	15.9	<LOD	<LOD	9.51	<LOD	<LOD	34.1	3304	<LOD	0.29	<LOD	<LOD	17.7	18.0
Mosaic 2020	0.013	<LOD	<LOD	<LOD	<LOD	<LOD	<LOD	117	<LOD	127	2592	<LOD	0.13	<LOD	<LOD	25.9	26.0
Mosaic 2021	<LOD	<LOD	<LOD	15.6	<LOD	<LOD	<LOD	65.9	<LOD	386	2688	0.12	0.56	<LOD	1.24	25.3	27.2
Nugget	<LOD	<LOD	<LOD	16.1	<LOD	<LOD	5.77	29.9	<LOD	62.2	4197	<LOD	0.95	<LOD	2.64	33.2	36.8
Premium	<LOD	<LOD	<LOD	<LOD	<LOD	<LOD	<LOD	80.2	<LOD	80.2	4032	<LOD	0.82	<LOD	1.41	43.2	45.4
Saaz	0.080	<LOD	<LOD	15.4	<LOD	<LOD	<LOD	125	<LOD	141	3707	<LOD	0.22	<LOD	<LOD	28.0	28.2
Saaz 2021	0.050	<LOD	<LOD	14.2	<LOD	<LOD	<LOD	66.4	<LOD	487	2187	<LOD	0.29	<LOD	<LOD	14.7	15.0
Sabro 2020	<LOD	<LOD	<LOD	<LOD	<LOD	<LOD	<LOD	<LOD	<LOD	582	4066	<LOD	0.13	<LOD	1.42	6.62	8.17
Simcoe 2020	<LOD	<LOD	<LOD	<LOD	<LOD	<LOD	<LOD	44.4	<LOD	53.6	2296	<LOD	0.54	<LOD	1.17	9.47	11.2
Sladek	<LOD	<LOD	<LOD	15.3	<LOD	<LOD	<LOD	120	3.22	138	4888	<LOD	1.93	<LOD	8.01	49.9	59.9
Staročeško Magnum	0.282	<LOD	<LOD	16.0	21.4	28.8	<LOD	170	0.80	303	3120	<LOD	1.22	<LOD	2.41	45.7	49.3
Styrian Aurora	<LOD	<LOD	<LOD	15.4	<LOD	<LOD	<LOD	90.9	<LOD	107	2746	<LOD	0.55	<LOD	0.91	34.4	35.8
Styrian Aurora 2021	0.051	<LOD	<LOD	<LOD	<LOD	<LOD	<LOD	88.2	<LOD	88.5	3472	<LOD	0.35	<LOD	1.72	39.1	41.2
Styrian Cardinal (Ale)	<LOD	<LOD	<LOD	15.8	<LOD	<LOD	<LOD	110	<LOD	156	3742	<LOD	0.68	<LOD	0.60	28.9	30.1
Styrian Fox 2021	0.048	<LOD	<LOD	15.1	<LOD	<LOD	<LOD	127	<LOD	149	2509	<LOD	<LOD	0.09	<LOD	8.82	8.91
Styrian Goldine	0.304	<LOD	<LOD	14.7	<LOD	<LOD	12.8	562	<LOD	591	2742	<LOD	0.43	<LOD	<LOD	12.3	12.7
Styrian Golding 2021	0.113	<LOD	<LOD	<LOD	<LOD	<LOD	<LOD	804	<LOD	805	3242	<LOD	<LOD	<LOD	<LOD	20.3	20.3
Styrian Golding Celeia	0.267	<LOD	<LOD	<LOD	<LOD	<LOD	<LOD	690	<LOD	691	2573	<LOD	<LOD	<LOD	<LOD	0.48	0.48
Styrian Wolf 2021	<LOD	<LOD	<LOD	<LOD	<LOD	<LOD	<LOD	<LOD	<LOD	0.01	1255	<LOD	0.51	<LOD	<LOD	47.3	47.9
Yakima Cascade	<LOD	<LOD	<LOD	18.0	<LOD	<LOD	<LOD	<LOD	<LOD	34.4	2258	<LOD	0.10	<LOD	0.98	14.0	15.1
Yakima Chinook	<LOD	<LOD	<LOD	<LOD	<LOD	<LOD	<LOD	<LOD	<LOD	51.0	1883	<LOD	0.15	<LOD	<LOD	25.3	25.5
Yakima Citra	<LOD	<LOD	<LOD	15.8	<LOD	<LOD	<LOD	6.51	<LOD	32.3	2677	<LOD	<LOD	<LOD	<LOD	54.4	54.4
Yakima Columbus	<LOD	<LOD	<LOD	18.2	<LOD	<LOD	<LOD	<LOD	<LOD	840	2766	0.74	0.64	<LOD	1.15	6.30	8.83
average	0.304	0.203	0.002	15.77	19.78	24.42	19.59	171.1	1.314	207.1	3057	0.724	0.508	0.196	2.113	26.24	28.08
sd	0.277	0.006	0	1.07	4.758	8.327	18.47	182.4	1.195	259.1	712	0.432	0.358	0.22	1.974	17.64	18.12
min	0.003	0.199	0.002	14.22	13.09	10.7	4.39	6.507	0.2	0.013	1255	0.124	0.058	0.045	0.087	0.485	0.485
max	0.97	0.208	0.002	19.2	24.27	30.42	55.06	803.7	3.218	1495	4888	1.272	1.928	0.675	9.35	87.84	88.09
median	0.267	0.203	0.002	15.78	20.87	28.97	12.8	117	0.759	107.1	3074	0.743	0.426	0.119	1.672	25.32	26.61
IQR	0.397	0.005	0	1.05	3.623	9.58	17.15	107.6	1.364	216.7	994	0.635	0.422	0.129	1.581	20.61	22.06
above LOD	29	2	1	35	4	6	7	39	6	61	62	7	50	7	39	62	62
prevalence	47%	3%	2%	56%	6%	10%	11%	63%	10%	98%	100%	11%	81%	11%	63%	100%	100%

Figure 1. Results of all detected toxins and metabolites in the hop samples. * Toxins mentioned in EC 553/2022 [13]; ** toxins from EFSA's annual call for continuous collection of chemical contaminants occurrence data in food and feed; *** toxins mentioned in EC 915/2023 [29]; **** toxins mentioned in EC 165/2013 [30].

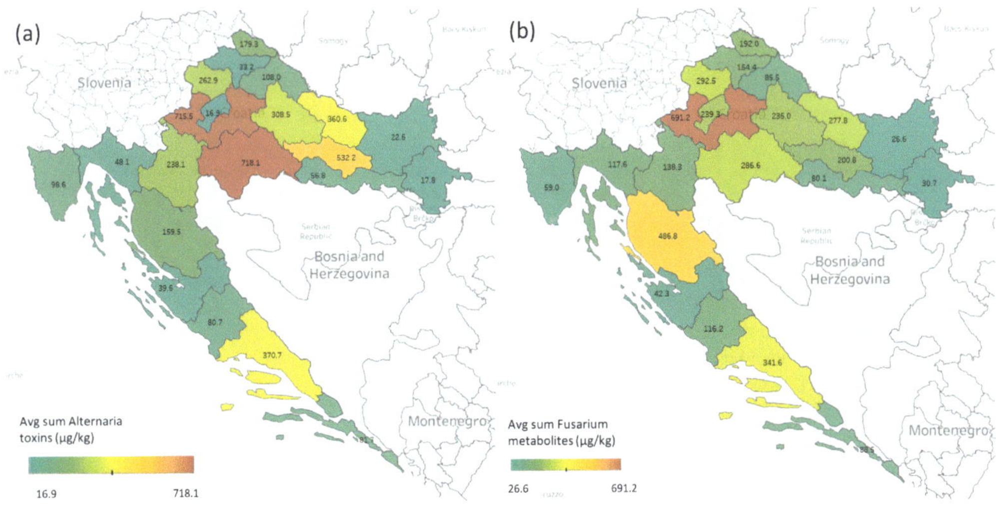

Figure 2. *Cont.*

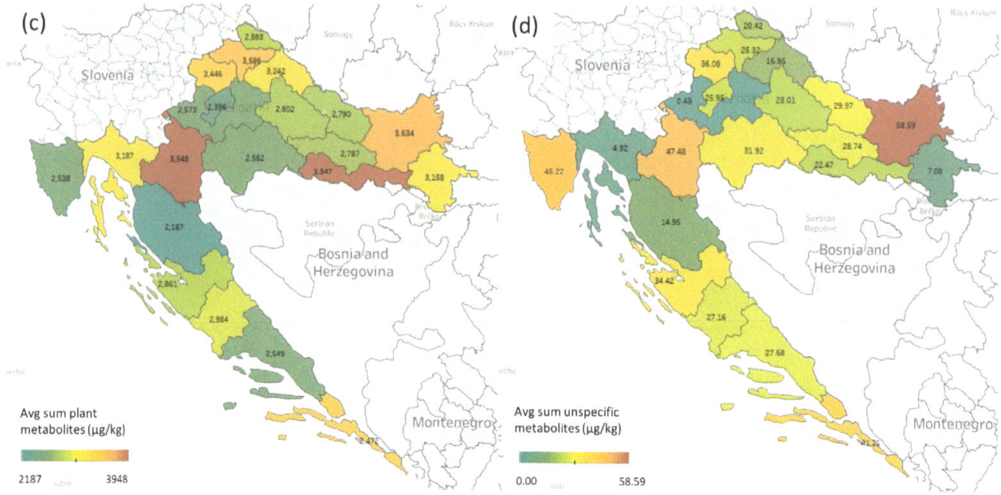

Figure 2. Heat map of the toxin distribution in hop samples from different Croatian counties. All concentrations are the average sum of the toxins in the selected county expressed in µg/kg. The division of the figure is (**a**) *Alternaria* toxins, (**b**) *Fusarium* toxins, (**c**) plant metabolites, (**d**) unspecific metabolites.

2.1. *Alternaria Toxins*

Alternaria alternata, a known producer of *Alternaria* toxins, is one of the fungi that contaminates hop plants [4,10,11,14]. The results in Figures 1 and 2a confirm the contamination of hops with *Alternaria* toxins. All tested samples confirmed the presence of at least one *Alternaria* toxin, with tenuazonic acid (TeA) showing the highest occurrence (100% of the samples), followed by altersetin (ALT) (98% of the samples), and alternariol methyl ether (AME) (quantified in 82% of the samples). The rest of the *Alternaria* toxins were present in less than 5% of the samples—infectopyrone, altenusin, tentoxin (TTX), and alternariol (AOH). A similar occurrence pattern of *Alternaria* toxins was found in similar matrices such as plants from India [31], tobacco [32], and pu-erh tea [33], while different occurrences were found in matrices such as figs, sunflower seeds, wine [34], barley [35], and millet [36].

The occurrence of TTX was low compared to other matrices, both in terms of occurrence data and measured concentrations, when compared to AOH, AME, or TeA. While some studies show higher AOH levels than AME in wheat-based products [37], TeA is always higher in concentration and occurrence compared to other *Alternaria* toxins. Overall, their occurrence data are well summarized in EFSA's dietary exposure assessment to *Alternaria* toxins in the European population [38]. It is well documented that *Alternaria* spp. produces host-specific mycotoxins [39–41], so every matrix will have its specific distribution of different *Alternaria* toxins. Since this is the first report, it is not possible to compare the results to other published results.

In the European Union, three *Alternaria* mycotoxins are recommended for monitoring (EC 553/2022) [13] but not in matrices such as hops, probably due to a lack of data on their occurrence. With these new data, they could be included in a monitoring plan, risk assessment by EFSA, or legislation due to a relatively high maximum concentration of TeA (1174 µg/kg), exceeding most of the proposed limits in the Commission recommendations (EC 553/2022) [13] on monitoring the presence of *Alternaria* toxins in food. The concentrations of the other two *Alternaria* toxins mentioned in the legislation do not exceed the proposed limits, although they should not be excluded from monitoring due to their toxicities [42]. When comparing the locations of the sampling, the highest average sums of all *Alternaria* toxins are in Zagreb County and Sisak-Moslavina County, while the lowest average sums of all *Alternaria* toxins were measured in samples collected in Zagreb City and Vukovar-Srijem County (Figure 2a). When comparing the results, there was a statistically significant difference in the individual and sum of all *Alternaria* toxin distributions between different Croatian counties ($p < 0.01$).

2.2. *Fusarium* Toxins

Fusarium spp. is one of the fungi occurring most often on hop plants, causing HF [16], FC (wilting, cankers in the crown, foliar necrosis, and death of infected plants) [17], and HW [18]. The frequent reports of *Fusarium* contamination and no mycotoxin occurrence data were a bit surprising. In this research, we confirmed the presence of 15 different *Fusarium* toxins, and on average 98% of the samples were contaminated by at least one of the *Fusarium* mycotoxins (Figures 1 and 2b). Out of the quantified mycotoxins, some are regulated in the European Union (deoxynivalenol (DON), fumonisin B1 (FB1) by commission regulation EC 915/2023 [29]); followed by some regulated by Commission recommendations on the presence of T-2 and HT-2 toxins in cereals and cereal products (EC 165/2013 [30]) (T-2 and HT-2 toxins); mycotoxins from EFSA's annual call for continuous collection of chemical contaminants occurrence data in food and feed (beauvericin (BEA), enniatin B (ENNB), enniatin B1 (ENNB1), enniatin B2 (ENNB2), enniatin B3 (ENNB3), moniliformin (MON), nivalenol (NIV)), and others are not mentioned in any official document. Interestingly, no modified or masked forms of mycotoxins were found, possibly due to relatively low contamination of specific metabolic pathways in hops, which do not include glucuronidation and sulfation. UDP-glucosyltransferase (UGT) is used in plants for modifying/masking of mycotoxins [9], but in hops it can also interfere with the taste through the formation of flavorless glucosides of terpenoids [43]. Therefore, hop varieties used in brewing are usually ones with lower UDP activity, diminishing their capacity for defending against mycotoxins. Although we also tested four local wild varieties, they also did not show any ability of masking compared to the commercial varieties.

Out of the unregulated *Fusarium* mycotoxins, the presence of bikaverin (BKV), butenolide (BUT), culmorin (CUL), and siccanol (SIC) was confirmed. When comparing the occurrence of *Fusarium* mycotoxins, ENNB had the highest occurrence, detected in 66% of samples, followed by SIC (63%), FB1 (56%), CUL (55%), ENNB1 (47%), BUT (34%), BEA (16%), DON (16%), NIV (11%), MON (10%), T-2 (10%,), HT-2 (6%), BKV (3%), ENNB2 (3%), and ENN B3 (2%). When comparing the regulated mycotoxins to their legal limits, the highest detected DON level (768 μg/kg) exceeds limits for baby food and processed cereal-based food for infants and young children, bread, pastries, biscuits, cereal snacks and breakfast cereals, cereals placed on the market for the final consumer, cereal flour, semolina, bran and germ as final products placed on the market for the final consumer, and other milling products of maize not placed on the market for the final consumer. With this detected level it is advisable to perform widespread screening and additional risk assessment, due to hops' usage in mainly water-soluble products (beer, infused water, and pharmaceutical products). Comparing the T-2 and HT-2 levels to the legislation, all samples with detected HT-2 toxin (with a maximum of 3.22 μg/kg and 24.3 μg/kg, respectively, and a median of 0.76 μg/kg and 15.8 μg/kg) exceeded the indicative level for cereal-based foods for infants and young children, and one sample exceeded the level for bread (including small bakery wares), pastries, biscuits, cereal snacks, and pasta. The FB1 levels were quite low compared to the legal limits, with a maximum detected level of 19.2 μg/kg, and median level of 15.8 μg/kg. Out of the other mycotoxins, the presence of enniatins (ENNs) was expected, and they were present in low concentrations, although their co-occurrence was high (all ENNs and BEA). The high occurrence of CUL (55%), and its high detected concentration of up to 1053 μg/kg, a *Fusarium* mycotoxin that suppress the in vitro glucuronidation of DON, can be concerning, due to lowering humans' detoxification capacity towards DON [44]. The other two unregulated, yet highly occurring, mycotoxins were SIC and BUT, with 63% and 55% occurrence in the tested samples. Their highest detected concentrations were 804 μg/kg and 51.8 μg/kg, but since there are not enough toxicological data at these concentrations, their toxicological relevance is yet to be investigated. Similar co-occurrence was not detected in other matrices, leaving the possibility of a similar host-dependent pattern of mycotoxins as found in *Alternaria* toxins [39,41], and also for *Fusarium* mycotoxins [45,46]. When comparing the regional distribution of the *Fusarium* toxins in hops, the same counties (and samples) that were highly contaminated by *Alternaria* toxins were also highly contaminated by *Fusarium* toxins. This is possible due to the phytotoxic effect of TeA, as it helps *Alternaria* spp. spread on the plants [47]. When compared, there was a statistically significant ($p < 0.05$) correlation between TeA and the sum of *Alternaria* toxins (r = 0.98), and TeA and the sum of *Fusarium* toxins (r = 0.61). Zagreb county had the highest average contamination by the sum of *Fusarium* toxins (691 μg/kg), while Osijek-Baranja and Vukovar-Srijem counties had the lowest average measured concentrations of *Fusarium* toxins (Figure 2b), similarly to *Alternaria* toxins (Figure 2a).

2.3. Plant Metabolites

The only plant metabolite that was detected in quantifiable concentrations was abscisic acid (ABA), as shown in Figure 1. It was detected in all of the samples at relatively high concentrations, ranging from 1255 μg/kg to 4888 μg/kg. As one of the drought stress indicators, ABA was surprisingly not correlated with a lot of the mycotoxins, although mycotoxin production is correlated with drought during growth [6,48,49]. The only statistically relevant correlations ($p < 0.05$) were between ABA and NIV (r = 0.26), and ABA and ALT (r = 0.41). Also, in regional distribution (Figure 2(c)), the average ABA concentrations were not similar to other mycotoxins, with the highest values in Karlovac and Slavonski Brod Posavina County (3948 μg/kg and 3947 μg/kg), and the lowest in Lika-Senj County and City of Zagreb (2187 μg/kg and 2396 μg/kg). The detected ABA concentrations are affected by the growth conditions and are not expected to change significantly during storage.

2.4. Unspecific Metabolites

The unspecific metabolites are all the other metabolites that can be produced by several fungal, bacterial, or plant sources. The most common was tryptophol, often found in yeast as a quorum-sensing molecule [50], which can be produced by plants, bacteria, fungi, and sponges [51] (Figure 1). In the tested hop samples, it was the most frequently occurring unspecific metabolite, with a range from 0.48 μg/kg to 88.1 μg/kg, and a median of 25.3 μg/kg. The second most frequently occurring metabolite was brevianamide F or cyclo-(L-Trp-L-Pro), found in different microorganisms as a precursor of tryptophan-proline 2,5-diketopiperazines, a large group of primary and secondary metabolites in microbes, that explains the high occurrence of 81%, with relatively low concentrations, from 0.06 μg/kg to 1.93 μg/kg. Also, in addition to this cyclopeptide, cyclo-(L-Pro-L-Val) was also often detected (in 63% of the samples), but also at similar low concentrations, from 0.09 μg/kg to 9.35 μg/kg. The last two of the unspecific metabolites were 3-nitro propionic acid and citreorosein, both in 11% of the samples, with relatively low ranges (0.12–1.27 μg/kg and 0.04–0.67 μg/kg, respectively). Both of these metabolites can be produced by a wide variety of fungi, and at the detected concentrations they do not pose a risk to human health. Interestingly, when comparing the regional distribution, the highest average concentration of the sum of unspecific metabolites was in Osijek-Baranja County, while the lowest was in Zagreb County, the reverse of the other fungal metabolites (*Alternaria* and *Fusarium*) (Figure 2d).

It is interesting to note that no *Aspergillus* or *Penicillium* metabolites were detected. Both fungal genera are considered storage fungi, suggesting that all craft brewers that were storing the hops for brewing were careful about the storage conditions. It is recommended by the producers that hops are stored in a cold, dark, and dry environment, such as a freezer, where all chemical reactions that degrade hops' quality are slower. On the other hand, it was expected, according to the literature data, that there should not be any *Aspergillus* spp. and a small amount of *Penicillium* spp., and this was confirmed by the metabolite scan.

2.5. Transfer of Mycotoxins in Hop Products

The major usage of hops is in the brewing industry, where they can be used in three forms: whole dried cone, pellets, and hop extract. The largest worldwide producer is the European Union, with an annual production of 50,000 tons, where the main producers are Germany, Czechia, Poland, and Slovenia [52]; and other high producing countries are the United States, China, and Australia, according to Statista data for the production volume of hops worldwide in 2022 by country [53]. Currently, more than 98% of all produced hops are used in the brewing industry, where the main product used for brewing is hop pellets [1], which have been investigated in this research. Outside the brewing industry, the rest of the usage is in the pharmaceutical industry, where they can use fresh hops or spent hops—a by-product of the brewing industry. As summarized by Korpelainen, et al. (2021), traditionally, hops were used in beer flavoring, for preserving and clarifying, as a vegetable, in bread making (to cultivate yeast), as a preservative in sausages, for flavoring water, in baked goods, in tobacco, as cattle fodder, for manure preparation, as a hair rinse for brunettes, as a deodorant (antimicrobial, fragrance), in perfumes, in skin lotions, as oil, in pharmacy as an antibiotic, as an anti–inflammatory, as a sedative, for sleep disturbances, headache, restlessness, tenderness of limbs, bleary eyes, gastric problems, indigestion, appetite, toothache, earache, neuralgia, for treating leprosy, tuberculosis, asbestosis, and silicosis, as an anthelmintic, as an antiparasitic drug, for cough, spasms, fever, and anxiety, for clearing blood, flatulence, delirium tremens, irritable bladder, aches, and diuresis, and in liver disorders (porphyria) [1].

The effects of processing on mycotoxins are diverse based on the type of processing. Generally, mycotoxins are considered thermostable, and that is why they are one of the most critical of the chemical contaminants in processed food [5]. Ninety-eight percent of produced hops are used in the brewing industry. The first step in hop processing is drying, which has to be performed gently at low temperatures (below 60 °C) to prevent color and flavor changes [1]. At these temperatures mycotoxins are stable, and no mycotoxin losses are expected [5]. After drying, the next step is milling and palletization, where low temperature extruders are used. The palletization temperature should not exceed 50 °C to reduce color and flavor changes [1]. There are no data on mycotoxin loss in the extrusion process at such low temperatures, but Janić Hajnal et al. (2022) have investigated the effects of the extrusion process on *Fusarium* and *Alternaria* mycotoxins (the two most relevant mycotoxin groups found in this research) in triticale flour [54] and concluded that due to complex interaction of various parameters, the effect of the extrusion process on the investigated mycotoxins still needs to be determined in detail for each combination of ingredients as well as for the applied parameters. Use of the optimal parameters determined for lowering the concentrations of the investigated mycotoxins gave a 9.5–85.7% reduction. Based on those results, it can be expected that raw hops have a higher mycotoxin content prior to palletization, but this presumption should be confirmed in a separate study. Finally, mycotoxin transfer from hops to beer would be expected to be highly dependent on the type of hopping during brewing and the chemical characteristics of the transferred mycotoxin. It is expected that all mycotoxins that are polar (e.g., DON) should be transferred to the final product, with a reduction by binding to yeast cells during fermentation, or masking due to the metabolic activity of yeast [9]. Hops can be added early during boiling, and all the way towards the end of fermentation. The contact time and temperature would change the mycotoxin transfer from the hops to the beer. This transfer has been noted in some research, where the authors had new, previously undetected mycotoxins in beer that were not found in cereals used in brewing or in wort, but were detected after hopping [55]. It is highly possible that those mycotoxins were extracted from hops. Mastanjević et al., (2019) detected statistically significant higher DON concentrations in beer samples than in wort [56], while in other research a decrease in the mycotoxin content between the wort and the beer was noted, except for zearalenone and tentoxin, the concentrations of which increased without explanation [57]. None of the published research prior to this paper considered hops as the source of the mycotoxins. Still brewing is a complex process that can have unexpected and unpredicted changes in the mycotoxin mixture from the basic ingredients to the final beer, depending on numerous variables—contamination of the ingredients, parameters used during brewing, yeast used during fermentation, used equipment, and type of beer [55]. It is not possible to find mycotoxin-free samples of cereals and hops to be able to eliminate one variable in mycotoxins transfer to the final product, and the only possible reliable way the transfer could be calculated would be by adding a known amount of mycotoxins by spiking samples and comparing them to unspiked ingredients under the same brewing conditions, or even more precisely, the usage of stable-isotope-labeled mycotoxin standards, that are currently not available for all of the toxins that have been detected in cereals and hops.

2.6. Health Perspective, Limitations, and Outlook

Out of the detected mycotoxins, the most concerning for health are those that are mentioned in the legislation such as AOH, AME, TeA, DON, FB1, T-2, and HT-2, and those that EFSA recognized in the call for continuous collection of chemical contaminant occurrence data in food and feed such as TTX, ENNs MON, and NIV. There are many reports on the toxic effects of the mentioned mycotoxins. AOH, AME, TeA, and TTX were found in relatively low concentrations compared to those that were investigated in toxicity testing studies [38,58]. Aichinger et al. (2021) have reviewed the toxic effects of *Alternaria* toxins with the available data, and AOH and AME, based on EFSA's toxicological threshold of concern (TTC) values, showed increased risk compared to TeA. TeA has shown moderate acute toxicity in in vivo and in vitro tests, while AOH showed immunosuppression and topoisomerase poisoning in in vitro tests, and estrogenicity and androgenicity (endocrine disruption) in in silico and in vitro tests [58]. The toxic effects of *Fusarium* toxins are well documented, where trichothecenes have been studied in more detail, due to their higher occurrence compared to other *Fusarium* mycotoxins [5,9]. There are proposed values of TDI for DON (1 μg/kg b.w./day) based on the reduced body weight gain of mice, and an acute reference dose (ARfD) of 8 μg/kg b.w. per meal was calculated [59]; the value for FB1 was 1 μg/kg b.w./day based on the increased incidence of megalocytic hepatocytes in the chronic study in mice [60]; as well as a combined temporary TDI for the sum of T-2 and HT-2 of 0.06 μg/kg b.w./day based on the general toxicity, hematotoxicity, and immunotoxicity of T-2 toxins [61]. Since, currently, hops are mainly used for beer flavoring, the detected concentrations should not pose a risk to human health if ingested in beer, and there are other ingredients that pose a greater threat (malt, cereals) to the mycotoxin burden of the final beer. The potential hazard could be in the application of the hops as a pharmaceutically active substance, where they are applied differently and can be absorbed by the body through the skin or intravenously. Additionally, risks should be calculated for the farmers and hop processing plant workers that could inhale the hop dust containing microbes and mycotoxins. The additional concerns are due to co-occurring mycotoxins that can change the total toxic impact; one of these highly occurring combinations that was noted in hops is DON-CUL. Culmorin can suppress the in vitro glucuronidation of DON, a main route of detoxification of DON in animals [44]. In combination, CUL can decrease animals' potential to detoxify DON and increase its toxicity. There are still numerous uninvestigated combinations that could have antagonistic, additive, or synergistic effects.

The limitations of this study are the single year of monitoring, where only one season of weather that could affect the mycotoxin occurrence is captured; for further perspectives, it would be good to continue with regular monitoring to confirm the occurrence pattern. Additionally, it would be interesting to check if the same mycotoxin pattern is observed worldwide, or only in European hop samples. This first-report study, while contributing new insights, also underscores the need for further research to corroborate these initial results. In further research, the extraction patterns should also be evaluated due to differences in hop usage in the brewing industry (there are different times and temperatures of hop addition, that can affect the extraction of mycotoxins), or when used in the pharmaceutical industry different solvents can be used (water, vaseline). In brewing, bittering hops can be added early in the boil, flavor hops are added in the middle towards the end of the boil, aroma hops are added at the very end of the boil, or just after the boil, dry-hopping is performed during or after fermentation, usually cold, and whirlpool hopping is performed during the whirlpooling process. With an increased number of reports, a proper risk assessment can be calculated and it can be assessed whether changes in legislation are needed to ensure human health.

3. Conclusions

This study is the first one to confirm the presence of multiple mycotoxins associated with fungi in hop samples, indicating the need for additional risk assessment. Notably, Alternaria toxins were prevalent, with tenuazonic acid showing the highest occurrence (in all tested samples). Fusarium toxins were also widespread, posing concerns due to their regulatory limits, especially DON, which exceeded limits for various food products. The correlation between the distribution of Alternaria and Fusarium toxins suggests a shared contamination pattern in the field. These results highlight the importance of monitoring and regulating mycotoxin levels in hops to ensure consumer safety in craft brewery products.

4. Materials and Methods

A total of 62 hop samples used by Croatian craft breweries were collected, with at least one sample per county, so the whole Croatian market was included. The samples were collected in winter 2022 and spring 2023 in accordance with the Commission Regulation (EC) No 401/2006 of 23 February 2006, laying down the methods of sampling and analysis for the official control of the levels of mycotoxins in foodstuffs, with the help of official inspectors and trained scientists for the sampling. All of the hop samples collected were from the harvesting season of 2021 and 2022, and were in the form of pellets. Brewers (where samples were collected) kept the hop samples tightly closed or in the original packaging with zip locks and stored them in freezers (−18 °C) or refrigerators (+4 to +8 °C) to keep the aroma profile of the hops. The samples were stored and transported to Austria at −20 °C. For the extraction and quantitation, the LC-MS/MS multi-mycotoxin method developed by Sulyok et al. (2020) [62] was used.

4.1. Chemicals and Reagents

Acetonitrile, methanol, and glacial acetic acid (MS grade) were purchased from Merck (Darmstadt, Germany), while ammonium acetate was purchased from Sigma-Aldrich (Vienna, Austria). Ultra-pure water was produced by Elga Purelab ultra from Veolia Water (Bucks, UK). The standards used for calibration of the multi-toxin LC-MS/MS method were obtained either as gifts from various research groups, or from commercial sources including AnalytiCon Discovery (Potsdam, Germany), Axxora Europe (Lausanne, Switzerland), BioAustralis (Smithfield, Australia), Enzo Life Sciences (Lausen, Switzerland), Iris Biotech GmbH (Marktredwitz, Germany), LGC Promochem GmbH (Wesel, Germany), RomerLabs® (Tulln Austria), Sigma-Aldrich (Vienna, Austria), and Toronto Research Chemicals (Toronto, ON, Canada). Stock solutions were purchased or prepared by dissolving a solid standard in water, acetonitrile, methanol, or their mixtures. All stock solutions were stored at −20 °C and allowed to reach room temperature prior to usage.

4.2. Sample Extraction

All samples were weighed in a 50 mL Falcon tube (5.00 g each) and extracted with 40 mL of extraction solvent (acetonitrile/water/acetic acid 79:20:1, *V*/*V*/*V*). After the addition of the extraction solvent, the tubes were vortexed to allow separation of the hop pellets and ensure proper extraction. The extraction was carried out on a GFL 3017 rotary shaker (GFL; Burgwedel, Germany) for 90 min, at 180 rpm at room temperature. After extraction, the tubes were allowed to settle and 1 mL of the extract was transferred to an Eppendorf tube where the fatty layer was separated by pipetting 500 µL of the bottom layer to an autosampler vial and diluted with 500 µL of dilution solvent (acetonitrile/water/acetic acid 20:79:1, *V*/*V*/*V*). The samples were analyzed without any further manipulation according to the state-of-the-art LC-MS/MS multi-toxin method described by Sulyok et al. (2020) [62]. The samples were analyzed on an Agilent 1290 UHPLC system (Agilent Technologies, Waldbronn, Germany), coupled to a Sciex QTrap 5500 MS/MS equipped with an ESI source (TurboV) (Sciex, Foster City, CA, USA). The chromatographic separation was performed on a Gemini C18 column (150 × 4.6 mm i.d., 5 μm filling particle size), and a C18 security guard cartridge (4 × 3 mm i.d., 5 μm filling particle size) (Phenomenex, Torrance, CA, USA). The used eluents were as follows: eluent A: methanol:water:acetic acid (10:89:1, *V*/*V*/*V*); eluent B: methanol:water:acetic acid (97:2:1, *V*/*V*/*V*). All of the further method details are described in the method paper published by Sulyok et al., (2020) [62].

4.3. Statistical Analysis

For the statistical analysis, Statistica 14.1.0.8. (Cloud Software Group Inc.) and Tableau Desktop 2023.3.1 (Tableau Software) were used. For the normality of the data distribution the Shapiro–Wilk W test was used, and for further comparison of the data distribution non-parametric tests were used (Mann–Whitney U test, and Kruskal–Wallis ANOVA), while for the correlations the Spearman rank-order test was used. Tableau was used to generate heat maps based on the average concentrations in the samples from different Croatian counties. All data below LOD values were substituted with 0 in accordance with the lower bound EFSA guide on the management of left-censored data [63].

Author Contributions: Conceptualization, I.D.Š., N.V., K.M. and B.Š.; methodology, I.D.Š., N.V., K.M., M.S., R.K. and B.Š.; software, K.H.; validation, M.S., R.K. and B.Š.; formal analysis, I.D.Š., J.H., N.U.-T., K.H., M.S. and B.Š.; resources, I.D.Š., N.U.-T., R.K. and B.Š.; writing—original draft preparation, I.D.Š., N.V., K.M. and B.Š.; writing—review and editing, J.H., N.U.-T., K.H., M.S. and R.K.; visualization, K.H. and B.Š.; supervision, N.V., K.M. and R.K. All authors have read and agreed to the published version of the manuscript.

Funding: This research and APC was funded by Brico (UNIN-BIOTEH-21-1-2), VIP BEER (UNIN-BIOTEH-22-1-2), and MycoWIN (UNIN-BIOTEH-24-1-2).

Institutional Review Board Statement: Not applicable.

Informed Consent Statement: Not applicable.

Data Availability Statement: Dataset available on request from the authors.

Acknowledgments: We wish to thank all of the craft breweries in Croatia that participated in this survey, and Hannes Puntscher for all of the help in the laboratory.

Conflicts of Interest: The authors declare no conflicts of interest.

References

1. Korpelainen, H.; Pietiläinen, M. Hop (*Humulus lupulus* L.): Traditional and Present Use, and Future Potential. *Econ. Bot.* **2021**, *75*, 302–322. [CrossRef]
2. Karabín, M.; Hudcová, T.; Jelínek, L.; Dostálek, P. Biologically Active Compounds from Hops and Prospects for Their Use. *Compr. Rev. Food Sci. Food Saf.* **2016**, *15*, 542–567. [CrossRef] [PubMed]
3. Kolenc, Z.; Langerholc, T.; Hostnik, G.; Ocvirk, M.; Štumpf, S.; Pintarič, M.; Košir, I.J.; Čerenak, A.; Garmut, A.; Bren, U. Antimicrobial Properties of Different Hop (*Humulus lupulus*) Genotypes. *Plants* **2023**, *12*, 120. [CrossRef] [PubMed]
4. Phalip, V.; Hatsch, D.; Laugel, B.; Jeltsch, J.M. An Overview of Fungal Community Diversity in Diseased Hop Plantations. *FEMS Microbiol. Ecol.* **2006**, *56*, 321–329. [CrossRef] [PubMed]
5. Šarkanj, B.; Šarkanj Dodlek, I.; Shamtsyan, M. Mycotoxins in Food—How to Prevent and What to Do When Things Go Bad. *E3S Web Conf.* **2020**, *215*, 01004. [CrossRef]
6. Medina, Á.; Rodríguez, A.; Magan, N. Climate Change and Mycotoxigenic Fungi: Impacts on Mycotoxin Production. *Curr. Opin. Food Sci.* **2015**, *5*, 99–104. [CrossRef]
7. Gwinn, K.D.; Leung, M.C.K.; Stephens, A.B.; Punja, Z.K. Fungal and Mycotoxin Contaminants in Cannabis and Hemp Flowers: Implications for Consumer Health and Directions for Further Research. *Front. Microbiol.* **2023**, *14*, 1278189. [CrossRef] [PubMed]
8. Lattab, N.; Kalai, S.; Bensoussan, M.; Dantigny, P. Effect of Storage Conditions (Relative Humidity, Duration, and Temperature) on the Germination Time of *Aspergillus carbonarius* and *Penicillium chrysogenum*. *Int. J. Food Microbiol.* **2012**, *160*, 80–84. [CrossRef] [PubMed]

9. Kovač, M.; Šubarić, D.; Bulaić, M.; Kovač, T.; Šarkanj, B. Yesterday Masked, Today Modified; What Do Mycotoxins Bring Next? *Arh. Hig. Rada Toksikol.* **2018**, *69*, 196–214. [CrossRef]
10. Liu, H.; Park, S.; Sang, H. Identification and Fungicide Control of Bipolaris Sorokiniana Causing Leaf Spot and Blight on Common Hop (*Humulus lupulus*) in Korea. *Plant Dis.* **2023**, *107*, 2929–2943. [CrossRef]
11. Gargani, E.; Faggioli, F.; Haegi, A. A Survey on Pests and Diseases of Italian Hop Crops. *Italus Hortus* **2018**, *24*, 1–17. [CrossRef]
12. Krofta, K.; Mravcová, L.; Kolek, D.; Patáková, P.; Patzak, J.; Henychová, A.; Dostálek, P. Microbial Community of Hop (*Humulus lupulus* L.) and Its Impact on the Quality of Hop Products. *Acta Hortic.* **2021**, *1328*, 109–114. [CrossRef]
13. European Commission. Commission Recommendation (EU) 2022/553 of 5 April 2022 on monitoring the presence of *Alternaria* toxins in food. *Off. J. Eur. Union* **2022**, *L 107/90*, 90–92.
14. Darby, P. *Alternaria alternata* Infection of Hop (*Humulus lupulus*) Cones. *Trans. Br. Mycol. Soc.* **1988**, *90*, 650–653. [CrossRef]
15. Alonso-Esteban, J.I.; Pinela, J.; Barros, L.; Ćirić, A.; Soković, M.; Calhelha, R.C.; Torija-Isasa, E.; de Cortes Sánchez-Mata, M.; Ferreira, I.C.F.R. Phenolic Composition and Antioxidant, Antimicrobial and Cytotoxic Properties of Hop (*Humulus lupulus* L.) Seeds. *Ind. Crops Prod.* **2019**, *134*, 154–159. [CrossRef]
16. Nionelli, L.; Pontonio, E.; Gobbetti, M.; Rizzello, C.G. Use of Hop Extract as Antifungal Ingredient for Bread Making and Selection of Autochthonous Resistant Starters for Sourdough Fermentation. *Int. J. Food Microbiol.* **2018**, *266*, 173–182. [CrossRef] [PubMed]
17. Solarska, E. Fusariosis of Hop. *Mycotoxin Res.* **1991**, *7*, 184. [CrossRef] [PubMed]
18. Augusto Moretti Ferreira Pinto, F.; Araujo, L.; de Andrade, C.C.L.; Mendes Fagherazzi, M.; Fontanella Brighenti, A.; Schlichting de Martin, M.; Gomes, L.B.; Fernandes, J.; Duarte, V.; João Arioli, C.; et al. First Report of *Fusarium* Meridionale Causing Canker in Hop Plants. *Australas. Plant Dis. Notes* **2022**, *17*, 1–4. [CrossRef]
19. Sabo, J.; Đurić, T.; Jasnić, S. *Fusarium* Fungi as a Pathogen Causing Hop Wilt. *Plant Prot. Sci.* **2002**, *38*, 308–310. [CrossRef]
20. Shi, W.; Tan, Y.; Wang, S.; Gardiner, D.M.; De Saeger, S.; Liao, Y.; Wang, C.; Fan, Y.; Wang, Z.; Wu, A. Mycotoxigenic Potentials of *Fusarium* Species in Various Culture Matrices Revealed by Mycotoxin Profiling. *Toxins* **2016**, *9*, 6. [CrossRef]
21. Bottalico, A.; Perrone, G. Toxigenic *Fusarium* Species and Mycotoxins Associated with Head Blight in Small-Grain Cereals in Europe. *Mycotoxins Plant Dis.* **2002**, *108*, 611–624. [CrossRef]
22. Thomas, W.J.; Borland, T.G.; Bergl, D.D.; Claassen, B.J.; Flodquist, T.A.; Montgomery, A.S.; Rivedal, H.M.; Woodhall, J.; Ocamb, C.M.; Gent, D.H. A Quantitative PCR Assay for Detection and Quantification of *Fusarium* Sambucinum. *Plant Dis.* **2022**, *106*, 2601–2606. [CrossRef]
23. Srinivasan, V.; Goldberg, D.; Haas, G.J. Contributions to the Antimicrobial Spectrum of Hop Constituents. *Econ. Bot.* **2004**, *58*, S230–S238. [CrossRef]
24. Cottrell, M.T. A Search for Diastatic Enzymes Endogenous to *Humulus lupulus* and Produced by Microbes Associated with Pellet Hops Driving "Hop Creep" of Dry Hopped Beer. *J. Am. Soc. Brew. Chem.* **2023**, *81*, 435–447. [CrossRef]
25. Young, J.; Oakley, W.R.M.; Fox, G. *Humulus lupulus* and Microbes: Exploring Biotic Causes for Hop Creep. *Food Microbiol.* **2023**, *114*, 104298. [CrossRef]
26. Okorska, S.B.; Dąbrowska, J.A.; Głowacka, K.; Pszczółkowska, A.; Jankowski, K.J.; Jastrzębski, J.P.; Oszako, T.; Okorski, A. The Fungicidal Effect of Essential Oils of Fennel and Hops against *Fusarium* Disease of Pea. *Appl. Sci.* **2023**, *13*, 6282. [CrossRef]
27. Yan, Y.F.; Wu, T.L.; Du, S.S.; Wu, Z.R.; Hu, Y.M.; Zhang, Z.J.; Zhao, W.B.; Yang, C.J.; Liu, Y.Q. The Antifungal Mechanism of Isoxanthohumol from *Humulus lupulus* Linn. *Int. J. Mol. Sci.* **2021**, *22*, 10853. [CrossRef]
28. Bartmańska, A.; Wałecka-Zacharska, E.; Tronina, T.; Popłoński, J.; Sordon, S.; Brzezowska, E.; Bania, J.; Huszcza, E. Antimicrobial Properties of Spent Hops Extracts, Flavonoids Isolated Therefrom, and Their Derivatives. *Mol. A J. Synth. Chem. Nat. Prod. Chem.* **2018**, *23*, 2059. [CrossRef]
29. European Commission. Commission Regulation (EU) 2023/915 of 25 April 2023 on maximum levels for certain contaminants in food and repealing Regulation (EC) No 1881/2006. *Off. J. Eur. Union* **2023**, *L119/103*, 103–157.
30. European Commission. Recomendations on the presence of T-2 and HT-2 toxin in cereals and cereal products (EC) No 165/2013. *Off. J. Eur. Union* **2013**, *L91/12*, 12–15.
31. Meena, M.; Swapnil, P.; Upadhyay, R.S. Isolation, Characterization and Toxicological Potential of *Alternaria*-Mycotoxins (TeA, AOH and AME) in Different *Alternaria* Species from Various Regions of India. *Sci. Rep.* **2017**, *7*, 1–19. [CrossRef]
32. Pero, R.W.; Harvan, D.; Lucas, G.B.; Snow, J.P. Analysis of Tobacco for the *Alternaria* Toxins, Alternariol and Alternariol Monomethyl Ether. *J. Agric. Food Chem.* **1971**, *19*, 1274–1275. [CrossRef]
33. Wang, Q.; Šarkanj, B.; Jurasovic, J.; Chisti, Y.; Sulyok, M.; Gong, J.; Sirisansaneeyakul, S.; Komes, D. Evaluation of Microbial Toxins, Trace Elements and Sensory Properties of a High-Theabrownins Instant Pu-Erh Tea Produced Using *Aspergillus* Tubingensis via Submerged Fermentation. *Int. J. Food Sci. Technol.* **2019**, *54*, 1541–1549. [CrossRef]
34. López, P.; Venema, D.; de Rijk, T.; de Kok, A.; Scholten, J.M.; Mol, H.G.J.; de Nijs, M. Occurrence of *Alternaria* Toxins in Food Products in The Netherlands. *Food Control* **2016**, *60*, 196–204. [CrossRef]
35. Habschied, K.; Krska, R.; Sulyok, M.; Šarkanj, B.; Krstanović, V.; Lalić, A.; Šimić, G.; Mastanjević, K. Screening of Various Metabolites in Six Barley Varieties Grown under Natural Climatic Conditions (2016–2018). *Microorganisms* **2019**, *7*, 532. [CrossRef]
36. Houissa, H.; Lasram, S.; Sulyok, M.; Šarkanj, B.; Fontana, A.; Strub, C.; Krska, R.; Galindo, S.; Ghorbel, A. Multimycotoxin LC-MS/MS Analysis in Pearl Millet (*Pennisetum Glaucum*) from Tunisia. *Food Control* **2019**, *106*, 106738. [CrossRef]
37. Zhao, K.; Shao, B.; Yang, D.; Li, F.; Zhu, J. Natural Occurrence of *Alternaria* Toxins in Wheat-Based Products and Their Dietary Exposure in China. *PLoS ONE* **2015**, *10*, e0132019. [CrossRef]

38. Food, E.; Authority, S.; Arcella, D.; Eskola, M.; Angel, J.; Omez Ruiz, G. Dietary Exposure Assessment to *Alternaria* Toxins in the European Population. *EFSA J.* **2016**, *14*, e04654. [CrossRef]
39. Meena, M.; Samal, S. *Alternaria* Host-Specific (HSTs) Toxins: An Overview of Chemical Characterization, Target Sites, Regulation and Their Toxic Effects. *Toxicol. Rep.* **2019**, *6*, 745–758. [CrossRef]
40. Logrieco, A.; Moretti, A.; Solfrizzo, M. *Alternaria* Toxins and Plant Diseases: An Overview of Origin, Occurrence and Risks. *World Mycotoxin J.* **2009**, *2*, 129–140. [CrossRef]
41. Tsuge, T.; Harimoto, Y.; Akimitsu, K.; Ohtani, K.; Kodama, M.; Akagi, Y.; Egusa, M.; Yamamoto, M.; Otani, H. Host-Selective Toxins Produced by the Plant Pathogenic Fungus *Alternaria alternata*. *FEMS Microbiol. Rev.* **2013**, *37*, 44–66. [CrossRef]
42. Louro, H.; Vettorazzi, A.; López de Cerain, A.; Spyropoulou, A.; Solhaug, A.; Straumfors, A.; Behr, A.C.; Mertens, B.; Žegura, B.; Fæste, C.K.; et al. Hazard Characterization of *Alternaria* Toxins to Identify Data Gaps and Improve Risk Assessment for Human Health. *Arch. Toxicol.* **2023**, *98*, 425–469. [CrossRef]
43. Holt, S.; Miks, M.H.; De Carvalho, B.T.; Foulquié-Moreno, M.R.; Thevelein, J.M. The Molecular Biology of Fruity and Floral Aromas in Beer and Other Alcoholic Beverages. *FEMS Microbiol. Rev.* **2019**, *43*, 193–222. [CrossRef]
44. Woelflingseder, L.; Warth, B.; Vierheilig, I.; Schwartz-Zimmermann, H.; Hametner, C.; Nagl, V.; Novak, B.; Šarkanj, B.; Berthiller, F.; Adam, G.; et al. The *Fusarium* Metabolite Culmorin Suppresses the in vitro Glucuronidation of Deoxynivalenol. *Arch. Toxicol.* **2019**, *93*, 1729–1743. [CrossRef]
45. Drakulic, J.; Bruce, T.J.A.; Ray, R.V. Direct and Host-Mediated Interactions between *Fusarium* Pathogens and Herbivorous Arthropods in Cereals. *Plant Pathol.* **2017**, *66*, 3–13. [CrossRef]
46. Gao, X.; Kolomiets, M.V. Host-Derived Lipids and Oxylipins Are Crucial Signals in Modulating Mycotoxin Production by Fungi. *Toxin Rev.* **2009**, *28*, 79–88. [CrossRef]
47. Shi, J.; Zhang, M.; Gao, L.; Yang, Q.; Kalaji, H.M.; Qiang, S.; Strasser, R.J.; Chen, S. Tenuazonic Acid-Triggered Cell Death Is the Essential Prerequisite for *Alternaria alternata* (Fr.) Keissler to Infect Successfully Host Ageratina Adenophora. *Cells* **2021**, *10*, 1010. [CrossRef]
48. Magan, N.; Medina, A. Integrating Gene Expression, Ecology and Mycotoxin Production by *Fusarium* and *Aspergillus* Species in Relation to Interacting Environmental Factors. *World Mycotoxin J.* **2016**, *9*, 673–684. [CrossRef]
49. Medina, A.; Akbar, A.; Baazeem, A.; Rodriguez, A.; Magan, N. Climate Change, Food Security and Mycotoxins: Do We Know Enough? *Fungal Biol. Rev.* **2017**, *31*, 143–154. [CrossRef]
50. Jagtap, S.S.; Bedekar, A.A.; Rao, C.V. Quorum Sensing in Yeast. *ACS Symp. Ser.* **2020**, *1374*, 235–250. [CrossRef]
51. Palmieri, A.; Petrini, M. Tryptophol and Derivatives: Natural Occurrence and Applications to the Synthesis of Bioactive Compounds. *Nat. Prod. Rep.* **2019**, *36*, 490–530. [CrossRef]
52. European Commission Crop productions and Plant-Based Products—Hops. Available online: https://agriculture.ec.europa.eu/farming/crop-productions-and-plant-based-products/hops_en (accessed on 16 June 2024).
53. Statista Production Volume of Hops Worldwide in 2022, by Country. Available online: https://www.statista.com/statistics/757722/hop-production-global-by-country/#statisticContainer (accessed on 15 June 2024).
54. Janić Hajnal, E.; Babič, J.; Pezo, L.; Banjac, V.; Čolović, R.; Kos, J.; Krulj, J.; Pavšič-Vrtač, K.; Jakovac-Strajn, B. Effects of extrusion process on Fusarium and Alternaria mycotoxins in whole grain triticale flour. *LWT* **2022**, *155*, 112926. [CrossRef]
55. Mastanjević, K.; Krstanović, V.; Mastanjević, K.; Šarkanj, B. Malting and brewing industries encounter Fusarium spp. related problems. *Fermentation* **2018**, *4*, 3. [CrossRef]
56. Mastanjević, K.; Šarkanj, B.; Mastanjević, K.; Šantek, B.; Krstanović, V. Fusarium culmorum mycotoxin transfer from wheat to malting and brewing products and by-products. *World Mycotoxin J.* **2019**, *12*, 55–66. [CrossRef]
57. Mastanjević, K.; Šarkanj, B.; Krska, R.; Sulyok, M.; Warth, B.; Mastanjević, K.; Šantek, B.; Krstanović, V. From malt to wheat beer: A comprehensive multi-toxin screening, transfer assessment and its influence on basic fermentation parameters. *Food Chem.* **2018**, *254*, 115–121. [CrossRef]
58. Aichinger, G.; Del Favero, G.; Warth, B.; Marko, D. Alternaria toxins-Still emerging? *Compr. Rev. Food Sci. Food Saf.* **2021**, *20*, 4390–4406. [CrossRef]
59. EFSA CONTAM Panel (EFSA Panel on Contaminants in the Food Chain); Knutsen, H.K.; Alexander, J.; Barregård, L.; Bignami, M.; Brüschweiler, B.; Ceccatelli, S.; Cottrill, B.; Dinovi, M.; Grasl-Kraupp, B.; et al. Scientific Opinion on the risks to human and animal health related to the presence of deoxynivalenol and its acetylated and modified forms in food and feed. *EFSA J.* **2017**, *15*, 4718. [CrossRef]
60. EFSA CONTAM Panel (EFSA Panel on Contaminants in the Food Chain); Knutsen, H.-K.; Barregård, L.; Bignami, M.; Brüschweiler, B.; Ceccatelli, S.; Cottrill, B.; Dinovi, M.; Edler, L.; Grasl-Kraupp, B.; et al. Scientific opinion on the appropriateness to set a group health-based guidance value for fumonisins and their modified forms. *EFSA J.* **2018**, *16*, 5172. [CrossRef]
61. EFSA CONTAM Panel (EFSA Panel on Contaminants in the Food Chain); Knutsen, H.-K.; Barregård, L.; Bignami, M.; Brüschweiler, B.; Ceccatelli, S.; Cottrill, B.; Dinovi, M.; Edler, L.; Grasl-Kraupp, B.; et al. Scientific opinion on the appropriateness to set a group health based guidance value for T2 and HT2 toxin and its modified forms. *EFSA J.* **2017**, *15*, 4655. [CrossRef]

62. Sulyok, M.; Stadler, D.; Steiner, D.; Krska, R. Validation of an LC-MS/MS-Based Dilute-and-Shoot Approach for the Quantification of >500 Mycotoxins and Other Secondary Metabolites in Food Crops: Challenges and Solutions. *Anal. Bioanal. Chem.* **2020**, *412*, 2607–2620. [CrossRef]
63. Management of Left-Censored Data in Dietary Exposure Assessment of Chemical Substances. *EFSA J.* **2010**, *8*. [CrossRef]

toxins

MDPI

Article

Mechanism of Fumonisin Self-Resistance: *Fusarium verticillioides* Contains Four Fumonisin B$_1$-Insensitive-Ceramide Synthases

Tamara Krska [1,2,†], Krisztian Twaruschek [1,2,†], Gerlinde Wiesenberger [1,3], Franz Berthiller [3] and Gerhard Adam [1,*]

[1] Institute of Microbial Genetics, Department of Applied Genetics and Cell Biology, BOKU University, Konrad-Lorenz-Strasse 24, 3430 Tulln, Austria; tamara.krska@boku.ac.at (T.K.); krisztian.twaruschek@boku.ac.at (K.T.); gerlinde.wiesenberger@boku.ac.at (G.W.)

[2] Austrian Competence Centre for Feed and Food Quality, Safety and Innovation FFoQSI GmbH, Konrad-Lorenz-Strasse 20, 3430 Tulln, Austria

[3] Institute of Bioanalytics and Agro-Metabolomics, Department of Agrobiotechnology (IFA-Tulln), BOKU University, Konrad Lorenz Strasse 20, 3430 Tulln, Austria; franz.berthiller@boku.ac.at

* Correspondence: gerhard.adam@boku.ac.at

† These authors contributed equally to this work.

Abstract: *Fusarium verticillioides* produces fumonisins, which are mycotoxins inhibiting sphingolipid biosynthesis in humans, animals, and other eukaryotes. Fumonisins are presumed virulence factors of plant pathogens, but may also play a role in interactions between competing fungi. We observed higher resistance to added fumonisin B$_1$ (FB$_1$) in fumonisin-producing *Fusarium verticillioides* than in nonproducing *F. graminearum*, and likewise between isolates of *Aspergillus* and *Alternaria* differing in production of sphinganine-analog toxins. It has been reported that in *F. verticillioides*, ceramide synthase encoded in the fumonisin biosynthetic gene cluster is responsible for self-resistance. We reinvestigated the role of *FUM17* and *FUM18* by generating a double mutant strain in a *fum1* background. Nearly unchanged resistance to added FB$_1$ was observed compared to the parental *fum1* strain. A recently developed fumonisin-sensitive baker's yeast strain allowed for the testing of candidate ceramide synthases by heterologous expression. The overexpression of the yeast *LAC1* gene, but not *LAG1*, increased fumonisin resistance. High-level resistance was conferred by *FUM18*, but not by *FUM17*. Likewise, strong resistance to FB$_1$ was caused by overexpression of the presumed *F. verticillioides* "housekeeping" ceramide synthases *CER1*, *CER2*, and *CER3*, located outside the fumonisin cluster, indicating that *F. verticillioides* possesses a redundant set of insensitive targets as a self-resistance mechanism.

Keywords: fumonisin; self-resistance; ceramide synthase; gene disruption; heterologous expression; target insensitivity

Key Contribution: Using a recently described fumonisin-sensitive *Saccharomyces cerevisiae* strain, evidence has been obtained that not only one FUM cluster-encoded ceramide synthase gene (*FUM18*), but also *CER1*, *CER2*, and *CER3* of *F. verticillioides* encode insensitive enzymes involved in fumonisin self-resistance.

1. Introduction

Sphingolipids are abundant in the membranes of eukaryotes but also exist in some prokaryotes [1]. In eukaryotes, they are involved in processes like membrane trafficking, cell signaling, apoptosis, and others. Furthermore, disturbances in sphingolipid metabolism have been implicated in a variety of human diseases [2]. The sphingolipid core structure consists of a long acyl chain amide, which is linked to a fatty acid by ceramide synthase [3]. The long chain base in animal ceramides is sphingosine, while in plants and fungi, sphingolipid biosynthesis starts by the condensation of the tri-hydroxylated long chain base phytosphingosine with an alpha-hydroxylated very long chain fatty acid [4]. Phosphosphingolipids have a polar headgroup linked to ceramide via a phosphodiester bond. Highly complex structures [5,6] exist in different organisms with different roles due to the attachment of inositol(-phosphates) and different sugar moieties.

Fumonisins are the major group of "sphinganine analog mycotoxins" [7], alongside the AAL toxin produced by *Alternaria alternata* f.sp. *lycopersici*. Fumonisin B_1 (FB_1) in particular is known to efficiently inhibit ceramide synthase in plants [8,9] and animals [10] by competing with sphinganine and acyl-coenzyme A [11,12]. Disturbances of sphingolipid biosynthesis have many effects: FB_1 is a potential human carcinogen (group 2B according to the International Agency for Research on Cancer), further implicated in esophageal cancer and neural tube defects in humans, and known to cause animal diseases such as equine leukoencephalomalacia, porcine pulmonary edema and cancer. Also, teratogenic, mutagenic, cytotoxic, nephrotoxic, neurotoxic, and immunotoxic effects have been described [13–15].

The main producers of different fumonisins are plant pathogenic fungi, such as different species of *Fusarium*, several species of black *Aspergilli* and also *Verticillium* and some *Alternaria* strains [16]. Yet, *Alternaria alternata* f.sp. *lycopersici* typically produces the structurally related AAL toxin (see [7] for review). The gene clusters for fumonisin biosynthesis in different fungi have been elucidated [7,17–19].

Whether fumonisin production is a virulence factor of plant pathogenic fungi is a controversial issue. Fumonisin-deficient *fum1* mutants of *F. verticillioides* were still able to cause Fusarium ear rot in maize [20]. An *F. verticillioides* strain from banana (now *F. musae*) containing a large deletion of the FUM cluster was not pathogenic to seedlings of maize. Yet, when the FUM cluster was added back by transformation and fumonisin biosynthesis was restored, it gained virulence [21]. Also, inactivation of *fum1* in several strains led to reduced stunting of seedlings, indicating that it is a virulence factor in seedlings at least in some sensitive maize cultivars. Maize can have highly variable resistance to FB_1 in a seed germination assay [22]. For *F. proliferatum*, which causes rice spikelet rot disease, it was shown that the disruption of several genes leading to loss of fumonisin production caused reduced virulence [23]. Also, in *Verticillium dahliae* causing wilting disease in cotton, fumonisin-deficient knockout strains were less virulent [24]. In the case of *Alternaria alternata* f.sp. *lycopersici*, which causes stem canker on susceptible tomato cultivars, resistance to the AAL toxin leads to resistance against the fungal pathogen (host selective toxin) [25]. Tomatoes with a homozygous loss of function of *Asc1*, encoding a ceramide synthase, are susceptible to the toxin and to the fungus [26]. Similarly, in *Arabidopsis*, inactivation of one of three ceramide synthase genes in this species, *LOH2*, leads to toxin sensitivity and breakdown of non-host resistance against an AAL-producing *Alternaria alternata* [27].

F. graminearum and *F. verticillioides* can co-occur and compete in infected maize ears. In a recent study [28], no significant difference between wild-type and *fum1* mutants in disease severity or amount of fungal DNA in the inoculated maize line was found. Yet, it was demonstrated that wild-type *F. verticillioides* could suppress the growth of *F. graminearum* in a co-culture on autoclaved kernels more strongly than a fumonisin-nonproducing strain. The authors hypothesized that fumonisin production in seeds suppresses colonization by other fungi after the seeds have been shed and that the main function of fumonisins thereby is to increase saprophytic fitness.

Data on fumonisin resistance or susceptibility in different fungi are scarce. It has been reported that FB_1 in very high concentrations (200 μL of up to 40 mM—corresponding to mg amounts per well in the agar) produced large growth inhibition zones with isolates of *Botrytis cinerea* and (not AAL-toxin-producing) *A. alternata* from a South African collection, while *F. graminearum* showed much higher resistance [29]. Conversely, Dawidziuk et al. [30] reported that a *Fusarium graminearum* isolate from Poland showed strong growth retardation by fumonisin already at the low concentration of 3 mg/L FB_1 mixed into the agar medium, while *F. oxysporum* and *F. proliferatum* isolates were unaffected by this concentration.

In principle, very high concentrations of fumonisins and also AAL toxin can be produced in fungal cultures and some mechanism of self-resistance must exist in toxin-producing fungi. Recently it has been reported that in the case of *Fusarium verticillioides*, self-protection against FB_1 is conferred by a FUM cluster-encoded ceramide synthase [31].

The aim of our study was to test whether *Fusarium*, *Aspergillus*, and *Alternaria* strains producing sphinganine-analog mycotoxins have higher levels of FB_1 resistance than related non-producers. Testing by gene disruption revealed that the cluster-encoded ceramide synthases of *F. verticillioides* are unexpectedly NOT necessary for high-level resistance. This result is explained by our finding that three presumed housekeeping ceramide synthases, when expressed in a sensitive yeast strain, are sufficient to confer high-level FB_1 resistance.

2. Results

2.1. Sphinganine-Analog Producing Fungal Species Are More Resistant to Fumonisin B_1 Than Non-Producers

To investigate whether the production of fumonisins or the related AAL toxin is associated with increased toxin resistance, we compared the growth of various fungal strains (see Table 1) in the presence of FB_1. First, we compared the growth of a well-studied *F. verticillioides* strain (FGSC 7600), which had been previously utilized for elucidation of the FUM cluster and for determination of the first genome sequence [32], with the growth of the likewise relevant fumonisin-nonproducer *F. graminearum* (strain PH-1, [33]) at different temperatures and different levels of fumonisins added to minimal medium. Since very high concentrations were needed for full inhibition, a crude concentrated extract containing fumonisins B_1, B_2, and B_3 was used as previously described [34], which contained 3.18 g/L FB_1. Without added toxin at 20 °C, *F. graminearum* (red pigmented, on the right half of the plates shown in Figure 1) grew more vigorously and covered a larger portion of the medium than *F. verticillioides*. At 30 °C, *F. verticillioides* grew better, and after two weeks, both strains covered about half of the plate. When increasing amounts of fumonisin were added to the medium, *F. graminearum* was increasingly inhibited, while *F. verticillioides* continued to grow. At the highest concentration tested (176 μM FB_1), growth of *F. graminearum* was completely inhibited, while *F. verticillioides* showed only marginally reduced radial growth after 7 days at 30 °C (Figure 1). We conclude that the fumonisin-producing *F. verticillioides* has clearly higher resistance to fumonisin than *F. graminearum*.

Table 1. Fungal strains used in this study.

Species	Strain Designation (Other Collection)	Genotype
Fusarium verticillioides	FGSC 7600; (FRC M-3125, NRRL 20956)	wt [1]
Fusarium graminearum	PH-1 (NRRL 31084)	wt
Alternaria alternata f.sp. *lycopersici*	AS27-12	wt
Alternaria alternata (mali)	MA 304 (CBS 106.24, ATCC 13963)	wt
Alternaria alternata	MA 308 (CBS 150.24)	wt
Aspergillus niger	ATCC 11414	wt
Aspergillus nidulans	FGSC A4 (ATCC 38163)	wt
F. verticillioides	GfA2364	*fum1::hygB*
F. verticillioides	KTFD1 KTFD4	*fum1::hygB* *fum17-18Δ::HSVtk-nptII* (this study)

[1] wt (wild-type).

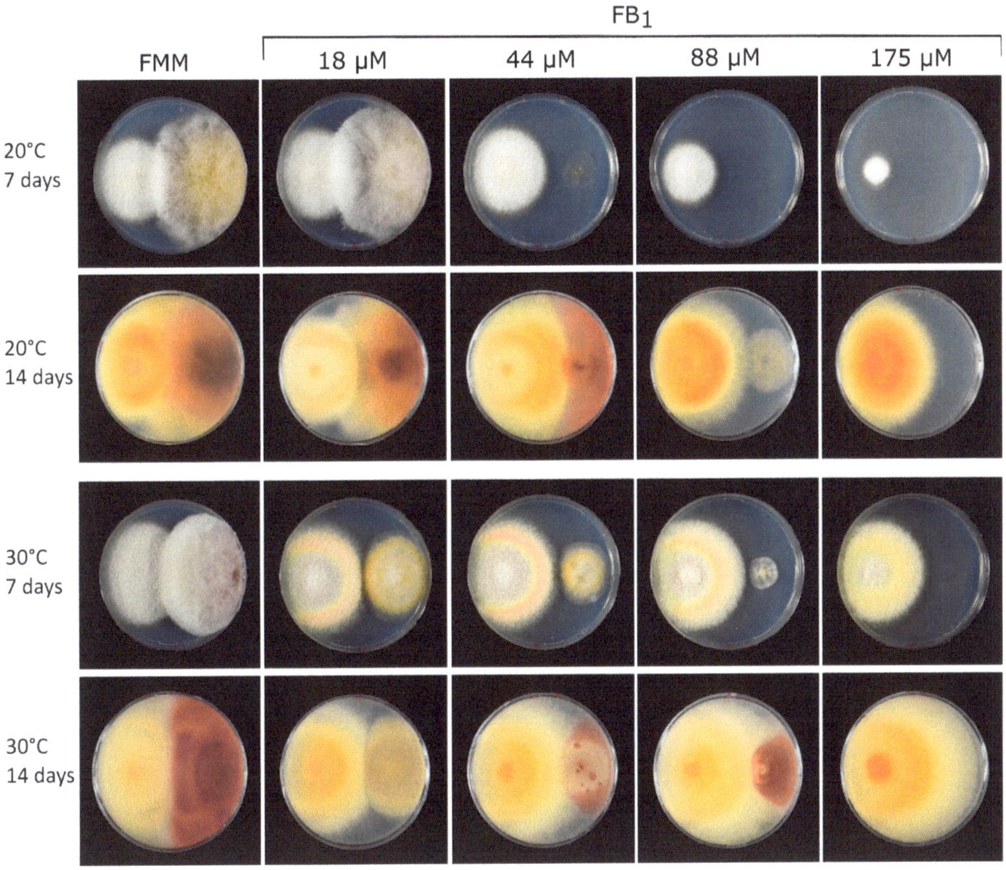

Figure 1. Growth of *F. verticillioides* and *F. graminearum* on FMM medium containing FB_1 (crude extract) at different temperatures. Pictures were taken after the indicated incubation time (on day 7 from above and on day 14 taken from below for better visualization of the red *F. graminearum* pigment).

Next, we compared various *Alternaria* strains (Figure 2A) producing or not producing sphinganine-analog toxins. The *A. alternata* f.sp. *lycopersici* strain AS27-12 is a well-known producer of AAL toxin and related derivatives [35]. Its resistance level was compared to two *A. alternata* isolates from our local university collection (Austrian Center for Biological Resources (https://acbr-database.boku.ac.at/, accessed on 21 May 2024)). The strain MA 304 was originally isolated from apple in the USA, whereas MA 308 caused leaf spot in *Solanum tuberosum*. Both strains do not produce AAL toxin. Already, at the low concentration of 10 µM (about 7.2 mg/L), the growth of both nonproducing strains was strongly reduced to about 20% of the diameter, while the AAL-producing strain had 78% of its diameter on the no-toxin control. At 50 µM FB_1, the AAL strain had an about 50% reduced diameter, while the two nonproducer strains were almost completely inhibited.

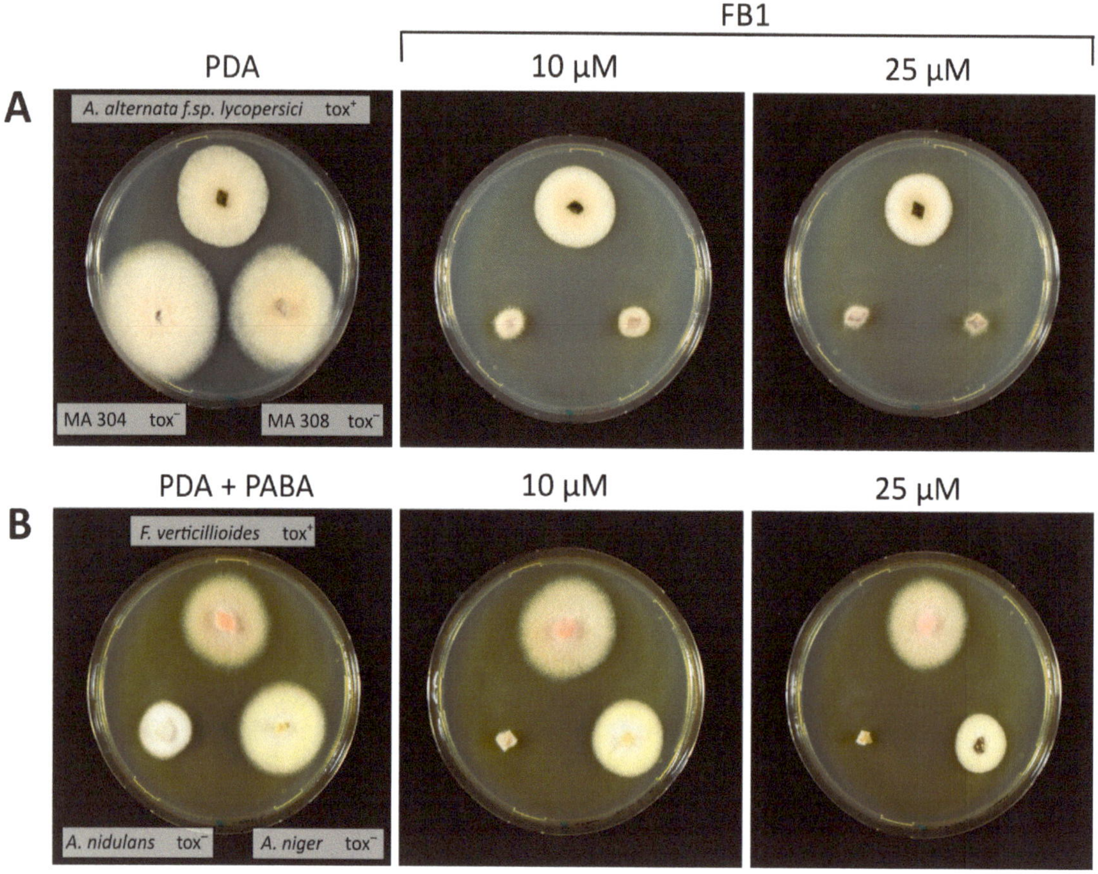

Figure 2. (**A**) Growth of *Alternaria* strains on FB_1 containing PDA medium. Small agar blocks of the indicated *Alternaria* strains (AAL toxin producer (tox^+) on top, the two nonproducers (tox^-) below) were transferred to PDA plates containing the indicated amount of FB_1. (**B**) Growth of *A. nidulans* (not fumonisin producing) and *A. niger* on PDA plates supplemented with PABA (p-aminobenzoic acid (PABA), 1.0 mg/L) and containing the indicated concentration of FB_1.

We also tested (Figure 2B) whether an *Aspergillus niger* wild-type strain, for which fumonisin production had been demonstrated (ATCC 11414, [36], see Table S1 therein), is more resistant than a wild-type *A. nidulans* strain (FGSC A4, [37]). *F. verticillioides* was added on top of the plates as a control (Figure 2B). At 10 µM FB_1, the *A. nidulans* strains were already fully inhibited, while *A. niger* was only slightly inhibited (compared to no toxin 86% diameter at 10 µM and 66% at 50 µM). Seemingly, a mechanism of protection exists in sphinganine-analog toxin producers. We set out to test whether target insensitivity is involved.

2.2. Generation and Characterization of a fum17-18 Deletion Strain in a fum1 Background

To be able to study the effects of added toxin undisturbed by endogenously synthesized fumonisin, we generated a *fum17-fum18* double mutant in the background of a *fum1::hygB* mutant. The previously described *fum1* mutant strain GfA2364, which is derived from the wild-type strain FGSC 7600 [20,38] by insertion of the *hygB* resistance gene into the *FUM1* PKS, was transformed with a construct that allows for simultaneous deletion of both genes using a *nptII* (G418) resistance cassette (see Section 4). Two transformants, designated

KTFD1 and KTFD4, were obtained and used in the fumonisin resistance tests: their growth was compared to the growth of the parental *fum1* mutant strain GfA2364. As evident from Figure 3—even on the highest concentration tested—both, the wild-type and the knockout strains were still able to grow. For unknown reasons, stronger and earlier pigmentation occurred in the wild type. We conclude that the cluster-encoded ceramide synthase genes *FUM17* and *FUM18* are not necessary for high-level resistance to fumonisin B_1.

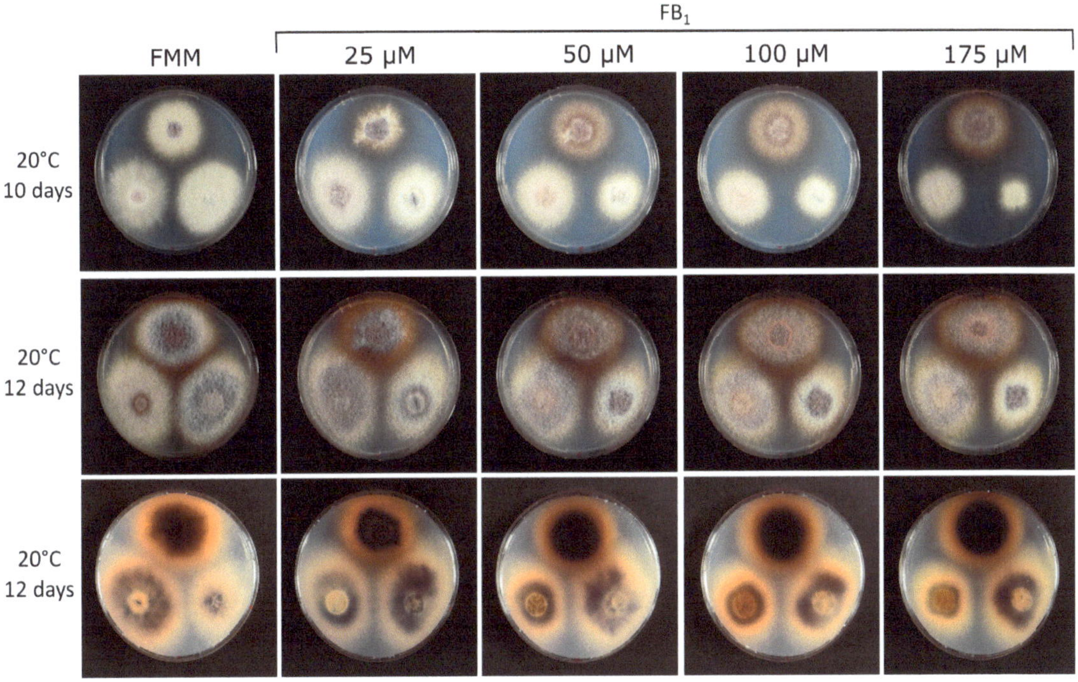

Figure 3. Growth of Δ*fum1* and two Δ*fum1* Δ*fum17-18* (double mutants, KTFD1 and KTFD4) mutants on FB_1-containing plates. The *fum17-fum18* (**bottom**) were inoculated onto FMM plates containing different concentrations of crude FB_1 together with the parental *fum1* (**top**). Strains were grown for 12 days with pictures taken after 10 and 12 days. The bottom row shows the backside of the plates after 12 days (note, that plates are mirrored).

2.3. *Testing Fumonisin Resistance of Ceramide Synthase Genes by Heterologous Expression in Yeast*

The finding that the *FUM17* and *FUM18* genes are not necessary for self-resistance against FB_1 indicates possible redundancy. *F. verticillioides* has three additional predicted ceramide synthase genes, *CER1*, *CER2*, and *CER3* [31], encoded outside the FUM cluster. We have recently reported the construction of a fumonisin-sensitive *Saccharomyces cerevisiae* strain [34]. We transformed this strain with the plasmids described by Janevska et al. [31] for the expression of the *F. verticillioides* ceramide synthase cDNAs behind the constitutive *TEF1* promoter. As controls, the yeast ceramide synthases *LAG1* ("longevity assurance gene", [39]) and its paralog ("longevity assurance cognate") *LAC1* [40] were also overexpressed. Yeast transformants were spotted onto SC-URA plates supplemented with increasing amounts of FB_1. Overexpression of *LAC1*, but not of *LAG1*, conferred low-level resistance at concentrations that were inhibitory for the empty vector controls. At higher concentrations, the yeast genes did not confer resistance, in contrast to *F. verticillioides* *FUM18*, *CER1*, *CER2*, and *CER3*.

As shown in Figure 4, the yeast host (containing functional endogenous *LAG1* and *LAC1* genes) transformed with the empty vector was already sensitive to 2.5 μM FB_1. Overexpression of *LAC1* but not *LAG1* in the 2 μ multicopy plasmid behind the strong *TEF1* promoter conferred a low level of increased resistance. On the other hand, high-level resistance (highest concentration tested 150 μM) was conferred by the expression of *FUM18* but not *FUM17*, and equally well by all three *F. verticillioides* ceramide synthases, *CER1*, *CER2*, and *CER3*.

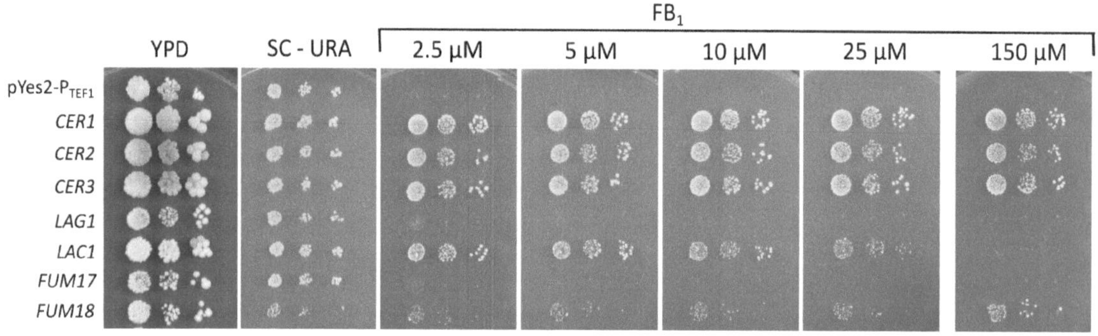

Figure 4. Growth of transformants of the FB_1-sensitive *Saccharomyces cerevisiae* strain YTKT33 on URA-dropout SC agar media containing increasing concentrations of FB_1. For the highest concentration, 75% pure FB_1 was used. YTKT33 was transformed with the empty expression vector, pYes2-P_{TEF1} (negative control), or expression vectors containing: *F. verticillioides* ceramide synthase *CER1*, *CER2*, *CER3*, the two *S. cerevisiae* ceramide synthases *LAG1* and *LAC1*, and two putative ceramide synthase genes from the *F. verticillioides* fumonisin cluster, *FUM17* and *FUM18*.

3. Discussion

The *FUM* cluster of *F. verticillioides* contains two genes, *FUM17* (FVEG_00327) and *FUM18* (FVEG_00328), which have sequence similarity to ceramide synthases. It has previously been reported [41] that both *FUM17* and *FUM18* expression were upregulated by FB_1 addition to the medium. We used a *fum1* background to avoid contribution by differences in endogenous FB_1 production to the overall effect. Both genes, which are located next to each other (with overlapping 3′ ends of the mRNAs), had been inactivated simultaneously by the insertion of a hygromycin resistance gene, which did not lead to a significant reduction of fumonisin production [12]. More recently, it was reported that "self-protection against the sphingolipid biosynthesis inhibitor fumonisin B_1 is conferred by a FUM cluster-encoded ceramide synthase" [31]. Using an assay supposedly reflecting fungal biomass, which is based on the activation of resazurin (a dye that is converted into the fluorescent derivative resorufin by respiratory activity), more relative inhibition (about 120% compared to wild-type) in liquid culture was observed upon addition of FB_1 [31]. No direct evidence for reduced growth of the knock-out strain on a solid medium was shown.

Our results confirm that the *FUM18* gene is sufficient to confer fumonisin resistance when expressed in fumonisin-sensitive yeast. The double mutant *lag1 lac1* is lethal in most yeast strains. The *FUM17* plasmid did not complement the conditional yeast mutant [31] when doxycycline was added in order to switch off the expression of the integrated tetracycline-regulated promoter P_{TET}-*LAG1* gene in the Δ*lac1* background. In agreement with this finding, we did not observe increased resistance compared to the empty vector in our sensitive yeast strain. It has been reported that *FUM17* is obviously non-functional in two other strains of the *F. fujikuroi* species complex [41], and more subtle mutations may also lead to the inactivity of the *F. verticillioides FUM17* gene product. While *FUM18* is sufficient to confer resistance in yeast, it is surprisingly not necessary for the high-level resistance to FB_1 in *F. verticillioides*. The knockout mutants (*fum17-18* double mutant, similar to that described but with a different selection marker, *fum17-18Δ::HSVtk-nptII*) in the *fum1::hygB* background are hardly inhibited in growth on solid medium by added FB_1 (Figure 3). The observed effect of inactivating the cluster ceramide synthases on FB_1-mediated growth inhibition is minor, but unexpectedly, differences in the amount and timing of pigment formation were observed between independent *fum1 fum17-18* and *fum1* mutants. Since both strains are *fum1* mutants, this should not be due to an alteration of the metabolic flux from fumonisin into a different metabolite that is responsible for this phenotype. Further research would be necessary to elucidate which changes occur at the level of the transcriptome or metabolome. The result that *FUM18* is not necessary for fumonisin resistance can be explained by our finding that other ceramide synthases of *F. verticillioides* also confer high-level resistance in yeast. A surprising result is the high level of resistance conferred by *CER3* (FVEG_15375), as it was reported that this gene is not able to complement the yeast ceramide synthase's loss of function [41]. Janevska et al. [41] reported that in the resazurin assay, overexpression of *CER1* (FVEG_06971) and *CER2* (FVEG_06971) showed a slightly reduced growth inhibition compared to the empty vector. In our strain, the same overexpression plasmids conferred high-level resistance (no evident inhibition at 175 μM FB_1) and the transformants were growing at least as well as the *FUM18*-overexpressing strain (see Figure 4).

The finding that *FUM18* is not necessary for self-resistance is in agreement with results with an *alt7* knockout strain in *A. alternata* producing AAL toxin. *ALT7* is a ceramide synthase gene located in the cluster for AAL toxin biosynthesis. The knockout of *ALT7* had no deleterious effect on the AAL toxin-producing pathogen, so the authors concluded that the gene does not act as a resistance/self-tolerance factor [42].

For the fungi that we tested, the hypothesis holds true that fumonisin producers are more resistant to FB_1 than related non-producers. Several black *Aspergilli* can produce fumonisins B_2, B_4, and B_6, e.g., on grapes [43] or maize [44], although the levels are typically lower than in *Fusarium*. The non-producing model fungus *A. nidulans* turned out to be extremely sensitive to added FB_1; growth was already fully inhibited by 10 μM FB_1. The FUM cluster of *A. niger* does not contain a ceramide synthase [45], but it nevertheless showed higher resistance than *A. nidulans* and it potentially also has "housekeeping" ceramide synthase genes responsible for the higher FB_1 resistance. In agreement with the reported much higher resistance in *F. graminearum* [29] and in contrast to Dawidziuk et al. [30], we found that the *F. graminearum* strain PH-1, lacking a *FUM18* ortholog, displayed quite high FB_1 resistance (Figure 1). If the fungus–fungus competition hypothesis is meaningful, as obviously production of very high levels of fumonisins is needed in this scenario.

Numerous cases exist (for review see [46]) where duplicated housekeeping genes containing sequence alterations encode insensitive target enzymes. These are often associated with toxin or antibiotic biosynthetic gene clusters, which allow for the elucidation of the mode of action of some compounds [47,48]. The presence of a (duplicated) putative self-resistance gene in a secondary metabolite biosynthetic cluster has been successfully used to identify new compounds with a desired mode of action, for instance, in the case of aspterric acid of *A. terreus* targeting branched-chain amino-acid biosynthesis [49]. Yet, the conclusion that such enzymes are also necessary for self-resistance may not be correct. Our results show that in the case of *F. verticillioides*, the fumonisin-cluster-encoded ceramide synthase *FUM18* is not necessary for self-resistance due to redundancy in self-resistance genes, as three other ceramide synthases, *CER1–3*, are additionally sufficient to confer fumonisin resistance.

4. Materials and Methods

4.1. FB_1-Sensitivity of Growth of Fusarium and Other Fungi

F. verticillioides (FGSC 7600) and *F. graminearum* (PH-1) were activated on Fusarium minimal medium (FMM; 1 g/L KH_2PO_4, 0.5 g/L $MgSO_4{\cdot}7H_2O$, 0.5 g/L KCl, 2 g/L $NaNO_3$, 30 g/L sucrose, 20 g/L agar, 200 μl/L of a trace element solution that was added after autoclaving) plates. Conidia of the *Fusarium* strains were generated by inoculating 50 mL of mung bean extract (MBS, filtrate of 10 g mung beans per L water boiled for 20 min) in a 250 mL baffled flask with fungal mycelium. After 3 days of incubation on a shaker at 140 rpm at 20 °C in the dark, conidia were obtained by removing mycelium using sterilized glass wool and subsequent sedimentation overnight at 4 °C. Five hundred spores were spotted onto FMM plates containing different concentrations of FB_1. *Aspergillus* and *Alternaria* strains were grown on potato dextrose agar (PDA, Sigma-Aldrich, Vienna, Austria). Agar blocks from colonies grown on PDA were transferred onto plates containing different concentrations of FB_1. The plates were supplemented with different concentrations of a crude FB_1 extract that was previously used for yeast spottings [34]. Pure fumonisin was purchased from Fermentek (Jerusalem, Israel) and Fumizol Ltd. (Szeged, Hungary), and the 70% pure FB_1 was a gift from Romer Labs. The plates were incubated at 20 °C and pictures were taken after 5 days.

4.2. Generation of Δ*fum1*, Δ*fum17-18* Mutants

The fumonisin-nonproducing Δ*fum1* mutant GfA2364 containing a hygromycin resistance cassette disrupting the coding region of *FUM1* polyketide synthetase was kindly provided by Dr. Robert Proctor. *FUM1* disruption was confirmed by using primers hyg-FW and hyg-RV to amplify an internal 861 bp hygromycin fragment, as well as by implementing primers flanking the insertion site (primers GfA2364_fum1test_fw and GfA2364_fum1test_rv), leading to a 2.9 kb fragment (Table 2).

Table 2. Primers used in this study.

Name	Sequence
Δfum1 confirmation	
GfA2364_fum1test_fw	AGAAGCCTTGATGCTGCCTA
GfA2364_fum1test_rv	GAGTGATGTCCCATGGCAGA
hyg-FW	GCTTTCAGCTTCGATGTAGGAGG
hyg-RV	CTACACAGCCATCGGTCCAGAC
Δfum17,18 disruption	
Fw_Fum327KO	ACTAGTCACGACAGTAAGAAGCAA
Rv_Fum327KO	GACTTGACGGGGATCGGTTC
Fw_Fum328KO	GGATTTGGAGACAAGTACGA
Rv_Fum328KO	GTCGACATCCTTCTCGAAGGCCAG
P#926	TGCTCCAACTCAGGCGATGCTG
P#940	CCGTCTAGCGCTGTTGATTGTATT
FUM1718_upstream_PCRtest	GCCTTCAAAGTTCATCATGGC
FUM1718_downstr_PCRtest	TAAGCGTGTCGTAACCTGTG

A double knock-out of putative self-protection genes *FUM17* (FVEG_00327) and *FUM18* (FVEG_00328) was performed in strain GfA2364 by replacing them with a geneticin resistance marker, *nptII* (G418). The 5′ UTR upstream of the FVEG_00328 promoter was amplified from *F. verticillioides* genomic DNA using the primers Fw_Fum328KO and Rv_Fum328KO, while the downstream UTR of FVEG_00327 was obtained using the primers Fw_Fum327KO and Rv_Fum327KO. The 5′ UTR was digested with BcuI and EcoRI and ligated into vector pKT300 containing a fusion gene between HSV-thymidine kinase and nptII [50]. Likewise, the 3′ UTR, was also cloned into pKT300 using HindIII and SalI. Finally, they were cloned into the same disruption plasmid, named pKT314. GfA2364 was transformed using a standard transformation protocol [50]. The knockout was confirmed by using the primers FUM1718_downstr_PCRtest, located downstream of *FUM17*, in combination with #940 (inside terminator region of disruption plasmid), located inside the disruption plasmid, as well as FUM1718_upstream_PCRtest, upstream of *FUM18* together with #926 (inside promoter region of disruption plasmid). Both of these amplifications lead to the expected 1 kb fragment, while the control (GfA2364) did not give a band. Primary transformants were obtained and purified to generate homokaryotic transformants. To this end, conidiospores were re-isolated from single colonies for two rounds while maintaining selection pressure to generate second-generation transformants. Primer sequences and purposes are given in Table 2.

4.3. Expression of Putative Self-Protection Genes in an FB_1-Sensitive Baker's Yeast Strain

Plasmid pYes2-P_{TEF1} [31] was used to express predicted fumonisin self-protection genes. It contains *URA3* as a selection marker, with genes expressed under the constitutive yeast *TEF1* promoter. Plasmids containing *CER1*, *CER2*, *CER3*, *LAG1*, *LAC1*, *FUM17*, and *FUM18* were described in [31] and kindly provided by Dr. Vito Valiante (Leibniz Institute for Natural Product Research and Infection Biology, Hans Knöll Institute, Jena, Germany). The fumonisin-sensitive baker's yeast YTKT33 [34] was transformed with these plasmids using the lithium transformation protocol and selected on synthetic complete media lacking uracil (SC-URA). For plate assays, liquid overnight cultures were diluted back to an OD_{600nm} of 0.1. After reaching an OD_{600nm} of about 0.3, they were diluted to an OD_{600nm} of 0.1, 0.01, and 0.001, and 3 µL of these suspensions was spotted on the agar plates. Photographs were taken after a 5-day incubation period at 30 °C.

Author Contributions: Conceptualization, G.A. and T.K.; validation, G.A.; investigation, T.K., K.T. and F.B.; resources, G.A. and F.B.; writing—original draft preparation, T.K.; writing—review and editing, G.A., T.K., G.W., F.B. and K.T.; supervision, G.A. and G.W.; project administration, G.A.; funding acquisition, G.A. and F.B. All authors have read and agreed to the published version of the manuscript.

Funding: This research was funded by FFoQSI GmbH (Austrian Competence Centre for Feed and Food Quality, Safety & Innovation, project C30-P12-W03: Toxin Inactivation). T.K. and K.T. were employed by FFoQSI. BOKU is a 35% co-owner of FFoQSI. Additional funding for analytics and toxin preparation was obtained by F.B. from FWF (Austrian Science Fund) project P33011 (Toxicological significance of modified fumonisins).

Institutional Review Board Statement: Not applicable.

Informed Consent Statement: Not applicable.

Data Availability Statement: The original contributions presented in the study are included in the article, further inquiries can be directed to the corresponding author.

Acknowledgments: *F. graminearum* PH-1 was kindly provided by Frances Trail (Michigan State University, East Lansing, MI, USA), *F. verticillioides* strains FGSC 7600 and the *fum1* mutant GfA2364 by Robert Proctor (USDA ARS NCAUR, Peoria, IL, USA). The AAL-producing strain AS27-12 was provided by David G. Gilchrist (University of California, Davies, CA, USA) on the basis of an MTA. We thank Christian Voitl (BOKU) for activating ACBR strains. The *Aspergillus nidulans* and *A. niger* strains were kindly provided by Christian P. Kubicek (TU Wien). We especially thank Vito Valiante (Leibniz Institute for Natural Product Research and Infection Biology, Hans Knöll Institut, Jena, Germany) for generously providing the *Fusarium verticillioides* and yeast ceramide synthase overexpression plasmids and the empty vector. We thank Guenther Jaunecker (Romer Labs) for a generous gift of partially purified FB_1. We also acknowledge the support of Marco Reiter during the extraction and purification of FB_1.

Conflicts of Interest: The funders had no role in the design of the study; in the collection, analyses, or interpretation of data; in the writing of the manuscript; or in the decision to publish the results. For the fumonisin-sensitive strain YTKT33, a patent application has been filed by FFoQSI (PCT/EP2023/062973), requests (on the basis of a material transfer agreement) should be directed to FFoQSI (juergen.marchart@ffoqsi.at).

References

1. Stankeviciute, G.; Tang, P.; Ashley, B.; Chamberlain, J.D.; Hansen, M.E.B.; Coleman, A.; D'Emilia, R.; Fu, L.; Mohan, E.C.; Nguyen, H.; et al. Convergent Evolution of Bacterial Ceramide Synthesis. *Nat. Chem. Biol.* **2022**, *18*, 305–312. [CrossRef] [PubMed]

2. Hannun, Y.A.; Obeid, L.M. Sphingolipids and Their Metabolism in Physiology and Disease. *Nat. Rev. Mol. Cell Biol.* **2018**, *19*, 175–191. [CrossRef]
3. Michaelson, L.V.; Napier, J.A.; Molino, D.; Faure, J.-D. Plant Sphingolipids: Their Importance in Cellular Organization and Adaption. *Biochim. Biophys. Acta* **2016**, *1861*, 1329–1335. [CrossRef] [PubMed]
4. Fougère, L.; Mongrand, S.; Boutté, Y. The Function of Sphingolipids in Membrane Trafficking and Cell Signaling in Plants, in Comparison with Yeast and Animal Cells. *Biochim. Biophys. Acta Mol. Cell Biol. Lipids* **2024**, *1869*, 159463. [CrossRef] [PubMed]
5. Haslam, T.M.; Feussner, I. Diversity in Sphingolipid Metabolism across Land Plants. *J. Exp. Bot.* **2022**, *73*, 2785–2798. [CrossRef] [PubMed]
6. Santos, T.C.B.; Dingjan, T.; Futerman, A.H. The Sphingolipid Anteome: Implications for Evolution of the Sphingolipid Metabolic Pathway. *FEBS Lett.* **2022**, *596*, 2345–2363. [CrossRef]
7. Chen, J.; Li, Z.; Cheng, Y.; Gao, C.; Guo, L.; Wang, T.; Xu, J. Sphinganine-Analog Mycotoxins (SAMs): Chemical Structures, Bioactivities, and Genetic Controls. *J. Fungi* **2020**, *6*, 312. [CrossRef] [PubMed]
8. Berkey, R.; Bendigeri, D.; Xiao, S. Sphingolipids and Plant Defense/Disease: The "Death" Connection and Beyond. *Front. Plant Sci.* **2012**, *3*, 68. [CrossRef] [PubMed]
9. Luttgeharm, K.D.; Cahoon, E.B.; Markham, J.E. Substrate Specificity, Kinetic Properties and Inhibition by Fumonisin B_1 of Ceramide Synthase Isoforms from Arabidopsis. *Biochem. J.* **2016**, *473*, 593–603. [CrossRef] [PubMed]
10. Merrill, A.H.; Wang, E.; Vales, T.R.; Smith, E.R.; Schroeder, J.J.; Menaldino, D.S.; Alexander, C.; Crane, H.M.; Xia, J.; Liotta, D.C.; et al. Fumonisin Toxicity and Sphingolipid Biosynthesis. In *Fumonisins in Food*; Jackson, L.S., DeVries, J.W., Bullerman, L.B., Eds.; Advances in Experimental Medicine and Biology; Springer: Boston, MA, USA, 1996; pp. 297–306; ISBN 978-1-4899-1379-1.
11. Merrill, A.H.; van Echten, G.; Wang, E.; Sandhoff, K. Fumonisin B_1 Inhibits Sphingosine (Sphinganine) N-Acyltransferase and de Novo Sphingolipid Biosynthesis in Cultured Neurons in Situ. *J. Biol. Chem.* **1993**, *268*, 27299–27306. [CrossRef] [PubMed]
12. Proctor, R.H.; Brown, D.W.; Plattner, R.D.; Desjardins, A.E. Co-Expression of 15 Contiguous Genes Delineates a Fumonisin Biosynthetic Gene Cluster in Gibberella Moniliformis. *Fungal Genet. Biol.* **2003**, *38*, 237–249. [CrossRef] [PubMed]
13. EFSA Panel on Contaminants in the Food Chain (CONTAM); Schrenk, D.; Bignami, M.; Bodin, L.; Chipman, J.K.; Del Mazo, J.; Grasl-Kraupp, B.; Hogstrand, C.; Leblanc, J.-C.; Nielsen, E.; et al. Assessment of Information as Regards the Toxicity of Fumonisins for Pigs, Poultry and Horses. *EFSA J.* **2022**, *20*, e07534. [CrossRef] [PubMed]
14. International Agency for Research on Cancer (IARC). *IARC Monographs on the Evaluation of Carcinogenic Risks to Humans*; Fumonisin b1; IARC Press: Lyon, France, 2002; pp. 275–366.
15. Wangia-Dixon, R.N.; Nishimwe, K. Molecular Toxicology and Carcinogenesis of Fumonisins: A Review. *J. Environ. Sci. Health C Toxicol. Carcinog.* **2021**, *39*, 44–67. [CrossRef] [PubMed]
16. Mirocha, C.J.; Chen, J.; Xie, W.; Xu, Y.; Abbas, H.K.; Hogge, L.R. Biosynthesis of Fumonisin and Aal Derivatives by Alternaria and Fusarium in Laboratory Culture. *Adv. Exp. Med. Biol.* **1996**, *392*, 213–224. [CrossRef] [PubMed]
17. Kim, H.-S.; Lohmar, J.M.; Busman, M.; Brown, D.W.; Naumann, T.A.; Divon, H.H.; Lysøe, E.; Uhlig, S.; Proctor, R.H. Identification and Distribution of Gene Clusters Required for Synthesis of Sphingolipid Metabolism Inhibitors in Diverse Species of the Filamentous Fungus Fusarium. *BMC Genom.* **2020**, *21*, 510. [CrossRef] [PubMed]
18. Proctor, R.H.; Van Hove, F.; Susca, A.; Stea, G.; Busman, M.; van der Lee, T.; Waalwijk, C.; Moretti, A.; Ward, T.J. Birth, Death and Horizontal Transfer of the Fumonisin Biosynthetic Gene Cluster during the Evolutionary Diversification of Fusarium. *Mol. Microbiol.* **2013**, *90*, 290–306. [CrossRef] [PubMed]
19. Proctor, R.H.; Busman, M.; Seo, J.-A.; Lee, Y.W.; Plattner, R.D. A Fumonisin Biosynthetic Gene Cluster in Fusarium Oxysporum Strain O-1890 and the Genetic Basis for B versus C Fumonisin Production. *Fungal Genet. Biol.* **2008**, *45*, 1016–1026. [CrossRef]
20. Desjardins, A.E.; Munkvold, G.P.; Plattner, R.D.; Proctor, R.H. FUM1—A Gene Required for Fumonisin Biosynthesis but Not for Maize Ear Rot and Ear Infection by Gibberella Moniliformis in Field Tests. *Mol. Plant Microbe Interact.* **2002**, *15*, 1157–1164. [CrossRef] [PubMed]
21. Glenn, A.E.; Zitomer, N.C.; Zimeri, A.M.; Williams, L.D.; Riley, R.T.; Proctor, R.H. Transformation-Mediated Complementation of a FUM Gene Cluster Deletion in Fusarium Verticillioides Restores Both Fumonisin Production and Pathogenicity on Maize Seedlings. *Mol. Plant Microbe Interact.* **2008**, *21*, 87–97. [CrossRef] [PubMed]
22. Desjardins, A.E.; Plattner, R.D.; Stessman, R.J.; McCormick, S.P.; Millard, M.J. Identification and Heritability of Fumonisin Insensitivity in Zea Mays. *Phytochemistry* **2005**, *66*, 2474–2480. [CrossRef] [PubMed]
23. Sun, L.; Chen, X.; Gao, J.; Zhao, Y.; Liu, L.; Hou, Y.; Wang, L.; Huang, S. Effects of Disruption of Five FUM Genes on Fumonisin Biosynthesis and Pathogenicity in Fusarium Proliferatum. *Toxins* **2019**, *11*, 327. [CrossRef] [PubMed]
24. Xu, F.; Huang, L.; Wang, J.; Ma, C.; Tan, Y.; Wang, F.; Fan, Y.; Luo, M. Sphingolipid Synthesis Inhibitor Fumonisin B_1 Causes Verticillium Wilt in Cotton. *J. Integr. Plant Biol.* **2022**, *64*, 836–842. [CrossRef] [PubMed]
25. Akamatsu, H.; Itoh, Y.; Kodama, M.; Otani, H.; Kohmoto, K. AAL-Toxin-Deficient Mutants of Alternaria Alternata Tomato Pathotype by Restriction Enzyme-Mediated Integration. *Phytopathology* **1997**, *87*, 967–972. [CrossRef] [PubMed]
26. Spassieva, S.D.; Markham, J.E.; Hille, J. The Plant Disease Resistance Gene Asc-1 Prevents Disruption of Sphingolipid Metabolism during AAL-Toxin-Induced Programmed Cell Death. *Plant J.* **2002**, *32*, 561–572. [CrossRef] [PubMed]
27. Egusa, M.; Miwa, T.; Kaminaka, H.; Takano, Y.; Kodama, M. Nonhost Resistance of Arabidopsis Thaliana against Alternaria Alternata Involves Both Pre- and Postinvasive Defenses but Is Collapsed by AAL-Toxin in the Absence of LOH2. *Phytopathology* **2013**, *103*, 733–740. [CrossRef] [PubMed]

28. Sherif, M.; Kirsch, N.; Splivallo, R.; Pfohl, K.; Karlovsky, P. The Role of Mycotoxins in Interactions between Fusarium Graminearum and F. Verticillioides Growing in Saprophytic Cultures and Co-Infecting Maize Plants. *Toxins* **2023**, *15*, 575. [CrossRef] [PubMed]
29. Keyser, Z.; Vismer, H.F.; Klaasen, J.A.; Snijman, P.W.; Marasas, W.F.O. The Antifungal Effect of Fumonisin B_1 on Fusarium and Other Fungal Species. *S. Afr. J. Sci.* **1999**, *95*, 455–458. [PubMed]
30. Dawidziuk, A.; Koczyk, G.; Popiel, D. Adaptation and Response to Mycotoxin Presence in Pathogen-Pathogen Interactions within the Fusarium Genus. *World Mycotoxin J.* **2016**, *9*, 565–575. [CrossRef]
31. Janevska, S.; Ferling, I.; Jojić, K.; Rautschek, J.; Hoefgen, S.; Proctor, R.H.; Hillmann, F.; Valiante, V. Self-Protection against the Sphingolipid Biosynthesis Inhibitor Fumonisin B_1 Is Conferred by a FUM Cluster-Encoded Ceramide Synthase. *mBio* **2020**, *11*, e00455-20. [CrossRef]
32. Ma, L.-J.; van der Does, H.C.; Borkovich, K.A.; Coleman, J.J.; Daboussi, M.-J.; Di Pietro, A.; Dufresne, M.; Freitag, M.; Grabherr, M.; Henrissat, B.; et al. Comparative Genomics Reveals Mobile Pathogenicity Chromosomes in Fusarium. *Nature* **2010**, *464*, 367–373. [CrossRef] [PubMed]
33. Cuomo, C.A.; Güldener, U.; Xu, J.-R.; Trail, F.; Turgeon, B.G.; Di Pietro, A.; Walton, J.D.; Ma, L.-J.; Baker, S.E.; Rep, M.; et al. The Fusarium Graminearum Genome Reveals a Link between Localized Polymorphism and Pathogen Specialization. *Science* **2007**, *317*, 1400–1402. [CrossRef]
34. Krska, T.; Twaruschek, K.; Valente, N.; Mitterbauer, R.; Moll, D.; Wiesenberger, G.; Berthiller, F.; Adam, G. Development of a Fumonisin-Sensitive Saccharomyces Cerevisiae Indicator Strain and Utilization for Activity Testing of Candidate Detoxification Genes. *Appl. Environ. Microbiol.* **2023**, *89*, e0121123. [CrossRef] [PubMed]
35. Caldas, E.D.; Jones, A.D.; Ward, B.; Winter, C.K.; Gilchrist, D.G. Structural Characterization of Three New AAL Toxins Produced by *Alternaria alternata* f. Sp. Lycopersici. *J. Agric. Food Chem.* **1994**, *42*, 327–333. [CrossRef]
36. Frisvad, J.C.; Larsen, T.O.; Thrane, U.; Meijer, M.; Varga, J.; Samson, R.A.; Nielsen, K.F. Fumonisin and Ochratoxin Production in Industrial Aspergillus Niger Strains. *PLoS ONE* **2011**, *6*, e23496. [CrossRef] [PubMed]
37. Galagan, J.E.; Calvo, S.E.; Cuomo, C.; Ma, L.-J.; Wortman, J.R.; Batzoglou, S.; Lee, S.-I.; Baştürkmen, M.; Spevak, C.C.; Clutterbuck, J.; et al. Sequencing of Aspergillus Nidulans and Comparative Analysis with A. Fumigatus and A. Oryzae. *Nature* **2005**, *438*, 1105–1115. [CrossRef] [PubMed]
38. Proctor, R.H.; Desjardins, A.E.; Plattner, R.D.; Hohn, T.M. A Polyketide Synthase Gene Required for Biosynthesis of Fumonisin Mycotoxins in Gibberella Fujikuroi Mating Population A. *Fungal Genet. Biol.* **1999**, *27*, 100–112. [CrossRef] [PubMed]
39. D'mello, N.P.; Childress, A.M.; Franklin, D.S.; Kale, S.P.; Pinswasdi, C.; Jazwinski, S.M. Cloning and Characterization of LAG1, a Longevity-Assurance Gene in Yeast. *J. Biol. Chem.* **1994**, *269*, 15451–15459. [CrossRef] [PubMed]
40. Jiang, J.C.; Kirchman, P.A.; Zagulski, M.; Hunt, J.; Jazwinski, S.M. Homologs of the Yeast Longevity Gene LAG1 in Caenorhabditis Elegans and Human. *Genome Res.* **1998**, *8*, 1259–1272. [CrossRef] [PubMed]
41. Sultana, S.; Kitajima, M.; Kobayashi, H.; Nakagawa, H.; Shimizu, M.; Kageyama, K.; Suga, H. A Natural Variation of Fumonisin Gene Cluster Associated with Fumonisin Production Difference in *Fusarium Fujikuroi*. *Toxins* **2019**, *11*, 200. [CrossRef]
42. Kheder, A.; Akagi, Y.; Tsuge, T.; Kodama, M. Functional Analysis of the Ceramide Synthase Gene ALT7, a Homologue of the Plant Disease Resistant Gene Asc1, in a Plant Pathogenic Fungus *Alternaria alternata*. *Plant Pathol. Microbiol.* **2012**. [CrossRef]
43. Mogensen, J.M.; Frisvad, J.C.; Thrane, U.; Nielsen, K.F. Production of Fumonisin B2 and B4 by Aspergillus Niger on Grapes and Raisins. *J. Agric. Food Chem.* **2010**, *58*, 954–958. [CrossRef] [PubMed]
44. Susca, A.; Moretti, A.; Stea, G.; Villani, A.; Haidukowski, M.; Logrieco, A.; Munkvold, G. Comparison of Species Composition and Fumonisin Production in Aspergillus Section Nigri Populations in Maize Kernels from USA and Italy. *Int. J. Food Microbiol.* **2014**, *188*, 75–82. [CrossRef] [PubMed]
45. Susca, A.; Proctor, R.H.; Butchko, R.A.E.; Haidukowski, M.; Stea, G.; Logrieco, A.; Moretti, A. Variation in the Fumonisin Biosynthetic Gene Cluster in Fumonisin-Producing and Nonproducing Black Aspergilli. *Fungal Genet. Biol.* **2014**, *73*, 39–52. [CrossRef] [PubMed]
46. Yan, Y.; Liu, N.; Tang, Y. Recent Developments in Self-Resistance Gene Directed Natural Product Discovery. *Nat. Prod. Rep.* **2020**, *37*, 879–892. [CrossRef] [PubMed]
47. O'Neill, E.C.; Schorn, M.; Larson, C.B.; Millán-Aguiñaga, N. Targeted Antibiotic Discovery through Biosynthesis-Associated Resistance Determinants: Target Directed Genome Mining. *Crit. Rev. Microbiol.* **2019**, *45*, 255–277. [CrossRef] [PubMed]
48. Stahlecker, J.; Mingyar, E.; Ziemert, N.; Mungan, M.D. SYN-View: A Phylogeny-Based Synteny Exploration Tool for the Identification of Gene Clusters Linked to Antibiotic Resistance. *Molecules* **2020**, *26*, 144. [CrossRef] [PubMed]
49. Yan, Y.; Liu, Q.; Zang, X.; Yuan, S.; Bat-Erdene, U.; Nguyen, C.; Gan, J.; Zhou, J.; Jacobsen, S.E.; Tang, Y. Resistance-Gene-Directed Discovery of a Natural-Product Herbicide with a New Mode of Action. *Nature* **2018**, *559*, 415–418. [CrossRef] [PubMed]
50. Twaruschek, K.; Spörhase, P.; Michlmayr, H.; Wiesenberger, G.; Adam, G. New Plasmids for Fusarium Transformation Allowing Positive-Negative Selection and Efficient Cre-loxP Mediated Marker Recycling. *Front. Microbiol.* **2018**, *9*, 1954. [CrossRef]

toxins

MDPI

Article

Inhibition of Aflatoxin B1 Production by Procyanidins Present in *Annona muricata* and *Uncaria tomentosa* Aqueous Extracts

Laura F. Cadenillas [1], Guillaume Billerach [1], Christopher Hernandez [1], Vanessa Durrieu [1] and Jean-Denis Bailly [1,2,*]

[1] Laboratoire de Chimie Agro-Industrielle (LCA), Université de Toulouse, INRAE, INPT, 4 Allée Emile Monso, 31030 Toulouse, France; laura.cadenillas.s@gmail.com (L.F.C.); guillaume.billerach@toulouse-inp.fr (G.B.); hernandezhernandezchristopher@gmail.com (C.H.); vanessa.durrieu@ensiacet.fr (V.D.)

[2] École Nationale Vétérinaire de Toulouse, 23 Chemin des Capelles, CEDEX, 31076 Toulouse, France

[*] Correspondence: jean-denis.bailly@envt.fr; Tel.: +33-56-1193-229

Abstract: Aflatoxin B_1 (AFB_1), primarily produced by *Aspergillus flavus* and *A. parasiticus*, is the most dangerous mycotoxin for humans and contaminates a variety of crops. To limit fungal growth and aflatoxin production in food and feed, research has been increasingly focusing on alternatives to pesticides. Studies show that some aqueous plant extracts with strong antioxidant properties could significantly impact AFB_1 production, representing an eco-friendly and sustainable method to protect crops. The present study demonstrates that aqueous extracts of *Anonna muricata* (*AM*) and *Uncaria tomentosa* (*UT*) inhibit AFB_1 synthesis in a dose-dependent manner with a half-maximal inhibitory concentration of 0.25 and 0.28 mg dry matter per milliliter of culture medium, respectively. This effect correlates with the presence of polyphenols and, more precisely, with condensed tannins. It is also related to the subsequent antioxidant activity of both extracts. A bio-guided fractionation followed by high-performance liquid chromatography and mass spectrometry analysis of the active fractions identifies procyanidins and, more precisely, catechin (5.3% *w*/*w* for *AM* and 5.4% *w*/*w* for *UT*) and epicatechin (10.6% *w*/*w* for *AM* and 25.7% *w*/*w* for *UT*) as the major components in both extracts. The analysis of how pure standards of these molecules affect AFB_1 production demonstrates that catechin plays an essential role in the inhibition observed for both plant extracts, since the pure standard inhibits 45% of AFB_1 synthesis at a concentration close to that of the extracts.

Keywords: Aflatoxin B_1; *Annona muricata*; *Uncaria tomentosa*; antioxidant activity; procyanidins

Key Contribution: This study demonstrates the role of condensed tannins, mainly catechin and epicatechin, present in two different aqueous plant extracts, in the inhibition of AFB_1 in *Aspergillus flavus*.

1. Introduction

Aflatoxin B_1 (AFB_1) is a mycotoxin produced by several fungal species within the section *Flavi* of the genus *Aspergillus*. *Aspergillus flavus* and *A. parasiticus* are the two major aflatoxin-producing species and are consistently responsible for the contamination of various types of foods worldwide. AFB_1 is the most dangerous mycotoxin and is classified as a class 1 carcinogen for humans by the International Agency for Research on Cancer. It is responsible for hepatocellular carcinoma in humans and has other toxic effects such as growth suppression, immune system modulation, and malnutrition [1].

AFB_1 poses a major risk to public health, particularly in warm regions where environmental conditions favor the growth of toxigenic fungi. Aflatoxigenic species can grow on various foods such as cereals, groundnuts, and spices. Due to the physiological characteristics of producing species and their xerophilic character, AFB_1 can enter at different steps of the production chain: in the fields before harvest, during the peri-harvest period, before crops are sufficiently dried or later, during storage in the case of insufficient drying or remoistening. Given the stability of AFB_1, the most effective strategy to protect consumers is to limit the toxin synthesis and the subsequent contamination of food.

Various strategies have now been established to limit or prevent fungal development and subsequent AFB_1 synthesis in crops during pre-harvest, post-harvest, and storage of foods. Some of these strategies typically include good agricultural practices, although the key role of climate, which cannot be controlled, may strongly limit their efficacy. In addition, fungicides are used mainly to avoid the growth of toxigenic fungi, but their numerous harmful side effects and the increasing resistance of target organisms require alternative strategies to prevent aflatoxin contamination of crops [2,3].

Ensuring food safety for consumers while developing sustainable production tools with minimal environmental impact has triggered extensive studies of new eco-friendly approaches based on natural products, such as plant-derived substances. Such products could be used to limit contamination in the fields, during peri-harvest by coating grains, and even during storage, although it would require specific formulations to guarantee the homogenous distribution of the product in silos.

Various plant-derived molecules limit the synthesis of mycotoxins such as AFB_1 [4–9]. For example, polyphenols such as phenols, flavonoids, tannins, and terpenes are active molecules that inhibit fungal development and AFB_1 production [10–12]. Additionally, certain active molecules with antioxidant potential may inhibit AFB_1 production by regulating the oxidative stress around the fungus [13].

Current research suggests that condensed tannins, present in some plant extracts, may significantly inhibit some mycotoxins. For example, a study on the condensed tannins of *Dalea purpurea* reported that these compounds restrict the synthesis of deoxynivalenol and ochratoxin A [14]. Similarly, another study on the aqueous extract of *Mimosa tenuiflora* demonstrated that condensed tannins inhibit AFB_1 synthesis due to a down-regulation of the internal regulators (*aflR* and *aflS*) of AFB_1 cluster genes [15].

The leaves of the *Annona muricata* (*AM*) tree are widely used in infusions to treat diabetes, insomnia, cystitis, and rheumatic problems [16–18]. To date, a total of 212 bioactive compounds have been reported in *AM* leaves, the predominant ones being acetogenins, alkaloids, and polyphenols [19,20].

In the same way, *Uncaria tomentosa* (*UT*), a large woody vine native to the Amazon, has antioxidant, antimicrobial, and anti-inflammatory properties and can also be used to treat asthma, abscesses, and urinary tract infections [21–23]. More than 50 phytochemicals from the plant have been identified and isolated, with condensed tannins being the most abundant. Others include indole and oxindole alkaloids, quinic acid, and polyphenols [24,25].

Previous work demonstrated that *AM* and *UT* aqueous extracts are rich in condensed tannins, which contribute to the high antioxidant potential of these plants [6]. Considering the antiradical capacity of these plant extracts and the potential link between AFB_1 synthesis and oxidative stress levels, the present study seeks to determine whether polyphenols, and more precisely condensed tannins, present in the aqueous extracts of *AM* and *UT* are responsible for the anti-AFB_1 activity. We thus used bio-guided fractionation of the extracts followed by high-performance liquid chromatography (HPLC) and mass spectrometry analysis of the fractions to identify the condensed tannins responsible for AFB_1 inhibition. Additionally, we evaluated how the major condensed tannins present in extracts affect the inhibition of AFB_1.

2. Results and Discussion

2.1. Effect of AM and UT Aqueous Extracts on A. flavus Growth and AFB_1 Synthesis

The impacts of four increasing concentrations (from 0.04 to 0.30 mg of dry matter per milliliter of culture medium) of aqueous extracts of *AM* leaves and *UT* bark on both fungal growth and AFB_1 synthesis were evaluated after one week of incubation at 27 °C and compared with a control culture grown on an extract-free culture medium. Figure 1 shows the results.

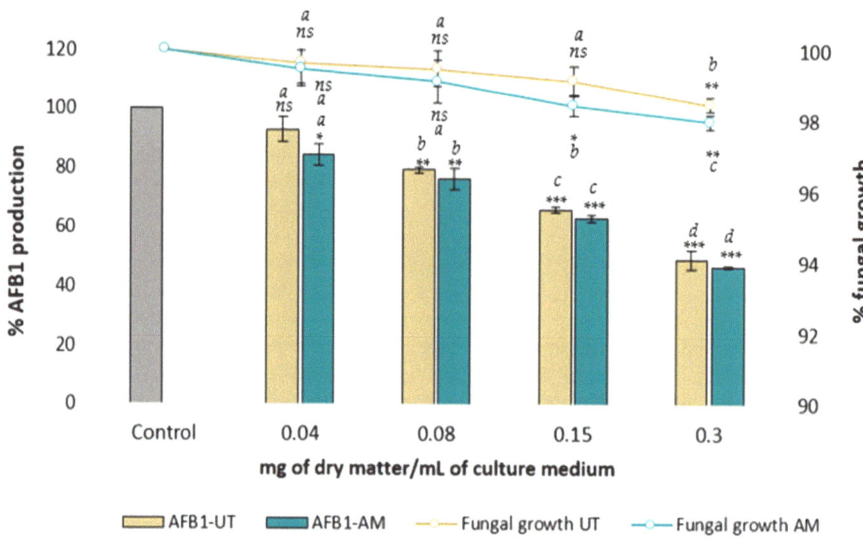

Figure 1. Dose effect of increasing concentrations of *UT* (brown) and *AM* (blue) aqueous extract on *A. flavus* NRRL 62,477 growth (lines) and AFB_1 production (bars). Results are presented as the percentage of AFB_1 production or colony diameter relative to untreated control cultures ± one standard deviation (n = 4). Control and treated cultures with statistically significant differences are indicated by * (*ns* = no statistically significant change, * p value < 0.05, ** p value < 0.01, *** p value < 0.001). Significant differences between two successive concentrations of the same sample are denoted by different letters (p value < 0.05).

The aqueous extracts of *AM* and *UT* did not strongly affect the growth of *A. flavus*. A slight reduction in growth occurred for concentrations of 0.15 and 0.3 mg dry matter per milliliter (DM/mL) for *AM* and 0.3 mg DM/mL for *UT*. At the highest extract concentrations, the maximal inhibition of growth was only 3% and 2.5% for *AM* and *UT*, respectively.

Both *AM* and *UT* aqueous extracts led to a very similar dose-dependent inhibition of AFB_1 synthesis, with half-maximal inhibitory concentrations on AFB_1 synthesis ($IC50_{AFB1}$) of 0.25 and 0.28 mg DM/mL, respectively. The maximal observed inhibitions were 53% for *AM* and 51% for *UT*, with both occurring at 0.3 mg DM/mL culture medium.

AM aqueous extract reduced AFB_1 production more than reported for other plants of the same family (*Annonaceae*). For example, *Polyalthea longifolia* and *Artabotrys odoratissimus* have $IC50_{AFB1}$ values of 0.9 and 0.5 mg DM/mL, respectively [26,27]. The results for *UT* were in the same range as those obtained in previous studies on bark extracts of other plants. For example, *Sapindus mucorossi* and *Mimosa tenuiflora* have an $IC50_{AFB1}$ of 0.4 and 0.15 mg DM/mL, respectively [15,26].

2.2. Characterization and Fractionation of AM and UT Extracts

2.2.1. Composition of Extracts and Fractions

A bio-guided fractionation of *AM* and *UT* extracts was performed to better characterize the role of polyphenols and, more particularly, condensed tannins in AFB_1 inhibition. The fractionation proceeded by separating the components using macroporous absorption resin (amberlite FPX66), leading to fraction F1 (aqueous fraction) and fraction F2 (ethanolic fraction). Next, F2 was further fractionated into a tannin-free fraction F2.1 using polyvinylpolypyrrolidone (PVPP) precipitation. The initial extracts and their fractions and sub-fractions were analyzed to determine their polyphenol content, condensed tannin concentration, and antioxidant activity. Table 1 summarizes the results.

Table 1. Characterization of *AM* and *UT* aqueous extract, aqueous (F1) and ethanolic (F2) fractions after separation on macroporous resin, and subfraction 2.1 (tannin-free fraction) obtained by PVPP precipitation.

Fraction	Dry Matter (g)	Polyphenols Content (mg GAE/g DM) *	Antioxidant Activity $IC50_{DPPH}$ (mg/L) **	Condensed Tannins (mg/g DM)
AM	3.91 ± 0.01^{a}	188 ± 9^{a}	35^{a}	137 ± 2^{a}
F1	1.70 ± 0.02^{b}	56 ± 5^{b}	310^{b}	Nd
F2	1.83 ± 0.04^{b}	320 ± 1^{b}	20^{b}	220 ± 4^{b}
F2.1	0.21 ± 0.01^{b}	10 ± 1^{b}	460^{b}	2 ± 1^{b}
UT	5.31 ± 0.03^{a}	226 ± 3^{a}	15^{a}	219 ± 5^{a}
F1	1.92 ± 0.01^{b}	18 ± 2^{b}	363^{b}	Nd
F2	2.24 ± 0.01^{b}	530 ± 7^{b}	8^{b}	502 ± 6^{b}
F2.1	0.18 ± 0.03^{b}	6 ± 1^{b}	$>500^{b}$	3 ± 1^{b}

* DM: dry matter; GAE: gallic acid equivalent; ** Concentration of extract or fraction reducing 50% of DPPH (2,2-Diphenyl-1-picrylhydrazyl), Nd: not detected. For each parameter, significant differences between the initial plant extract and its fractions are indicated by different letters (p value < 0.05).

The aqueous extract of *AM* had a polyphenol content of 188 mg GAE/g DM and an antioxidant activity equivalent to an $IC50_{DPPH}$ of 35 mg/L. To compare, Trolox and quercetin, two strong antioxidant molecules often used as standards, have $IC50_{DPPH}$ = 3.5 mg/L [28] and 8.1 mg/L [29], respectively, confirming that *AM* has a significant antioxidant activity. Both polyphenol content and antioxidant activity exceeded those previously reported for different *AM* leaf extracts, whether they were aqueous (polyphenol content = 155 mg GAE/g and $IC50_{DPPH}$ = 81 mg/L) [30] or methanolic ($IC50_{DPPH}$ of 63 mg/L) [31]. Antioxidant activity also exceeded that reported for plants belonging to the same family, such as *A. diversifolia* ($IC50_{DPPH}$ = 98 mg/L), *A. purpurea* ($IC50_{DPPH}$ = 151 mg/L), and *A. reticulata* ($IC50_{DPPH}$ = 126 mg/L) [32].

UT extract exhibited considerably greater antioxidant activity ($IC50_{DPPH}$ = 15 mg/L) than *AM*, which could be linked to *UT*'s greater polyphenolic compound content (226 mg GAE/g DM) and, more specifically, condensed tannins (219 mg/g DM). Note that the extract was prepared from *UT* bark, which is rich in proanthocyanidins (condensed tannins), as demonstrated by Gonçalves et al. [33]. In that work, chromatographic analysis (HPLC-diode array detection and thin-layer chromatography) revealed that the main constituents of a *Uncaria tomentosa* bark decoction were proanthocyanidins and phenolic acids. In addition, the authors demonstrated that proanthocyanidins correlate directly with the high antioxidant activity of the aqueous extract.

The fractions of *AM* and *UT* showed similar trends. As expected, F1 (aqueous fraction) had a low polyphenol content (56 mg GAE/g DM for *AM* and 18 mg GAE/g DM for *UT*), which explains the low antioxidant capacity of this fraction ($IC50_{DPPH}$ = 310 mg/L for *AM* and $IC50_{DPPH}$ = 363 mg/L for *UT*). In contrast, F2 (ethanolic fraction) concentrated the polyphenols: 320 mg GAE/g DM for *AM* and 530 mg GAE/g DM for *UT*. Condensed tannins represented a significant fraction of these polyphenols (220 mg/g for *AM* and 502 mg/g for *UT*), as highlighted by the low residual polyphenol content after tannin precipitation with PVPP. Fraction F2 of both plants also displayed high antioxidant activities ($IC50_{DPPH}$ of 20 and 8 mg/L for *AM* and *UT*, respectively).

These results are consistent with each other because the fractionation used solvents of different polarities to separate the compounds of interest in the extract. Fraction F1 used only water as an elution solvent, which favored the isolation of sugars and proteins. In contrast, fraction F2 was obtained using ethanol, which favored the elution of polyphenols from the column. These results are consistent with previous studies, demonstrating the high concentration of polyphenols in the ethanolic extracts of different plants [34,35]. The correlation of the polyphenol content of the plant extract with the antioxidant activity of the extract has also been shown [36,37], as in our study.

Previous reports state that condensed tannins such as procyanidin trimer C1 and dimer B1 participate strongly in the antioxidant activity of the *AM* extract [38]. Similarly, in *UT*, condensed tannins such as procyanidins are the most abundant polyphenols in the plant and are responsible for its high antioxidant activity [39]. Since both *AM* and *UT* tannin-free subfractions (F2.1) completely lost their antioxidant capacity, our results confirm the central role of condensed tannins in both *AM* and *UT* antioxidant activity.

2.2.2. Impact of *AM* and *UT* Fractions on AFB_1 Production

A well-established link exists between oxidative stress and AFB_1 synthesis. An increase in oxidative stress increases AFB_1 synthesis, whereas a decrease, obtained by the presence of antioxidant compounds, decreases AFB_1 production [13]. Different polyphenols with potent antioxidant capacity were reported to subsequently hinder AFB_1 synthesis [40–42]. Therefore, the AFB_1 inhibition observed with *AM* and *UT* extracts may be attributed to their high polyphenol content and the resulting significant antioxidant activity. To corroborate this claim, we tested on *A. flavus* the AFB_1-inhibiting potential of the different fractions by using various concentrations in polyphenols, condensed tannins, and antioxidative potential. Figure 2 presents the results.

The *AM* and *UT* fractions produced similar trends in AFB_1 inhibition.

Compared with the initial extracts, fractions F1 lost most of their ability to inhibit AFB_1 in *A. flavus*. At the highest concentration tested (0.3 mg DM/mL), *AM* and UT aqueous fractions 1 inhibited only 11% and 20% of AFB_1 synthesis, respectively. However, a slight bioactivity remained for both plants, which could be explained by the residual content of polyphenols found in these fractions (see Table 1). In addition, at the lowest concentrations tested (0.04 mg DM/mL), the F1 fractions of both plants induced a slight increase in AFB_1 synthesis, possibly due to the presence of molecules such as sugars, which may supply nutritional support for the growth of *A. flavus* and, therefore, indirectly promote the synthesis of AFB_1.

The ethanolic fractions F2 of both *AM* and *UT* produced a dose-dependent effect on AFB_1 inhibition. The F2 fractions of both *AM* and *UT* inhibited approximately 65% of AFB_1 synthesis at 0.3 mg DM/mL, the highest concentration tested. Since the F2 fractions were rich in polyphenolic compounds (32% *w/w* for *AM* and 53% *w/w* for *UT*), these results indicate that polyphenols may be what allows *AM* and *UT* extracts to inhibit AFB_1 synthesis.

Removing the condensed tannins by precipitation (producing fraction F2.1) significantly reduced the efficacy of the extract, with a residual maximum AFB_1 inhibition of 22% for *AM* and only 9% for *UT*. This result demonstrates that condensed tannins play a pivotal role in AFB_1 inhibition for both plants.

Further studies are needed to elucidate the precise mechanisms by which condensed tannins influence AFB_1 production. Current theories mainly propose that their antioxidant activity could be due to direct interference with the regulation of genes involved in aflatoxin biosynthesis. In fact, numerous studies have demonstrated a direct correlation between oxidative stress and the regulation of AFB_1 synthesis [12,13,15]. Some researchers even proposed that, in *A. flavus*, aflatoxin synthesis could be part of the response to oxidative stress [43]. For example, previous work showed that extracts from the bark of *Mimosa tenuiflora* and Pine Bark, which are rich in condensed tannins, inhibit AFB_1 synthesis by down-regulating the key regulatory genes *aflR* and *aflS* [15]. In addition, the extracts also interfere with the expression of various genes involved in the fungal stress response and antioxidant defense system (e.g., catA, sod1, mtfA, atfA, and msnA), suggesting a connection between the two processes.

Other studies suggested that extracts containing flavonoids could directly degrade AFB_1, most likely by targeting the furan double bond and the lactone ring of AFB_1, which are responsible for its toxic and carcinogenic properties. However, the reaction mechanisms were not fully elucidated and could involve additive or synergistic mechanisms with enzymes [12,44,45].

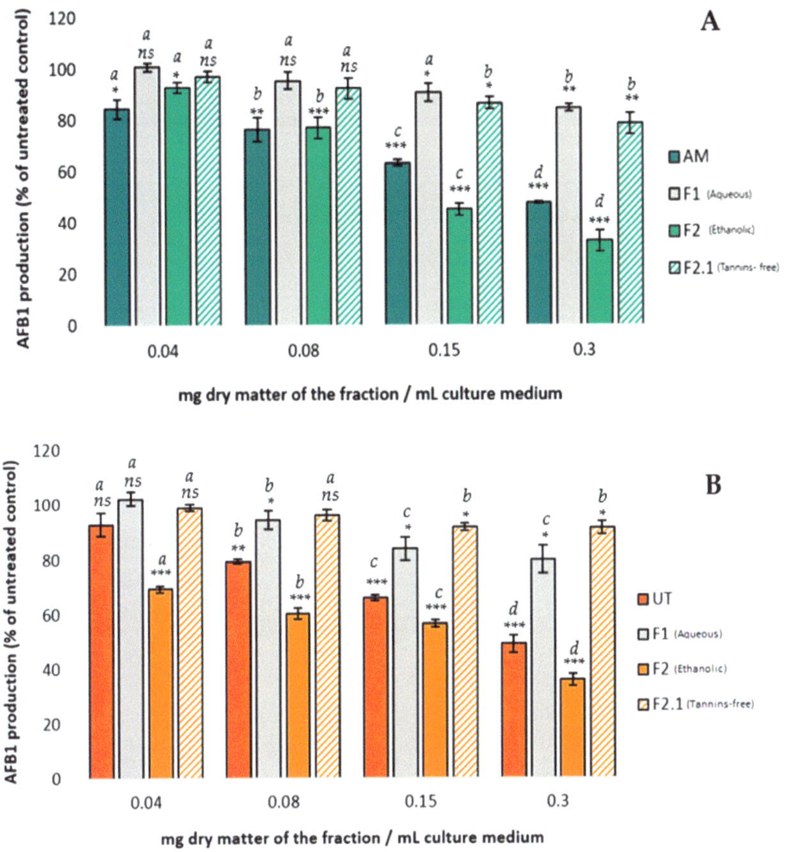

Figure 2. Comparison of the effect of increasing concentrations of *AM* (**A**) and *UT* (**B**) aqueous extracts and their fractions on AFB_1 synthesis in *A. flavus* NRRL 62,477. Results are expressed as the percentage of AFB_1 production relative to untreated control cultures ± one standard deviation ($n = 4$). Statistically significant differences from the control are indicated with * for the different concentrations of each treatment (*ns* = no statistically significant change between two concentrations, * p value < 0.05, ** p value < 0.01, *** p value < 0.001). Significant differences between two successive concentrations of the same extract and fraction are denoted by different letters (p value < 0.05).

2.2.3. Identification of the Primary Condensed Tannins in the Ethanolic Fractions of *AM* and *UT*

Research has yet to identify which type(s) of condensed tannins is (are) responsible for the anti-AFB_1 property. Thus, we further characterized the tannins present in our extracts.

Since condensed tannins are complex polymers, their characterization required depolymerization. Therefore, condensed tannins from fractions F2 of both plants were depolymerized for 120 min in a methanolic solution of hydrochloric acid and menthofuran. Next, the samples were analyzed using ultrahigh-performance liquid chromatography (UHPLC) with ultraviolet (UV) diode-array detection (DAD) coupled with mass spectrometry (MS). Figure 3 shows the UV chromatogram of the F2 fractions at 280 nm before and after depolymerization.

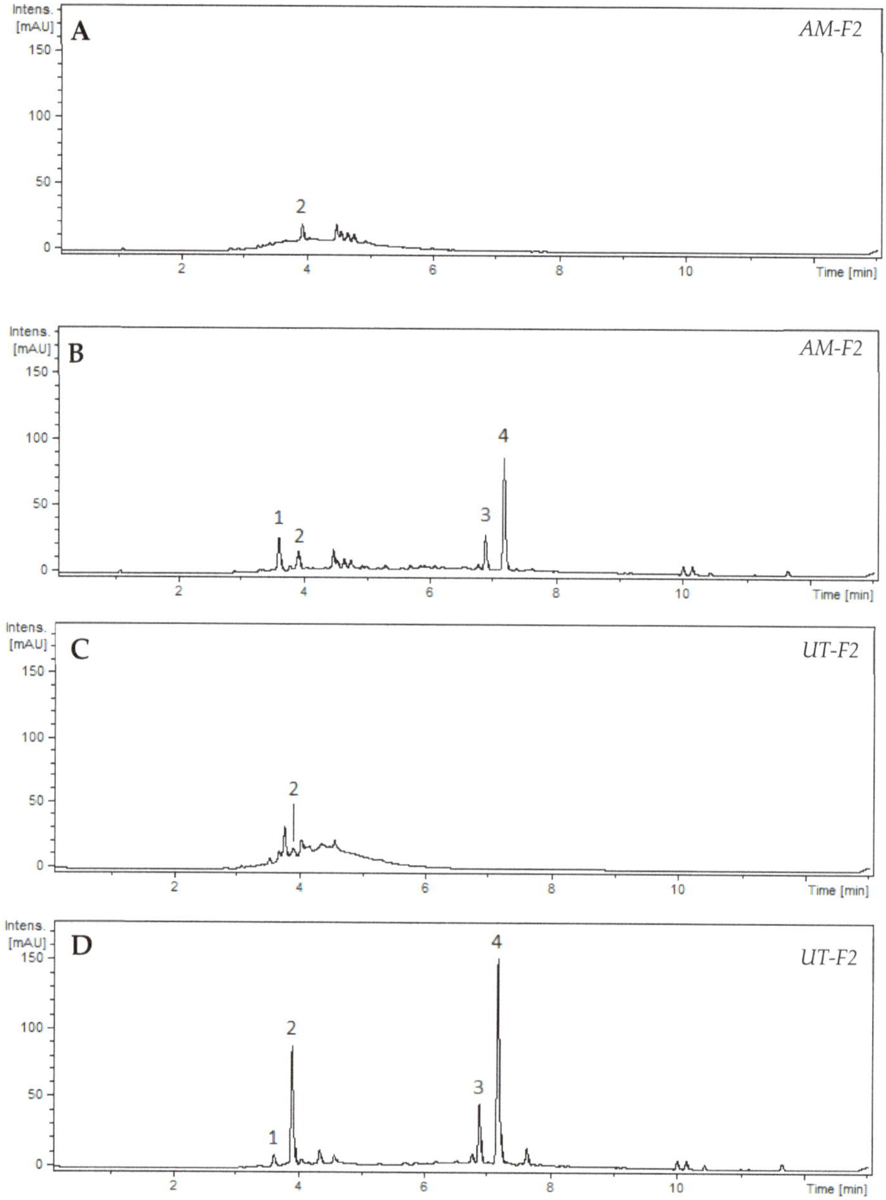

Figure 3. Chromatograms at 280 nm of condensed tannins present in fraction 2 of (**A**,**B**) *AM* and (**C**,**D**) *UT*, (**A**,**C**) before and (**B**,**D**) after depolymerization (1 g/L; 120 min reaction). Peak identification: **1**, Catechin (290.9 m/z); **2**, Epicatechin (290.9 m/z); **3**, Catechin-menthofuran (439.1 m/z); **4**, Epicatechin-menthofuran (439.1 m/z).

Table 2 details the nature of condensed tannins (% w/w) in *AM* and *UT* F2 fractions. The results show that, in the *AM* F2 fraction, the total procyanidin content represented 15.9% (w/w) of the extract and consisted of 5.3% of catechin and 10.6% of epicatechin. In contrast, *UT* had twice the content: 31.1% (w/w) total procyanidin content from the extract, with 5.4% catechin and 25.7% epicatechin.

Table 2. Primary condensed tannins present in F2 fractions of *AM* and *UT*.

	Composition in Condensed Tannins of the Tested Fractions (%*w*/*w*)				
Peak Number	**1**	**2**	**3**	**4**	
Detected compound	Catechin (terminal units)	Epicatechin (terminal units)	Catechin-Mf * (extension units)	Epicatechin-Mf * (extension units)	Total procyanidin content
Tested fractions					
AM-F2	2.8 ± 0.1	1.7 ± 0.1	2.5 ± 0.1	8.9 ± 0.1	15.9
UT-F2	1.1 ± 0.1	9.3 ± 0.1	4.3 ± 0.1	16.4 ± 0.2	31.1

* Mf: menthofuran.

The difference in the percentage of condensed tannins between *AM* and *UT* fractions is primarily attributed to the part of the plant used to prepare each extract. Whereas *UT* extract was prepared from bark, *AM* extract was prepared from leaves. Recall that bark is particularly rich in condensed tannins, more so than other plant parts. This fact was illustrated by Killedar and More in a study that quantified and characterized condensed tannins from different parts of *Memecylon umbellatum*. They demonstrated that bark contains the highest concentration of condensed tannins (24.1% *w*/*w*), followed by leaves (17.6% *w*/*w*) [46].

The nature of the condensed tannins identified from *AM* and *UT* is consistent with the composition of similar extracts reported in the literature. Besides catechin and epicatechin, procyanidins B2, B4, and C1 have been identified as the primary condensed tannins in ethanolic extracts of *UT* bark [47]. Propelargonidins reported in *UT* extracts [39] were also detected in our fractions [m/z = 423.1, tentatively identified as (epi)afzelechin-menthofuran]. However, they were not quantified because of their negligible contribution to the UV chromatogram compared with procyanidins. Furthermore, a study on an ethanolic fraction of *AM* reported (+)-catechin, gallocatechin, procyanidin dimer B1, and procyanidin trimer C1 as the main flavonoids [48,49]. All these compounds except gallocatechin were detected in our fractions, but at different concentrations, which can be explained by differences in species, age, and the part of the plant collected and its geographical origin. Additionally, considering that condensed tannins are synthesized by a defense mechanism against environmental stress, their production is closely related to the external conditions and cultivation in which the plants grow [50,51].

2.2.4. Effect of Catechin and Epicatechin on AFB_1 Synthesis

Catechin and epicatechin were identified as the primary condensed tannins in *UT* and *AM* extracts. When *AM* aqueous extract was cultured with *A. flavus* at the highest concentration (0.3 mg DM/mL of culture medium), 16.2 µg of catechin and 32.1 µg of epicatechin per milliliter of culture medium were quantified, achieving an AFB_1 inhibition of 53%. While using *UT* extract, a concentration of 15.9 µg of catechin and 77.1 µg of epicatechin per milliliter of culture medium resulted in a 51% AFB_1 inhibition. Therefore, to determine how these compounds affect the inhibition of AFB_1, we analyzed how AFB_1 synthesis is affected by increasing concentrations of pure catechin and epicatechin. Specifically, we used 5, 10, and 15 µg/mL of culture medium for catechin and 6.25, 12.5, and 25 µg/mL of culture medium for epicatechin. After a seven-day incubation at 27 °C, the highest concentration of catechin (epicatechin) inhibited AFB_1 synthesis by 45% (36%) (see Table 3). These results suggest that these two compounds play a significant role in inhibiting the synthesis of AFB_1.

Table 3. Effect of increasing concentrations of catechin and epicatechin standards on AFB_1 inhibition.

Standard	Concentration of the Standard (μg/mL of Culture Medium)	Inhibition of AFB_1 * (%)
Catechin	5.00	28 ± 1
	10.00	37 ± 2
	15.00	45 ± 1
Epicatechin	6.25	19 ± 1
	12.50	27 ± 2
	25.00	36 ± 2

* Results are expressed as the percent of AFB_1 inhibition with respect to untreated control cultures.

Currently, limited data are available to identify the molecules directly implicated in the inhibition of mycotoxin synthesis. Norton et al. studied how certain anthocyanidins and related flavonoids affected AFB_1 inhibition and showed that, at a concentration of 83 μg of catechin per milliliter culture medium, these anthocyanidins and flavonoids generated 30% AFB_1 inhibition [52]. Similarly, a microdilution study by Zhou et al. [53] of how a tea-derived polyphenol mixture affected AFB_1 synthesis showed that catechin at a concentration of 500 μg/mL of solvent generates 50% AFB_1 inhibition. Zhou et al. [53] also suggested that using a mixture of polyphenols, including epicatechin, epigallocatechin, and/or gallocatechin, could enhance the inhibition of AFB_1.

The present results indicate that catechin and epicatechin both inhibit AFB_1. However, considering the effects of these two compounds when used alone, one would naively expect a total AFB_1 inhibition of over 80% when the two compounds are used together. This expectation significantly exceeds the inhibition obtained when using both *AM* and *UT* aqueous extracts. This lower efficacy may be attributed to the fact that, in the plant extracts, these two compounds are present as polymers, whereas the standards were tested as monomers. Therefore, the effectiveness of the plant extracts may be strengthened by their depolymerization. Such a drop in efficacy could also result from an inhibition–saturation phenomenon for AFB_1 if both compounds target the same reaction pathway. Further studies are thus required to determine the exact reaction mechanism by which these molecules impact AFB_1 synthesis.

The results also indicate that catechin may affect other mycotoxins besides AFB_1. In fact, a chemical characterization of a methanolic extract of *Aloe vera* gel with high anti-ochratoxin A potential suggested that catechin is a primary contributor to this inhibitory process [54].

Combining these results allows us to propose catechin and epicatechin as compounds with promising anti-AFB_1 activity (and, more generally, antimycotoxin activity). However, more detailed studies are needed on the reaction mechanism(s) and on possible interactions between catechin and epicatechin.

3. Conclusions

The present results demonstrate that aqueous extracts of *AM* and *UT* inhibit AFB_1 synthesis without affecting the growth of *A. flavus*. The bio-guided fractionation demonstrates the importance of polyphenols, especially condensed tannins, in inhibiting AFB_1 synthesis. Although *AM* and *UT* are unrelated plants, the results show that catechin and epicatechin are the primary condensed tannins present in aqueous extracts from each plant. The results show that these compounds play an essential role in AFB_1 inhibition, opening the door to the valorization of certain agricultural by-products rich in condensed tannins. These by-products could be used during or after harvest to protect crops from AFB_1 contamination. The formulation of such a product is a key point to ensure the correct distribution of the active compounds throughout the grains. Finally, more research is required to determine the precise reaction mechanism(s) of these compounds and to verify that the inhibitory effect is reproduced in other *A. flavus* strains and other aflatoxigenic species.

4. Materials and Methods

4.1. Chemicals and Reagents

All analytical solvents and chemicals were purchased from Sigma-Aldrich (St Quentin-Fallavier, France). (+)− epicatechin and (−)− catechin were purchased from Extrasynthese (Genay, France). Menthofuran (3,6-dimethyl-4,5,6,7-tetrahydro-1-benzofuran, ≥95%) was purchased from Thermo-Fisher (Strasbourg, France). Ethanol (96%) and sodium carbonate were purchased from VWR International (Fontenay sous Bois, France).

4.2. Plant Material

Dried *Annona muricata leaves* and *Uncaria tomentosa bark* were purchased from a local market in Lima, Peru. The samples were ground using a mill with a 2 mm grid and stored until needed at 4 °C in plastic bags.

4.3. Preparation of Aqueous Extracts

The aqueous extraction was performed by maceration in a 5 L reactor. Ninety grams of ground material from each plant was combined with 3 L of ultrapure water and stirred for 15 h at room temperature. The mixture was then centrifuged for 15 min at 15,000 rpm (Sigma 6-16 K Centrifuge, Osterode AM Harz, Germany) and filtered under vacuum in a Whatman No. 1 paper (Cytiva). The extracts were subsequently sterilized at 121 °C for 20 min and stored at 4 °C until required.

4.4. Fractionation of Extracts

The fractionation method was adapted from Hernandez et al. [15] and involved macroporous adsorption resin amberlite FPX66 (Rohm and Haas, Philadelphia, PA, USA). First, 80 g of resin was preconditioned with 250 mL of ultrapure water (UHQ) stirred by a magnetic stirrer for 10 h. Subsequently, the solution was packed into a cylindrical glass column (inner diameter × length = 3 cm × 30 cm) fitted with a fritted disk. Three hundred mL of each extract was deposited into the column containing the resin. A first elution with 600 mL of ultrapure water (UHQ) was performed to remove sugars and proteins (fraction 1), followed by desorption using 375 mL of 96% ethanol to elute the compounds contained in the resin (fraction 2). The entire fractionation was conducted three times, and the fractions obtained with each solvent were pooled.

4.5. PVPP Fractionation

The condensed tannins in the ethanolic fraction (fraction 2) were precipitated using the nonprotein polymer polyvinylpolypyrrolidone (PVPP) to obtain a tannin-free fraction. Briefly, 5 mL of fraction 2 from each extract was combined with 5 mL of distilled water and 500 mg of PVPP. The samples were vortexed for 30 s, then stored at 4 °C for 15 min, and then centrifuged at 4000 rpm (3000 g) for 10 min. The supernatant was collected for subsequent analysis (fraction 2.1).

4.6. Characterization of Extracts and Fractions

4.6.1. Dry Matter Content

The DM content was determined by weighing 20 mL of each sample before and after drying in a Memmert oven (Schwabach, Germany) at 100 °C for 24 h. This procedure was performed in triplicate.

4.6.2. Dosage of Total Phenolic Content

The total phenolic content was assessed using the Folin–Ciocalteu method from Singleton and Rossi in a 96-well microplate [55]. The results are expressed in mg of gallic acid equivalent (GAE) per gram of dry extract.

4.6.3. Condensed Tannin Content

The condensed tannin content of each sample was quantified by using the Waterman and Mole method [56]. The following equation was used to calculate the tannin concentration:

$$\text{Tannin concentration (mg/g)} = [0.3866 \times (\text{sample absorbance} - \text{control absorbance}) \times \text{dilution factor}]/(\text{dry matter})$$

All tests were performed three times, and the results are expressed in milligrams of condensed tannins per gram of dry matter (DM) $\pm$ one standard deviation.

4.6.4. Dosage of Free Radical Scavenging Activity

The antioxidant activity was assessed using the DPPH (2,2-diphenyl-1-picrylhydrazyl) method from Brand-Williams et al. [57]. The radical-scavenging activity was reported as the half-maximal inhibitory concentration (IC50), which corresponds to the concentration of the extract (in mg/L) required to reduce 50% of the DPPH radicals. IC50 was calculated as follows:

$$\text{IC50} = 0.5 - \frac{a}{b}$$

The results are expressed in milligrams of extract per liter (mg/L).

4.7. Characterization of Condensed Tannins in the Ethanolic Fraction of AM and UT Extracts

4.7.1. Depolymerization of Condensed Tannins from Fraction 2

In a vial, 500 µL of methanolic solutions of *AM* and *UT* (2 g/L) previously prepared from a dried fraction 2 of both samples was mixed with 50 µL of menthofuran (13.8 mM) and 450 µL of HCl (0.1 M). The vials were sealed and incubated at 30 °C. After 120 min of reaction, the vials were immediately analyzed by UHPLC-DAD-MS (Thermo Scientific, Waltham, MA, USA). Each test was performed in triplicate per sample.

4.7.2. UHPLC-DAD-MS Analyses

The liquid chromatography system included an Ultimate 3000 ultrahigh-performance liquid chromatography (UHPLC) instrument equipped with a photodiode array detector (Thermo Fisher Scientific, Waltham, MA, USA). The column consisted of a Kinetex 2.6 µm EVO C18 100 Å, 150 mm $\times$ 4.6 mm (Phenomenex, Torrance, CA, USA) maintained at 40 °C. The flow rate was 1.40 mL/min and the gradient conditions were as follows: solvent A (H_2O–HCOOH, 999:1, v/v), solvent B (CH3CN–HCOOH, 999:1, v/v); 0–7 min, 1% to 60% B (linear gradient); 7–10 min, 60% to 99% B (linear); 10–11 min, 99% B (isocratic); 11–12 min, 99% to 1% B (linear); 12–13 min, 1% B (isocratic). The Ultimate 3000 UHPLC system was coupled online with an amaZon SL Ion-Trap mass spectrometer (Bruker Daltonics, Billerica, MA, USA), operating in positive-ion mode with electrospray ionization. In the source, the nebulizer pressure was set to 44 psi, the dry gas temperature was maintained at 200 °C with a flow rate of 10 L/min, and the capillary voltage was adjusted to 4.5 kV. Mass spectra were acquired in UltraScan mode over an m/z range of 100–1600 with a mass spectrum-acquisition speed of 8100 $(m/z)\,s^{-1}$.

4.7.3. Identification of Peaks and Quantification of Products

Peaks 1 and 2 from the UV chromatograms (280 nm) were attributed to (+)-catechin and (−)-epicatechin, respectively, by comparing their associated mass spectra and retention times with those obtained from pure standards. The products resulting from the trapping of extension units by menthofuran (i.e., (epi)catechin-(4 $\rightarrow$ 5)-menthofuran) are not commercially available. Peak 3 was attributed to catechin-(4 $\rightarrow$ 5)-menthofuran by comparing its associated mass spectrum and retention time with that obtained by depolymerizing a commercial standard of procyanidin B3 with menthofuran (this reaction yields only catechin and the targeted catechin-(4 $\rightarrow$ 5)-menthofuran). Peak 4 was attributed to epicatechin-(4 $\rightarrow$ 5)-menthofuran by using the same method applied to procyanidin B2. In a previous study, the procyanidin depolymerization products obtained using menthofuran were also characterized by NMR [58].

The molar responses of catechin and epicatechin at 280 nm were determined and found to be equal through calibration with their respective commercial standards. This molar response at 280 nm was applied to quantify the corresponding extension units because (epi)catechin-menthofuran has the same molar response at 280 nm as (epi)catechin [58].

4.8. Effect of U. tomentosa and A. muricata Extracts and Fractions on Aspergillus flavus Growth and Aflatoxin B1 Synthesis

4.8.1. Fungal Strain and Culture Conditions

Aspergillus flavus NRRL 62,477 strain was used for all analyses. First, a spore suspension in 0.05% Tween 80 was prepared from a seven-day-old culture maintained at 27 °C in malt extract agar (Biokar Diagnostics, Allone, France). The concentration of the spore suspension was adjusted to 10^5 spores/mL. Petri dishes were then prepared by adding 18 mL of malt extract agar medium and 2 mL of the extract or fraction at four concentrations (0.04, 0.08, 0.15, and 0.3 mg DM per milliliter culture medium) diluted in water. Control cultures were prepared by adding 2 mL of distilled water to the culture medium. Finally, 10 µL of the spore suspension was inoculated centrally onto each Petri dish and incubated for 7 days at 27 °C in the dark. After incubation, the colony diameter was measured to assess the fungal growth. Each measurement was performed in triplicate.

For catechin and epicatechin assays, the standards were dissolved in methanol to obtain an initial concentration of 1 mg/mL. Different concentrations were then prepared by dilution in distilled water: 300, 200, and 100 µg/mL for catechin, and 500, 250, and 125 µg/mL for epicatechin. Next, 2 mL of the standard solutions was added to 18 mL of malt extract agar; control cultures were prepared by adding 2 mL of a mixture of distilled water and methanol (3:1, *v:v*).

All cultures were incubated 7 days at 27 °C before analysis of AFB_1 production.

4.8.2. Aflatoxin Extraction and HPLC Quantification

AFB_1 extraction and quantification was performed as described by El Khoury et al. [5]. Aflatoxin B1 was quantified by using an Ultimate 3000 UHPLC system (Thermo Fisher Scientific) with an EvoC18 column (3 µm, 150 mm × 3.2 mm, Phenomenex) at 27 °C. AFB_1 was detected with a fluorescence detector at 365 nm excitation and 430 nm emission. Additionally, to confirm the nature of the molecules, the UV spectra were analyzed using a DAD integrated into the system.

We calculated the half-maximal inhibitory concentration to inhibit AFB_1 ($IC50_{AFB1}$), which corresponds to the concentration of extract (in mg/L) required to reduce AFB_1 synthesis by 50%. The equation shown in Section 4.6.4 was used for the calculation.

4.9. Statistics

The differences between control and treated samples were analyzed using the Student's *t*-test. Statistical significance was defined as p values less than 0.05. All errors were reported as the standard deviation of the mean. Data analyses were carried out using R studio software (version 1.4.1717).

Author Contributions: Formal analysis and experiments: L.F.C. Conceptualization: V.D., J.-D.B. and L.F.C. Methodology: V.D., J.-D.B., L.F.C., G.B. and C.H. Writing—original draft preparation: L.F.C. Writing—review: V.D., J.-D.B., G.B. and C.H. All authors have read and agreed to the published version of the manuscript.

Funding: The Ministère de l'Enseignement Supérieur et de la Recherche (MEST) funded the PhD grant of Laura F Cadenillas. This work has also benefited from a state grant managed by the National Research Agency under the "Investissements d'Avenir" programme with the reference ANR-18-EURE-0021.

Institutional Review Board Statement: Not applicable.

Informed Consent Statement: Not applicable.

Data Availability Statement: The data presented in this study are available on request from the corresponding author.

Conflicts of Interest: The authors declare no conflicts of interest.

References

1. IARC. *Fungi Producing Significant Mycotoxins*; IARC Scientific Publications: Lyon, France, 2012; pp. 1–30.
2. Nicolopoulou-Stamati, P.; Maipas, S.; Kotampasi, C.; Stamatis, P.; Hens, L. Chemical Pesticides and Human Health: The Urgent Need for a New Concept in Agriculture. *Front. Public Health* **2016**, *4*, 148. [CrossRef] [PubMed]
3. EFSA. The 2021 European Union Report on Pesticide Residues in Food. *EFSA J.* **2018**, *21*, e07939. [CrossRef]
4. Dikhoba, P.M.; Mongalo, N.I.; Elgorashi, E.E.; Makhafola, T.J. Antifungal and Anti-Mycotoxigenic Activity of Selected South African Medicinal Plants Species. *Heliyon* **2019**, *5*, e02668. [CrossRef] [PubMed]
5. El Khoury, R.; Caceres, I.; Puel, O.; Bailly, S.; Atoui, A.; Oswald, I.P.; El Khoury, A.; Bailly, J.-D. Identification of the Anti-Aflatoxinogenic Activity of *Micromeria graeca* and Elucidation of Its Molecular Mechanism in *Aspergillus flavus*. *Toxins* **2017**, *9*, 87. [CrossRef] [PubMed]
6. Cadenillas, L.F.; Hernandez, C.; Mathieu, C.; Bailly, J.-D.; Durrieu, V. Screening of the Anti-Aflatoxin B1 Activity of Peruvian Plant Extracts: Relation with Their Composition. *Food Bioprocess Technol.* **2023**, *16*, 1324–1334. [CrossRef]
7. Bluma, R.; Amaiden, M.R.; Etcheverry, M. Screening of Argentine Plant Extracts: Impact on Growth Parameters and Aflatoxin B1 Accumulation by *Aspergillus* Section *Flavi*. *Int. J. Food Microbiol.* **2008**, *122*, 114–125. [CrossRef]
8. Li, Q.; Zhao, Y.; Zhu, X.; Xie, Y. Antifungal Efficacy of Paeonol on *Aspergillus flavus* and Its Mode of Action on Cell Walls and Cell Membranes. *LWT* **2021**, *149*, 111985. [CrossRef]
9. Nobili, C.; De Acutis, A.; Reverberi, M.; Bello, C.; Leone, G.P.; Palumbo, D.; Natella, F.; Procacci, S.; Zjalic, S.; Brunori, A. Buckwheat Hull Extracts Inhibit *Aspergillus flavus* Growth and AFB1 Biosynthesis. *Front. Microbiol.* **2019**, *10*, 1997. [CrossRef]
10. Moon, Y.-S.; Lee, H.-S.; Lee, S.-E. Inhibitory Effects of Three Monoterpenes from Ginger Essential Oil on Growth and Aflatoxin Production of *Aspergillus flavus* and Their Gene Regulation in Aflatoxin Biosynthesis. *Appl. Biol. Chem.* **2018**, *61*, 243–250. [CrossRef]
11. Ribeiro, L.P.; Domingues, V.C.; Gonçalves, G.L.P.; Fernandes, J.B.; Glória, E.M.; Vendramim, J.D. Essential Oil from *Duguetia lanceolata* St.-Hil. (*Annonaceae*): Suppression of Spoilers of Stored-Grain. *Food Biosci.* **2020**, *36*, 100653. [CrossRef]
12. Loi, M.; Paciolla, C.; Logrieco, A.F.; Mulè, G. Plant Bioactive Compounds in Pre- and Postharvest Management for Aflatoxins Reduction. *Front. Microbiol.* **2020**, *11*, 243. [CrossRef] [PubMed]
13. Caceres, I.; Al Khoury, A.; El Khoury, R.; Lorber, S.; Oswald, I.P.; El Khoury, A.; Atoui, A.; Puel, O.; Bailly, J.-D. Aflatoxin Biosynthesis and Genetic Regulation: A Review. *Toxins* **2020**, *12*, 150. [CrossRef] [PubMed]
14. Peng, K.; Jin, L.; Niu, Y.D.; Huang, Q.; McAllister, T.A.; Yang, H.E.; Denise, H.; Xu, Z.; Acharya, S.; Wang, S.; et al. Condensed Tannins Affect Bacterial and Fungal Microbiomes and Mycotoxin Production during Ensiling and upon Aerobic Exposure. *Appl. Environ. Microbiol.* **2018**, *84*, e02274-17. [CrossRef] [PubMed]
15. Hernandez, C.; Cadenillas, L.; Maghubi, A.E.; Caceres, I.; Durrieu, V.; Mathieu, C.; Bailly, J.-D. *Mimosa tenuiflora* Aqueous Extract: Role of Condensed Tannins in Anti-Aflatoxin B1 Activity in *Aspergillus flavus*. *Toxins* **2021**, *13*, 391. [CrossRef]
16. Wahab, S.M.A.; Jantan, I.; Haque, M.A.; Arshad, L. Exploring the Leaves of *Annona muricata* L. as a Source of Potential Anti-Inflammatory and Anticancer Agents. *Front. Pharmacol.* **2018**, *9*, 661. [CrossRef]
17. Adewole, S.; Ojewole, J. Protective Effects of *Annona muricata* Linn. (*Annonaceae*) Leaf Aqueous Extract on Serum Lipid Profiles and Oxidative Stress in Hepatocytes of Streptozotocin-Treated Diabetic Rats. *Afr. J. Tradit. Complement. Altern. Med.* **2009**, *6*, 30–41. [CrossRef]
18. Hajdu, Z.; Hohmann, J. An Ethnopharmacological Survey of the Traditional Medicine Utilized in the Community of Porvenir, Bajo Paraguá Indian Reservation, Bolivia. *J. Ethnopharmacol.* **2012**, *139*, 838–857. [CrossRef]
19. Balderrama-Carmona, A.P.; Silva-Beltrán, N.P.; Gálvez-Ruiz, J.-C.; Ruíz-Cruz, S.; Chaidez-Quiroz, C.; Morán-Palacio, E.F. Antiviral, Antioxidant, and Antihemolytic Effect of *Annona muricata* L. Leaves Extracts. *Plants* **2020**, *9*, 1650. [CrossRef]
20. Santos, I.L.; Rodrigues, A.M.d.C.; Amante, E.R.; Silva, L.H.M. Soursop (*Annona muricata*) Properties and Perspectives for Integral Valorization. *Foods* **2023**, *12*, 1448. [CrossRef]
21. Azevedo, B.C.; Morel, L.J.F.; Carmona, F.; Cunha, T.M.; Contini, S.H.T.; Delprete, P.G.; Ramalho, F.S.; Crevelin, E.; Bertoni, B.W.; França, S.C.; et al. Aqueous Extracts from *Uncaria tomentosa* (Willd. Ex Schult.) DC. Reduce Bronchial Hyperresponsiveness and Inflammation in a Murine Model of Asthma. *J. Ethnopharmacol.* **2018**, *218*, 76–89. [CrossRef]
22. Ccahuana-Vasquez, R.A.; Santos, S.S.F.d.; Koga-Ito, C.Y.; Jorge, A.O.C. Antimicrobial Activity of *Uncaria tomentosa* against Oral Human Pathogens. *Braz. Oral Res.* **2007**, *21*, 46–50. [CrossRef] [PubMed]
23. Snow, A.D.; Castillo, G.M.; Nguyen, B.P.; Choi, P.Y.; Cummings, J.A.; Cam, J.; Hu, Q.; Lake, T.; Pan, W.; Kastin, A.J.; et al. The Amazon Rain Forest Plant *Uncaria tomentosa* (Cat's Claw) and Its Specific Proanthocyanidin Constituents Are Potent Inhibitors and Reducers of Both Brain Plaques and Tangles. *Sci. Rep.* **2019**, *9*, 561. [CrossRef] [PubMed]
24. Peñaloza, E.M.C.; Kaiser, S.; Resende, P.E.d.; Pittol, V.; Carvalho, Â.R.; Ortega, G.G. Chemical Composition Variability in the *Uncaria tomentosa* (Cat's Claw) Wild Population. *Quím. Nova* **2015**, *38*, 378–386. [CrossRef]

25. Batiha, G.E.-S.; Magdy Beshbishy, A.; Wasef, L.; Elewa, Y.H.A.; Abd El-Hack, M.E.; Taha, A.E.; Al-Sagheer, A.A.; Devkota, H.P.; Tufarelli, V. *Uncaria tomentosa* (Willd. Ex Schult.) DC.: A Review on Chemical Constituents and Biological Activities. *Appl. Sci.* **2020**, *10*, 2668. [CrossRef]
26. Shukla, R.; Singh, P.; Prakash, B.; Anuradha; Dubey, N.K. Antifungal, Aflatoxin Inhibitory and Free Radical-Scavenging Activities of Some Medicinal Plants Extracts. *J. Food Qual.* **2012**, *35*, 182–189. [CrossRef]
27. Srivastava, B.; Singh, P.; Srivastava, A.K.; Shukla, R.; Dubey, N.K. Efficacy of *Artabotrys odoratissimus* Oil as a Plant Based Antimicrobial against Storage Fungi and Aflatoxin B1 Secretion. *Int. J. Food Sci.* **2009**, *44*, 1909–1915. [CrossRef]
28. Muñoz-Acevedo, A.; Méndez, L.Y.V.; Stashenko, E.E.; Kouznetsov, V. Improved Trolox® Equivalent Antioxidant Capacity Assay for Efficient and Fast Search of New Antioxidant Agents. *Anal. Chem. Lett.* **2011**, *1*, 86–102. [CrossRef]
29. Choi, C.W.; Kim, S.C.; Hwang, S.S.; Choi, B.K.; Ahn, H.J.; Lee, M.Y.; Park, S.H.; Kim, S.K. Antioxidant Activity and Free Radical Scavenging Capacity between Korean Medicinal Plants and Flavonoids by Assay-Guided Comparison. *Plant Sci.* **2002**, *163*, 1161–1168. [CrossRef]
30. Nam, J.-S.; Park, S.-Y.; Jang, H.-L.; Rhee, Y.H. Phenolic Compounds in Different Parts of Young *Annona muricata* Cultivated in Korea and Their Antioxidant Activity. *Appl. Biol. Chem.* **2017**, *60*, 535–543. [CrossRef]
31. Orak, H.H.; Bahrisefit, I.S.; Sabudak, T. Antioxidant Activity of Extracts of Soursop (*Annona muricata* L.) Leaves, Fruit Pulps, Peels and Seeds. *Pol. J. Food Nutr. Sci.* **2019**, *69*, 359–366. [CrossRef]
32. Pineda-Ramírez, N.; Calzada, F.; Alquisiras-Burgos, I.; Medina-Campos, O.N.; Pedraza-Chaverri, J.; Ortiz-Plata, A.; Pinzón Estrada, E.; Torres, I.; Aguilera, P. Antioxidant Properties and Protective Effects of Some Species of the *Annonaceae*, *Lamiaceae*, and *Geraniaceae* Families against Neuronal Damage Induced by Excitotoxicity and Cerebral Ischemia. *Antioxidants* **2020**, *9*, 253. [CrossRef] [PubMed]
33. Gonçalves, C.; Dinis, T.; Batista, M.T. Antioxidant Properties of Proanthocyanidins of *Uncaria tomentosa* Bark Decoction: A Mechanism for Anti-Inflammatory Activity. *Phytochemistry* **2005**, *66*, 89–98. [CrossRef]
34. Tomsone, L.; Kruma, Z.; Galoburda, R. Comparison of Different Solvents and Extraction Methods for Isolation of Phenolic Compounds from Horseradish Roots (*Armoracia rusticana*). *J. Agric. Biol. Eng.* **2012**, *6*, 236–241.
35. Sharif, M.F.; Bennett, M.T. The Effect of Different Methods and Solvents on the Extraction of Polyphenols in Ginger (*Zingiber officinale*). *J. Teknol.* **2016**, *78*, 11-2. [CrossRef]
36. Du, G.; Li, M.; Ma, F.; Liang, D. Antioxidant Capacity and the Relationship with Polyphenol and Vitamin C in Actinidia Fruits. *Food Chem.* **2009**, *113*, 557–562. [CrossRef]
37. Kiselova, Y.; Ivanova, D.; Chervenkov, T.; Gerova, D.; Galunska, B.; Yankova, T. Correlation between the In Vitro Antioxidant Activity and Polyphenol Content of Aqueous Extracts from Bulgarian Herbs. *Phytother. Res.* **2006**, *20*, 961–965. [CrossRef]
38. Park, S.H.; Kim, J.M.; Kim, J.H.; Oh, Y.S.; Joo, D.H.; Lee, E.Y.; Shin, H.S.; Kim, A.R.; Lee, S.L.; Park, S.N. Antioxidative Effects and Component Analysis of Graviola (*Annona muricata*) Leaf Extract/Fractions. *J. Soc. Cosmet. Sci. Korea* **2017**, *43*, 309–320. [CrossRef]
39. Navarro-Hoyos, M.; Lebrón-Aguilar, R.; Quintanilla-López, J.E.; Cueva, C.; Hevia, D.; Quesada, S.; Azofeifa, G.; Moreno-Arribas, M.V.; Monagas, M.; Bartolomé, B. Proanthocyanidin Characterization and Bioactivity of Extracts from Different Parts of *Uncaria tomentosa* L. (Cat's Claw). *Antioxidants* **2017**, *6*, 12. [CrossRef]
40. Caceres, I.; El Khoury, R.; Bailly, S.; Oswald, I.P.; Puel, O.; Bailly, J.-D. Piperine Inhibits Aflatoxin B1 Production in *Aspergillus flavus* by Modulating Fungal Oxidative Stress Response. *Fungal Genet. Biol.* **2017**, *107*, 77–85. [CrossRef]
41. Prakash, B.; Shukla, R.; Singh, P.; Kumar, A.; Mishra, P.K.; Dubey, N.K. Efficacy of Chemically Characterized *Piper betle* L. Essential Oil against Fungal and Aflatoxin Contamination of Some Edible Commodities and Its Antioxidant Activity. *Int. J. Food Microbiol.* **2010**, *142*, 114–119. [CrossRef]
42. Ravinayagam, V.; Jaganathan, R.; Panchanadham, S.; Palanivelu, S. Potential Antioxidant Role of Tridham in Managing Oxidative Stress against Aflatoxin-B1-Induced Experimental Hepatocellular Carcinoma. *Int. J. Hepatol.* **2012**, *2012*, 428373. [CrossRef] [PubMed]
43. Narasaiah, K.V.; Sashidhar, R.B.; Subramanyam, C. Biochemical Analysis of Oxidative Stress in the Production of Aflatoxin and Its Precursor Intermediates. *Mycopathologia* **2006**, *162*, 179–189. [CrossRef] [PubMed]
44. Iram, W.; Anjum, T.; Iqbal, M.; Ghaffar, A.; Abbas, M. Structural Elucidation and Toxicity Assessment of Degraded Products of Aflatoxin B1 and B2 by Aqueous Extracts of *Trachyspermum ammi*. *Front. Microbiol.* **2016**, *7*, 346. [CrossRef] [PubMed]
45. Negera, M.; Washe, A. Use of Natural Dietary Spices for Reclamation of Food Quality Impairment by Aflatoxin. *J. Food Qual.* **2019**, *2019*, 4371206. [CrossRef]
46. Killedar, S.; More, H. Estimation of Tannins in Different Parts of *Memecylon umbellatum* Burm. *J. Pharm. Res.* **2010**, *3*, 554–556.
47. Navarro-Hoyos, M.; Alvarado-Corella, D.; Moreira-Gonzalez, I.; Arnaez-Serrano, E.; Monagas-Juan, M. Polyphenolic Composition and Antioxidant Activity of Aqueous and Ethanolic Extracts from *Uncaria tomentosa* Bark and Leaves. *Antioxidants* **2018**, *7*, 65. [CrossRef]
48. Nolasco-Gonzalez, Y.; Chacon-Lopez, M.; Ortiz-Basurto, R.; Aguilera-Aguirre, S.; Gonzalez-Aguilar, G.; Rodriguez-Aguayo, C.; Navarro-Cortez, M.; Garcia-Galindo, H.; Garcia-Magaña, M.d.L.; Meza-Espinoza, L.; et al. *Annona muricata* Leaves as a Source of Bioactive Compounds: Extraction and Quantification Using Ultrasound. *Horticulturae* **2022**, *8*, 560. [CrossRef]
49. Valdez-Guerrero, Y.D.; Esparza-González, S.; Morlett-Chávez, J.; Nery-Flores, S.; Flores-Gallegos, A.; Ascacio-Valdés, J.; Rodríguez-Herrera, R. Isolation of Polyphenols from Soursop (*Annona muricata* L.) Leaves Using Green Chemistry Techniques and Their Anticancer Effect. *Braz. Arch. Biol. Technol.* **2021**, *64*, e21200163. [CrossRef]

50. Rubert-Nason, K.F.; Yang, P.; Morrow, C.J.; Lindroth, R.L. Environment and Genotype Influence Quantitative and Qualitative Variation in Condensed Tannins in Aspen. *J. Chem. Ecol.* **2023**, *49*, 325–339. [CrossRef]
51. Top, S.M.; Preston, C.M.; Dukes, J.S.; Tharayil, N. Climate Influences the Content and Chemical Composition of Foliar Tannins in Green and Senesced Tissues of *Quercus rubra*. *Front. Plant Sci.* **2017**, *8*, 423. [CrossRef]
52. Norton, R.A. Inhibition of Aflatoxin B1 Biosynthesis in *Aspergillus flavus* by Anthocyanidins and Related Flavonoids. *J. Agric. Food Chem.* **1999**, *47*, 1230–1235. [CrossRef] [PubMed]
53. Zhou, W.; Hu, L.B.; Zhao, Y.; Wang, M.Y.; Zhang, H.; Mo, H.Z. Inhibition of Fungal Aflatoxin B1 Biosynthesis by Diverse Botanically-Derived Polyphenols. *Trop. J. Pharm. Res.* **2015**, *14*, 605–609. [CrossRef]
54. Dammak, I.; Lasram, S.; Hamdi, Z.; Ben Moussa, O.; Mkadmini Hammi, K.; Trigui, I.; Houissa, H.; Mliki, A.; Hassouna, M. In Vitro Antifungal and Anti-Ochratoxigenic Activities of *Aloe Vera* Gel against *Aspergillus carbonarius* Isolated from Grapes. *Ind. Crops* **2018**, *123*, 416–423. [CrossRef]
55. Singleton, V.L.; Rossi, J.A. Colorimetry of Total Phenolics with Phosphomolybdic-Phosphotungstic Acid Reagents. *Am. J. Enol. Vitic.* **1965**, *16*, 144–158. [CrossRef]
56. Waterman, P.G.; Mole, S. *Analysis of Phenolic Plant Metabolites*; Blackwell Scientific: Oxford, UK; Boston, MA, USA, 1994; ISBN 978-0-632-02969-3.
57. Brand-Williams, W.; Cuvelier, M.E.; Berset, C. Use of a Free Radical Method to Evaluate Antioxidant Activity. *LWT—Food Sci. Technol.* **1995**, *28*, 25–30. [CrossRef]
58. Billerach, G.; Rouméas, L.; Dubreucq, E.; Fulcrand, H. Furanolysis with Menthofuran: A New Depolymerization Method for Analyzing Condensed Tannins. *J. Agric. Food Chem.* **2020**, *68*, 2917–2926. [CrossRef]

toxins

MDPI

Article

High-Performance Liquid Chromatography–Fluorescence Detection Method for Ochratoxin A Quantification in Small Mice Sample Volumes: Versatile Application across Diverse Matrices Relevant for Neurodegeneration Research

Elba Beraza [1], Maria Serrano-Civantos [1], Maria Izco [2], Lydia Alvarez-Erviti [2], Elena Gonzalez-Peñas [1] and Ariane Vettorazzi [1,*]

1 MITOX Research Group, Department of Pharmaceutical Sciences, School of Pharmacy and Nutrition, Universidad de Navarra, 31008 Pamplona, Spain; eberaza@alumni.unav.es (E.B.); mserrano.14@alumni.unav.es (M.S.-C.); mgpenas@unav.es (E.G.-P.)

2 Laboratory of Molecular Neurobiology, Center for Biomedical Research of La Rioja (CIBIR), Piqueras 98, 26006 Logroño, Spain; mizco@riojasalud.es (M.I.); laerviti@riojasalud.es (L.A.-E.)

* Correspondence: avettora@unav.es

Abstract: Ochratoxin A (OTA) is a mycotoxin commonly found in various food products, which poses potential health risks to humans and animals. Recently, more attention has been directed towards its potential neurodegenerative effects. However, there are currently no fully validated HPLC analytical methods established for its quantification in mice, the primary animal model in this field, that include pivotal tissues in this area of research, such as the intestine and brain. To address this gap, we developed and validated a highly sensitive, rapid, and simple method using HPLC-FLD for OTA determination in mice tissues (kidney, liver, brain, and intestine) as well as plasma samples. The method was rigorously validated for selectivity, linearity, accuracy, precision, recovery, dilution integrity, carry-over effect, stability, and robustness, meeting the validation criteria outlined by FDA and EMA guidelines. Furthermore, the described method enables the quantification of OTA in each individual sample using minimal tissue mass while maintaining excellent recovery values. The applicability of the method was demonstrated in a repeated low-dose OTA study in Balb/c mice, which, together with the inclusion of relevant and less common tissues in the validation process, underscore its suitability for neurodegeneration-related research.

Keywords: ochratoxin A; HPLC-FLD; tissues; brain; plasma; mouse

Key Contribution: Data on the validation of a highly sensitive and rapid HPLC-FLD method for quantifying Ochratoxin A (OTA) in mice tissues and plasma are presented. The described method adheres to FDA and EMA guidelines, enabling accurate OTA quantification in individual samples of crucial tissues in neurodegeneration-related research including brain and intestine.

1. Introduction

Ochratoxin A (OTA) is a mycotoxin produced naturally as a secondary metabolite by some species of *Aspergillus* and *Penicillium* fungi that can contaminate a great variety of foods, such as cereals, beans, coffee nuts or species [1]. It is a thermostable compound that is not destroyed by common food preparation procedures and therefore enters the food chain via both raw and processed products [1]. In this regard, several human biomonitoring studies have demonstrated human exposure to this mycotoxin [2].

OTA is a potent renal carcinogen in rodents, which has been proven to cause renal tumors in rats and mice, with male rats being the most sensitive [1,3–5]. Considering this, but without having enough evidence in humans, the International Agency for Research on Cancer (IARC) classified it in 1993 as a possible human carcinogen (group 2B) [6], and maximum levels based on this effect have been established in several food commodities to reduce health risks in Europe [1]. However, even though its target organ is the kidney, OTA has also been described as immunotoxic, hepatotoxic, teratogenic, and neurotoxic [7–12].

In recent years, research has examined the impact of both acute and chronic exposure to OTA on the nervous system, revealing neurotoxicity as a sensitive endpoint for this mycotoxin [13,14]. In vivo, OTA has been proven to alter the proliferation and differentiation of cells in the hippocampus [15] and in the subventricular zone of adult mice [14]. Moreover, investigations have revealed the possibility of OTA causing parkinsonism in mice, demonstrating that the nigrostriatal pathway might be affected by OTA [7,16–18]. The relevance of these findings underlines the importance of carrying out studies involving OTA administration in mouse models, given that mice currently stand as the most widely used experimental model in neurodegenerative disease research. In order to further delve into this matter, it is crucial to develop methodologies able to determine OTA in biological samples of mice. In addition, in the context of neurodegenerative diseases, and more specifically Parkinson's disease (PD), it is crucial to quantify OTA not only in plasma, kidney, and liver, which are the most common organs reported in the literature when studying OTA, but also in brain and intestine tissue. The relevance of intestine tissue stems from Braak's hypothesis, which proposes that PD might initiate at the enteric nervous system and spread to the lower brainstem via the vagal nerve [19].

There are methods reported in the literature to measure OTA in biological samples, but they have either been developed for other species, use large sample volumes, do not include certain relevant tissues, or lack validation data. Indeed, considering that rats are the most sensitive to the toxin, most of the methods developed to quantify OTA for toxicology studies use rat biological samples [20], and adapting them for their use with mice presents unique challenges. An analytical method for OTA quantification in mouse samples requires the possibility of using smaller quantities of samples, approximately 10 times lower than that of a rat, due to their smaller size. Additionally, studies regarding neurodegeneration often include behavioral assessments, necessitating individual quantification of mycotoxin levels in each mouse to maintain traceability. Related to this, a higher number of animals are commonly used in these studies to be able to reach conclusions from behavioral studies, so achieving rapid OTA quantification is also essential to process the high number of samples efficiently. Furthermore, specific validation of the method for each species is strongly recommended by the reference Guidelines, both from the Food and Drug Administration (FDA) [21] and the European Medicines Agency (EMA) [22], to ensure the reliability of the data generated.

In this study, we validated a high-performance liquid chromatography–fluorescence detection (HPLC-FLD) analytical procedure suitable for the quantitative analysis of OTA in plasma, brain, kidney, liver, and intestine tissue in individual mice, making it possible to acknowledge interindividual variability. Guidelines from both the FDA [21] and the EMA [22] were consulted to ensure comprehensive validation of the method. Moreover, its effectiveness has been demonstrated via its successful application to the analysis of samples obtained after a 28-day OTA repeated intraperitoneal dose study in mice.

2. Results

2.1. Method Validation

2.1.1. Selectivity

The method was selective in all the matrices. The chromatograms of blank samples revealed no interfering signals at the retention time of OTA (6.7 min), and the retention time of OTA matches in both fortified plasma or tissues and real samples, as can be seen in Figure 1.

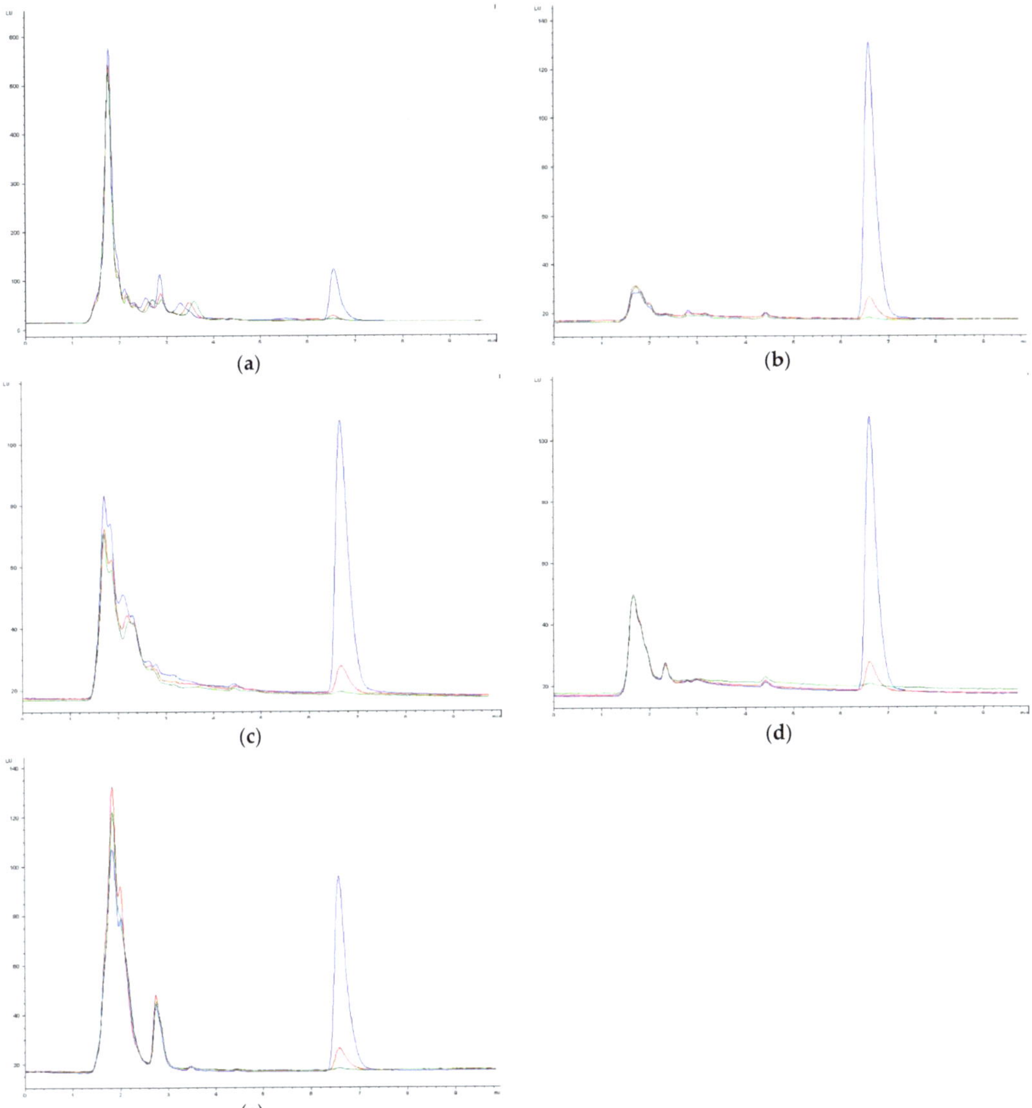

Figure 1. Chromatograms of blank samples of plasma (**a**), brain (**b**), kidney (**c**), intestine tissue (**d**), and liver (**e**) spiked with OTA. Green line = blank samples spiked with OTA as in the recovery study (2.35 ng/mL in plasma and 9.4 ng/g in brain, kidney, intestine tissue, and liver); red line = blank samples spiked with OTA as in the recovery study (22.83 ng/mL in plasma and 91.32 ng/g in brain, kidney, intestine tissue and liver); blue line = blank samples spiked with OTA as in the recovery study (228.33 ng/mL in plasma and 913.32 ng/g in brain, kidney, intestine tissue and liver).

2.1.2. Linearity

The calibration curves made on three different days for each one of the two concentration ranges (2.35–22.83 ng/mL and 22.83–228.33 ng/mL) showed a good linear relationship between peak areas and OTA concentrations. All the requirements for linearity have been met in all the cases (see Table S1 in the Supplementary Materials). Since the triplicated calibration curves for each range did not differ significantly, a global calibration curve for the 2.35–22.83 ng/mL range and another one for the 22.83–228.33 ng/mL range were made with the average of the triplicates. The results of the linearity study on the global calibration curves are presented in Table 1. The LLOQ for OTA was determined as the lowest calibration standard (2.35 ng/mL) as it showed acceptable accuracy and precision. Therefore, considering the dilution factors, the respective values of LLOQ in the different matrices are 2.35 ng/mL in plasma and 9.4 ng/g in brain, kidney, intestine tissue, and liver.

Table 1. Global calibration curves obtained in the linearity study in the following ranges: 2.35–22.83 ng/mL and 22.83–228.33 ng/mL. Eighteen data points were used for each calibration range.

	Range 2.35–22.83 ng/mL	Range 22.83–228.33 ng/mL
Curve equation [a]	y = 9.19x + 0.533	y = 8.50x + 15.33
r^2	0.997	0.999
Slope limits (p = 95%)	8.55; 9.83	8.19; 8.81
Intercept limits (p = 95%)	−6.96; 8.03	−20.61; 51.27
CV [b] of response factors (%)	3.72	3.47
Back-calculated RE [c] (%)	<5.5	<5.2

[a] y: peak area, x: concentration of OTA (ng/mL for plasma or ng/g for tissues). [b] Coefficient of variation. [c] Relative error.

2.1.3. Precision and Accuracy

Precision and accuracy of linearity showed adequate values according to FDA and EMA guidelines (Table 2). Raw data are available in the Supplementary Materials (Table S2).

Table 2. Results of the precision and accuracy study. The precision within days was studied by analyzing some calibrators (2.35, 22.83, 228.33 ng/mL) in triplicate each day. The precision between days was assured by analyzing calibrators of these levels in three different days.

	Within-Day Variability (n = 3)			Between-Day Variability (n = 9)		
$C_{nominal}$ (ng/mL)	$C_{measured}$ (ng/mL)	CV [a]	A [b]	$C_{measured}$ (ng/mL)	CV [a]	A [b]
Range 2.35–22.83 ng/mL [c,d]						
2.35	2.21	4.08	2.09	2.29	3.55	2.55
22.83	21.77	2.09	4.64	21.63	4.95	5.25
Range 22.83–228.33 ng/mL [e,f]						
22.83	22.43	2.13	1.77	22.28	4.95	2.42
228.33	235.82	0.49	3.28	226.25	4.99	0.91

[a] Coefficient of variation (%). [b] Accuracy (RE%). [c] Equivalent range in plasma: 2.35–22.83 ng/mL (no dilution factor). [d] Equivalent range in kidney, liver, brain, and intestine tissue: 9.4–91.32 ng/g (dilution factor: 4). [e] Equivalent range in plasma: 342.45–3424.95 ng/mL (dilution factor: 15). [f] Equivalent range in kidney, liver, brain, and intestine tissue: 91.32–913.32 ng/g (dilution factor: 4).

2.1.4. Recovery

The recovery values obtained for each matrix are summarized in Table 3. More data are provided in the Supplementary Materials (Table S3). Recovery was very efficient in all the matrices (74.8% for plasma, 79.7% for brain, 87.6% for kidney, 80.2% for intestine tissue and 76.2% for liver). Furthermore, the %CVs obtained in within- and between-day experiments were below 12% in each case, thus demonstrating the precision of the methodology.

Table 3. Results of the recovery study. The repeatability of the process was studied by carrying out the complete recovery experiment for each matrix on 1 day (within days) and on 3 different days (between days).

		Global Recovery (%)	CV [a] (%)
Plasma	Within-day	75.2 (n = 9)	6.8 (n = 9)
	Between day	74.8 (n = 27)	7.5 (n = 27)
Brain	Within-day	76.9 (n = 9)	12.8 (n = 9)
	Between day	79.7 (n = 27)	11.2 (n = 27)
Kidney	Within-day	88.8 (n = 9)	2.9 (n = 9)
	Between day	87.6 (n = 27)	4.5 (n = 27)
Intestine	Within-day	80.4 (n = 9)	11.8 (n = 9)
	Between day	80.2 (n = 27)	11.6 (n = 27)
Liver	Within-day	78.9 (n = 9)	5.5 (n = 9)
	Between day	76.2 (n = 27)	5.8 (n = 27)

[a] Coefficient of variation.

2.1.5. Dilution Integrity

The dilution integrity of OTA from plasma samples that needed a dilution of the supernatant before the HPLC injection was also studied for the dilution factor selected (1/15). Accuracy and precision were both below 15% (respectively, 2.93% (mean RE%) and 5.54% (CV%)). This demonstrates that acidified ACN is not saturated when extracting high OTA doses and that the dilution of the supernatant does not affect the recovery.

2.1.6. Carry-Over Effect

There was no carry-over effect when measuring blank samples after high-concentration calibration standards.

2.1.7. Stability

OTA in processed plasma, brain, intestine, liver, and kidney samples was stable for at least 6 h in the autosampler tray (Table 4). Considering these results, all samples were analyzed immediately after extraction and in less than 6 h, in order to assure OTA quantification. Thus, a maximum of 24 vials were measured in one run, including samples, calibrators, and QCs.

Table 4. Results of stability study. Time evolution of OTA concentrations in fortified blank samples at low and high concentrations.

	Plasma		Brain		Kidney		Intestine		Liver	
Time (h)	$C_{measured}$ (ng/mL)	A [a] (%)	$C_{measured}$ (ng/g)	A [a] (%)	$C_{measured}$ (ng/g)	A [a] (%)	$C_{measured}$ (ng/g)	A [a] (%)	$C_{measured}$ (ng/g)	A [a] (%)
$C_{nominal}$ 2.35 ng/mL										
0	2.44	3.99	2.14	9.10	2.43	3.44	2.24	4.74	2.61	10.92
2	2.36	0.52	2.46	4.88	2.25	4.06	2.30	2.26	2.41	2.59
6	2.52	7.28	2.61	11.05	2.56	9.11	2.60	10.66	2.12	9.99
12	5.70	142.64	3.48	48.17	3.17	34.81	1.62	30.92	2.05	12.77
$C_{nominal}$ 228.33 ng/mL										
0	237.81	4.15	208.54	8.67	202.43	11.34	213.38	6.55	213.77	6.38
2	230.46	0.93	229.75	0.62	197.13	13.67	206.65	9.50	216.87	5.02
6	247.86	8.55	256.83	12.48	228.21	0.05	227.34	0.43	195.83	14.23
12	462.34	102.49	357.30	56.48	279.62	22.46	232.49	1.82	202.84	11.16

[a] Accuracy (RE%).

2.1.8. Robustness

OTA retention time was very similar in both columns (6.7–6.9 min). When calibration standard areas obtained with column A were extrapolated with the global calibration curves obtained in column B and vice versa, the accuracy values obtained were acceptable according to guidelines (<15%). The results from the robustness study are summarized in Table 5. More data are available in the Supplementary Materials (Tables S4 and S5).

Table 5. Results of the robustness study. Precision and accuracy values of peak areas obtained in column A are quantified with calibration curves from column B. Precision between days (n = 9) is presented in the table.

	Areas from Column A Extrapolated with Calibration Curves Obtained in Column B			Areas from Column B Extrapolated with Calibration Curves Obtained in Column A		
$C_{nominal}$ (ng/mL)	$C_{measured}$ (ng/mL)	CV [a] (%)	A [b] (%)	$C_{measured}$ (ng/mL)	CV [a] (%)	A [b] (%)
Range 2.35–22.83 ng/mL [c,d]						
2.35	2.35	3.55	0.20	2.37	8.25	0.90
22.83	20.37	4.95	10.76	22.00	5.94	3.65
Range 22.83–228.33 ng/mL [e,f]						
22.83	22.26	4.95	2.50	24.01	5.94	5.16
228.33	216.82	4.99	5.04	231.80	5.55	1.52

[a] Coefficient of variation. [b] Accuracy (RE%). [c] Equivalent range in plasma: 2.35–22.83 ng/mL (no dilution factor). [d] Equivalent range in kidney, liver, brain, and intestine tissue: 9.4–91.32 ng/g (dilution factor: 4). [e] Equivalent range in plasma: 342.45–3424.95 ng/mL (dilution factor: 15). [f] Equivalent range in kidney, liver, brain, and intestine tissue: 91.32–913.32 ng/g (dilution factor: 4).

2.2. *Application to In Vivo Study*

This analytical procedure was applied to measure OTA in samples obtained after an OTA repeated intraperitoneal dose study. OTA levels in control samples were not detectable in any matrix, but they were found in all the matrices at both doses. Results are depicted in Table 6, and typical chromatograms obtained can be seen in Figure 2. The highest levels of OTA were found in plasma, followed by the liver. As explained above, all of the analytical results obtained were corrected by recovery.

Table 6. Concentrations of OTA in plasma, brain, kidney, intestine, and liver in control ($NaHCO_3$) and OTA-treated mice after 28 days of repeated intraperitoneal dose administration (0.21 and 0.5 mg/kg bw). Results are expressed as mean ± SD for each matrix and condition.

	OTA Concentration (ng/mL or ng/g)				
	Plasma	Brain	Kidney	Intestine	Liver
Control animals					
1	<LLOQ [a]	<LLOQ	<LLOQ	<LLOQ	<LLOQ
2	<LLOQ	<LLOQ	<LLOQ	<LLOQ	<LLOQ
3	<LLOQ	<LLOQ	<LLOQ	<LLOQ	<LLOQ
4	<LLOQ	<LLOQ	<LLOQ	<LLOQ	<LLOQ
5	<LLOQ	<LLOQ	<LLOQ	<LLOQ	<LLOQ
Mean	<LLOQ	<LLOQ	<LLOQ	<LLOQ	<LLOQ
Treated animals (0.21 mg/kg bw)					
1	835.66	*3.05*	47.51	39.20	58.99
2	1093.83	*4.28*	56.58	56.78	69.65
3	836.21	*1.94*	55.41	52.84	78.98
4	787.23	*3.00*	48.18	35.80	69.35
5	849.31	*3.69*	78.81	54.15	60.96
Mean	880.45 ± 121.61	*3.19 ± 0.87*	51.14 ± 12.70	47.82 ± 9.47	67.59 ± 7.98

Table 6. *Cont.*

	OTA Concentration (ng/mL or ng/g)				
	Plasma	**Brain**	**Kidney**	**Intestine**	**Liver**
Control animals					
Treated animals (0.5 mg/kg bw)					
1	3858.11	13.19	163.70	296.86	244.00
2	1785.37	*5.72*	107.15	115.02	183.08
3	2312.75	*7.27*	106.93	149.33	163.17
4	3247.83	10.28	140.32	164.76	251.95
5	3061.26	9.90	172.70	104.34	173.59
Mean	2853.06 ± 812.59	*9.27 ± 2.89*	138.16 ± 30.77	166.06 ± 77.14	203.16 ± 41.61

[a] Lower limit of quantification, Italics > LOD (limit of detection) < LLOQ.

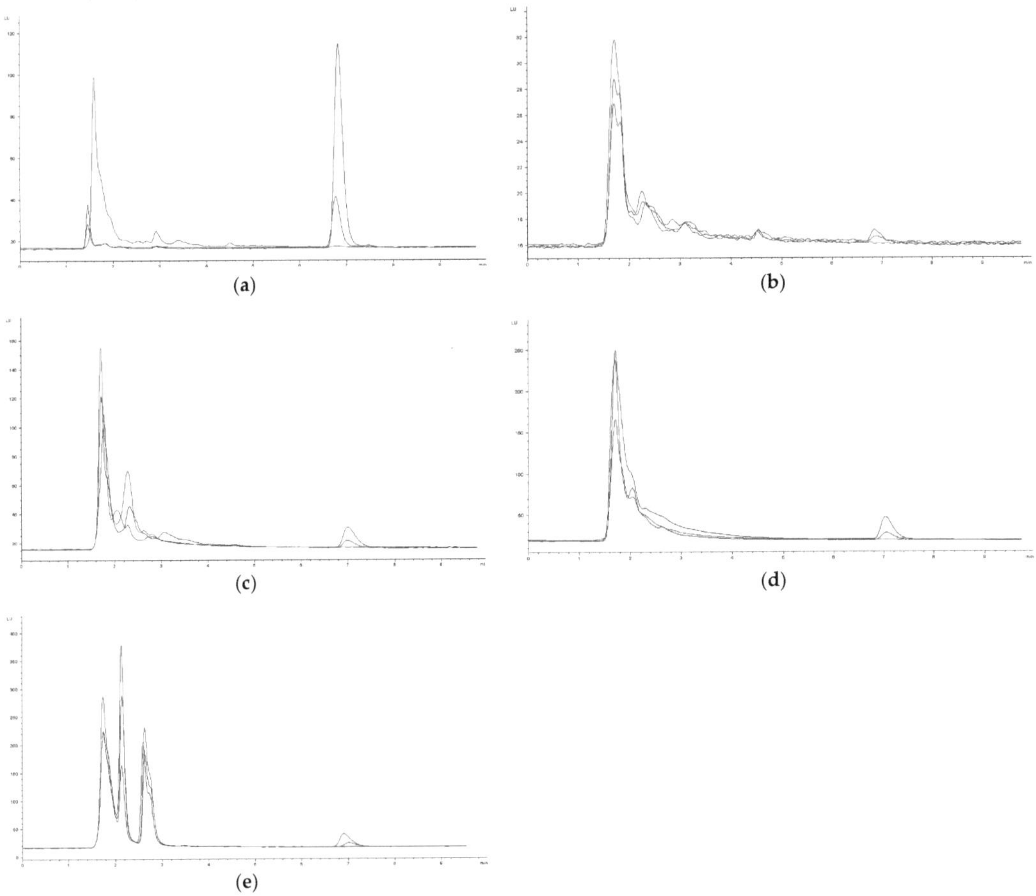

Figure 2. Representative chromatograms obtained in the analysis of samples from the in vivo study used in the application of the method in the different matrices: plasma (**a**), brain (**b**), kidney (**c**), intestine tissue (**d**), and liver (**e**). Green line = samples from control animals; red line = samples from treated animals (0.21 mg/kg bw); blue line = samples from treated animals (0.5 mg/kg bw). Plasma samples from treated animals (both 0.21 mg/kg bw and 0.5 mg/kg bw) were diluted 1/15.

3. Discussion

The method presented in this work is the first one published, to the best of our knowledge, that allows OTA quantification in five different matrices (plasma, kidney, liver, brain, and intestine tissue) in mice by means of HPLC-FLD. Indeed, there is no HPLC-FLD method to quantify OTA in mice brains or intestines. There have been published several analytical methods for OTA determination in biological samples. However, most of them are developed for other species, mainly rats, and therefore, they are not applicable to mice tissues, as they need a high quantity of samples in comparison with the sample amount that can be extracted from a mouse. Considering a mouse weighs around 20–25 g and a rat approximately 200–250 g, the sample size required in a method validated for OTA quantification in mice would have to be 10 times lower than if it were validated for rats.

Regarding methods used previously by other authors to quantify OTA in mice, except for the one published by Szöke et al. [23], they are all immunoassay methods, either enzyme immunoassays [24] or radioimmunoassays [25,26]. Immunological methods are typically less expensive but cannot distinguish between different ochratoxins, as reviewed by Meulenberg [27]. Moreover, the use of radiolabeled compounds has important disadvantages, as they present multiple health hazards and need specialized waste disposal. The method validated in this study allows to quantify OTA in five different tissues in each individual mouse by means of high-performance liquid chromatography coupled with a fluorescence detector. HPLC-FLD is one of the most sensitive, convenient, and widely used bioanalytical methods for OTA quantification in biological samples, as it is able to detect and distinguish between the various ochratoxin family members and metabolites. In fact, recently, Szöke et al. aimed to develop an immunoassay-based method that could be compared to HPLC-FLD (14). They do not present the full validation data for the chromatographic method used; however, they concluded that HPLC-FLD gave, indeed, the most reliable measurement at the lowest levels of OTA.

Apart from the high specificity, one important advantage of the present method is that only 50 µL of plasma and 12.5 mg of tissue are enough to obtain results in a wide range of concentrations. The low sample mass needed is extremely relevant in toxicology and toxicokinetic studies, as well as when studying neurodegeneration. In this last situation, the mice's brains must often be dissected to analyze different structures separately, and therefore, the sample mass available for HPLC analysis can be very low (15–20 mg), being this one of the critical factors. Many authors do not specify the tissue mass needed for the analysis [23], or they propose to use pooled tissues [24], which entails increasing the number of animals per group and losing the interindividual variability data. Additionally, working with such small volumes also accelerates the process, as the drying process is less time consuming (around 30–40 min), being able to process samples in less than 3 h.

Regarding the range of concentrations, we proved the linearity of the method between 2.35 and 228.33 ng/mL, defining 2.35 ng/mL as the LLOQ for plasma and 9.4 ng/g as the LLOQ for brain, kidney, liver, and intestine. In their publication, Szöke et al. [23] determined the linearity in a higher range of concentrations (between 4.0 and 403.8 ng/mL), although they defined the LLOQ as 2.4 ng/mL based on signal-to-noise ratio without checking the linearity, precision, or accuracy. Thus, the present method achieved a similar LLOQ while simultaneously demonstrating dilution integrity for expected high-concentration samples. This capability enabled the quantification of OTA across a broader concentration range, even after administering low doses of the mycotoxin.

The method presented allows OTA quantification in five different matrices (plasma, kidney, liver, brain, and intestine). Tissue analysis requires an additional step in sample preparation to turn biological samples into a liquid form. To this end, there are several techniques available, such as homogenization, digestion, or sonication. Homogenization is the most popular one [23]; however, the risk of cross-contamination is high, and the homogenizer should be thoroughly rinsed after each sample to avoid it. In this study, the validated method includes bead beating as the cell-disrupting procedure. Bead beating is a cell disruption technique that can be used to obtain DNA, RNA, proteins, metabolites, and small molecules from diverse samples like animal and human tissues, bacteria, plants, etc. [28]. This technique has become more popular lately, as it minimizes the risk of cross-contamination compared to traditional techniques while also simplifying, optimizing, and accelerating the sample preparation process. In this way, tissue samples are processed in individual tubes that contain grinding beads, and 40 s is enough to obtain the homogenates. Simultaneously, three samples can be homogenized with the model used, although the number can be up to 24 samples at the same time with a bigger apparatus. It is the first time, to our knowledge, that the bead beating technique has been used in mice tissue samples for OTA extraction, proving to be a useful and easy process.

Once tissues are homogenized, the extraction method is common for four out of the five matrices (plasma, kidney, brain, and intestine tissue) using a three-fold volume of ACN acidified with 0.4% formic acid. ACN is one of the most common organic precipitants for the pretreatment of tissue samples due to its strong precipitating ability [29]. However, the extraction of OTA has been demonstrated to be pH dependent, as at higher pHs (above pH 5.0), the deprotonation of the carboxyl and/or the phenolic hydroxyl group(s) can lower the efficacy of the extraction [30]. Thus, acidifying ACN with 0.4% formic acid allows the most effective extraction of OTA. The selection of the solvent was also made according to previous studies regarding the extraction of mycotoxins, which demonstrated that acidified ACN was the best solvent for the matter [31]. For protein precipitation, at least two volumes of organic solvent should be added to each unit weight of tissue; the more added, the more dilution factor but also, the more thorough the precipitation will be [29]. With the method developed, a three-fold volume is used to ensure protein precipitation as then the supernatant is evaporated, and no dilution is made in this step. In the case of the liver, ice-cold absolute ethanol and trichloroacetic acid 20% were used as ACN acidified with formic acid did not clean enough the samples. Amongst the options available to further clean the samples, the use of a salt like sodium acetate was discarded as it has been shown to reduce recovery values [31]. In the same line, cartridges could not be used due to the small volumes available. Ethanol is also widely used as a precipitation agent and is less toxic than other organic solvents such as chloroform or acetic acid. Again, the addition of a strong acid as trichloroacetic acid 20% lowers the pH of the extraction solvent making the extraction process more effective. However, it has to be noted that TCA is considered to be hazardous, and the Threshold Limit Value (TLV) established by the International Labour Organization (ILO) is much lower than those admitted for acetonitrile, formic acid and ethanol (0.5 ppm for TCA, 20 ppm for ACN, 5 ppm for formic acid and 1000 for ethanol) [32]. Therefore, to lower the risks, the extraction process was carried out with acidified ACN when possible. In both cases, the procedure is simple, fast, and economical because only one solvent step is needed, and the use of immunoaffinity columns is not necessary. Five minutes of vortexing was enough time with both extraction solvents to extract OTA from complex matrices such as the kidney, liver, plasma, intestines, and brain.

The recovery values have been studied and are presented for each matrix, which is important considering the differences in composition between the matrices that affect OTA binding and releasing. Recovery values are very efficient in all the matrices, ranging between 74.8 and 87.6%, with CV values below 15%. This good reproducibility validates the procedure of spiking blank samples and demonstrates the precision of the analytical procedure. Szöke et al. [23] obtained a higher recovery value, but they do not specify whether it is for plasma or for tissue samples nor present the within- and between-day variability. However, from their study it is clear that recovery values obtained from HPLC analysis are higher than those obtained from ELISA-based immunoassay or from flow cytometry measurements. These findings reinforce the idea of HPLC being an adequate technique to quantify OTA in mice tissues.

The fact that the method presented uses the same calibration curves and chromatographic conditions for all five matrices makes it easier to carry out the experimental work. The mobile phase used is also easy to prepare, as there is no need to adjust the pH. The LLOQ achieved in all the matrices was adequate for toxicological studies, where high doses are administered to small animals, but also could be applied to further studies in the neurodegeneration field, in which lower doses are administered for longer periods of time, as proven in the application of the method.

OTA concentrations were found to be similar in the kidney and liver, with slightly higher levels detected in the liver. While this finding may appear contradictory to OTA's primary nephrotoxic effect, it aligns with previous studies in rodents that reported similar or even higher OTA levels in the liver under various experimental conditions and administration routes [23,33–39]. In this regard, some of these studies have demonstrated that, despite similar accumulation in both target and non-target tissues, only the kidney exhibited biochemical and histopathological changes [36,37]. Moreover, the results also show OTA accumulation in the small intestine at levels similar to those in the kidney and liver and penetration via the blood–brain barrier, with concentrations in the brain falling between the LOD and the LLOQ. This is in agreement with Wang et al., who in 2020 reported similar biodistribution of OTA upon intravenous administration [25]. It is important to note that in this method tissue levels and not intestine content levels were measured, as in the latter, levels of OTA can be substantially higher [25]. However, considering that OTA-induced toxicity is not only driven by tissue distribution and kinetics but also by organ-specific toxicodynamics, the necessity for further delving into its possible role in neurodegenerative diseases is underscored.

4. Conclusions

In conclusion, a highly sensitive, rapid, and simple HPLC-FLD method for OTA determination in mice tissues (kidney, liver, brain, and intestine), as well as plasma samples, was developed and validated for selectivity, linearity, accuracy, precision, recovery, dilution integrity, carry-over effect, stability, and robustness. The applicability of the assay was evaluated in repeated low-dose OTA study in Balb/c mice.

The method described allows for the quantification of OTA in each individual needing a very low sample mass with good recovery values. In addition, it has been validated for novel and less common tissues in OTA-related research, such as the intestine and brain, which could, however, be essential for delving into its potential neurodegenerative effects. Moreover, the fact that immunoaffinity columns are not needed makes this also a simple, fast, and economical method. Following the FDA and EMA guidelines, it has been demonstrated that all the validation criteria were met; thus, the method has adequate characteristics to assure reliable results.

5. Material and Methods

5.1. Reagents

All the reagents used for the HPLC analysis were of LC-gradient grade. Acetonitrile (ACN) and formic acid were purchased from Merck (Darmstadt, Germany), whereas absolute ethanol UV–IR-HPLC and trichloroacetic acid 20% (w/v) (TCA) were both obtained from Panreac (Barcelona, Spain). Water used throughout the analysis was purified with a Milli-Q System (Millipore, Bedford, MA, USA).

For the tissue homogenates, sodium phosphate buffer (0.05 M, pH 6.50) was prepared by adding 6.90 g of $NaH_2PO_4 \cdot H_2O$ (Merck, Darmstadt, Germany) to 900 mL of distilled water. The pH of the dissolution was adjusted with NaOH 3 M, and the volume was adjusted to one liter. The buffer was kept at 4 °C until use.

OTA was obtained in powder from Sigma-Aldrich (Steinheim, Germany) (REF O1877, lot 0000149830). For animals' administration, OTA was dissolved in 0.1 M $NaHCO_3$ (sodium bicarbonate powder, Sigma-Aldrich, Steinheim, Germany), adjusted to pH 7.4 with HCl, and kept at −20 °C until use. To prepare the standard solutions, OTA was dissolved in methanol 99.9% (Panreac, Barcelona, Spain) and kept at −20 °C until use.

5.2. Application of the Method: Animals and Samples Collection

5.2.1. Animals and Experimental Design

The animals used were eight- to nine-week-old male Balb/cByJ (ref. 627) mice purchased from Charles River. On the day of arrival, animals were weighted and distributed randomly into individual polycarbonate cages with stainless steel covers, with a maximum of six mice per cage. Mice were maintained in constant environmental conditions of humidity (55 ± 10%) and temperature (22 ± 2 °C) on a 12 h light/dark cycle and allowed ad libitum access to standard pellet diet (Special Diet Service, Essex, UK) and normal tap water.

For the study involved in the application of the method, animals were distributed into 3 groups (n = 10 per group) and, after one week of acclimatization, received repeated OTA administrations (0.21 or 0.5 mg/kg) or vehicle ($NaHCO_3$) daily for 28 days, intraperitoneally. For blank samples, male Balb/cByJ (ref. 627) (n = 10) were purchased, left for the acclimatization period, and sacrificed without receiving any administrations.

These experiments were approved by the Ethics Committee on Animal Experimentation of the Universidad de Navarra (CEEA 049-19), and they were conducted according to the National Institute of Health (NIH) Guide for the Care and Use of Laboratory Animals.

5.2.2. Plasma and Samples Collection

Animals were sacrificed 24 h after the last OTA administration. Blood was obtained at sacrifice from cardiac puncture and collected in Sarstedt (Nümbrecht, Germany) Multivette® 600 EDTA K3 tubes for then to be centrifuged at 2000× g for 10 min. The obtained plasma was stored at −20 °C until HPLC analysis. The brain, liver, kidney, and small intestine were rapidly removed from the animals and washed with saline buffer until the external blood was removed. Intestines were also washed internally. Organs were carefully dried using filter paper and weighed; then, the brain was dissected, and the rest of the organs were carved. During the necropsies, all the dissection material was cleaned with saline buffer and rinsed with ethanol after each animal to prevent sample contamination.

For other purposes, the brain was dissected, and the different structures were collected as needed. After dissection, the remaining portion of the right hemisphere was promptly frozen in liquid nitrogen for OTA quantification. The kidneys, liver, and intestine were carved according to established laboratory protocols: the kidney was transversely cut to include both cortex and medulla, one lobule of the liver was longitudinally cut, and the intestine was transversely cut. Thereafter, smaller portions of these tissue pieces were immediately frozen in liquid nitrogen and kept at −80 °C until analysis. The weight of these tissue pieces varied between organs, ranging from 15 to 20 mg for the brain and 18 to 40 mg for the other tissues. Blank samples for method validation were obtained from non-treated animals and processed as described above.

5.3. Apparatus and Chromatographic Conditions

The analytical method was based on the one validated for rat plasma, kidney, and liver by our group [40], with some modifications in order to adapt the sample treatment to a different animal species and to new tissues (brain and intestine), reducing the sample mass needed, and also, to simplify both: chromatographic separation conditions (mobile phase) and sample preparation procedure. OTA was quantified by HPLC-FLD in an 1100 series LC (Agilent Technologies, Waldbronn, Germany) with a fluorescence detector (λ excitation 225 nm and λ emission 461 nm). The chromatographic system was equipped with a Tracer Extrasil ODS column (25 cm × 0.4 cm, 5µm particle size) from Teknokroma (Spain) preceded by a (4 mm i.d.) Tracer Extrasil ODS2 safeguard column and working at 40 °C. The mobile phase was a mixture of acetonitrile and an aqueous solution of formic acid (0.4%) (50:50) in isocratic conditions. The aqueous phase was filtered through a 0.45 µm nylon membrane filter (Teknokroma, Barcelona, Spain). The injection volume was 20 µL, and the flow rate was 1 mL/min. The retention time for OTA under these conditions was 6.7 min, and the total analysis time was 10 min.

5.4. Preparation of Stock and Working Solutions

The stock solution of OTA (1 mg/mL) was prepared by dissolving 1 mg of OTA in methanol, and its concentration was verified by spectrophotometry at 333 nm (MW = 403.8; ε = 5500 M^{-1} cm^{-1}). OTA working standard solutions were prepared by diluting the stock solution with methanol and stored at −20 °C until use. OTA stability was previously confirmed in these conditions [41]. Thirty minutes before using them, an aliquot of each standard was tempered in darkness.

5.5. Preparation of Calibration and Quality Control (QC) Samples

The calibrators used in the validation of the method were prepared in 1.5 mL Eppendorffs by evaporating 50 µL of the corresponding working standard solution and dissolving them in 50 µL of the mobile phase.

QC samples were prepared similarly to calibrators at low, medium, and high concentration levels (2.35 ng/mL, 22.83 ng/mL, and 228.33 ng/mL) and were included in every analytical run while the analysis of study samples, according to the recommendations in FDA end EMA guidelines [21,22].

During sample analysis, according to the guidelines, each analytical run contained the three QC levels and at least two replicates per QC level. In total, at least six QCs were used in each analytical run, or QCs were enough to accomplish 5% of the total number of the analyzed samples, whichever number was greater. The preparation of all calibrators and QC samples was just before the analysis.

5.6. Sample Treatment

5.6.1. Homogenization of Solid Tissues

In the case of solid tissues (brain, liver, kidney, and intestine), frozen samples were thawed, weighed, and mixed with a sodium dihydrogen phosphate-buffered solution at pH 6.5 (4 µL per mg of tissue, dilution factor of kidney, liver, and intestines 1/4). After that, the mixture was homogenized with a bead beater (BeadBug™ 3 Position Bead Homogenizer, Gentaur, Kampenhout, Belgium) using 2 mL screw cap microtubes and 3 mm glass beads. Homogenates were then transferred to another Eppendorf in order to be frozen without the presence of the beads at −80 °C. Homogenates were frozen for at least 24 h hours until the extraction was carried out.

5.6.2. OTA Extraction

Plasma, brain, kidney, and intestine. For OTA quantification in plasma, brain, kidney, and intestine tissue homogenates, frozen samples were kept at room temperature in darkness for 30 min before the extraction process. Following that, 50 µL of the sample was mixed with 150 µL of ACN acidified with formic acid (0.4%) for protein precipitation and OTA release. After vortexing for 5 min, the sample was centrifuged at 12,000× *g* for 10 min at room temperature. Then, 150 µL of the supernatant were dried in an evaporator (GeneVac, SP Scientific, Ipswich, England) under vacuum at 60 °C and reconstituted with 50 µL of mobile phase (no dilution factor for plasma, dilution factor of kidney, liver, and intestines 1/4). For plasma samples with an expected high OTA concentration, only 10 µL of the supernatant was evaporated before being reconstituted with 50 µL of mobile phase (dilution factor 1/15).

Liver. For OTA extraction in the liver, homogenates were tempered in darkness for 30 min at room temperature. After that, 50 µL of the sample was mixed with 120 µL of ice-cold absolute ethanol and 15 µL of TCA 20%. Then, samples were vortexed for 5 min, centrifuged at 12,000× *g* for 10 min at room temperature and treated further as previously described for plasma, brain, kidney, and intestine tissue samples.

5.7. Validation of the Method

Following the FDA end EMA guidelines [21,22], the analytical method was validated according to the following parameters: selectivity, linearity, accuracy, and precision (within and between days), recovery, dilution integrity, carry-over effect, stability, and robustness. The acceptance criteria, based on the guidelines, are summarized in Table 7.

Table 7. Summary of acceptance criteria used for the validation of the method (adapted from [21,22].)

Parameters	Criteria of Acceptance Based on EMA and FDA Criteria
Selectivity	Absence of interfering components is accepted where the response is not more than 20% of the analyte response at the LLOQ [a] for the analyte.
Linearity/calibration curve	At least 6 concentration levels. Back-calculated concentrations of the calibration standards should be within 15% of the nominal value (20% at LLOQ) for at least 75%.
Accuracy and precision	RE% [b] and CV% [c] (within runs and between runs): ±15% of nominal concentrations, except ± 20% at LLOQ.
Dilution integrity	RE% and CV%: ±15%.
Carry-over effect	Blank response after a calibrator at ULOQ [d] should not exceed 20% of the analyte response at LLOQ.
Stability	RE% at each level (LLOQ and ULOQ): ±15%.
Robustness	Column batches (2).

[a] LLOQ: Lower limit of quantification. [b] RE%: relative error. [c] CV%: coefficient of variation. [d] ULOQ: upper limit of quantification.

5.7.1. Selectivity

The ability of the method to distinguish OTA from other endogenous components present in the samples was evaluated for the five matrices individually. The selectivity of the method was assessed by measuring and comparing blank plasma or tissues from 6 different individuals (not treated with any substance), with blank plasma or tissues spiked with OTA and with samples from OTA-treated mice.

5.7.2. Linearity (Calibration Curves) and LLOQ

The linearity was tested with OTA calibrators prepared as described above in 5.5. A calibration curve was made in triplicate on three different days in each one of the following ranges: 2.35–22.83 ng/mL and 22.83–228.33 ng/mL. Each one of the ranges included six points. With regard to linearity, the following criteria were considered: correlation coefficient ($r^2 > 0.99$), coefficient of variation (CV) between response factors (<5%), slope interval not having to include zero (p = 95%), and intercept interval having to include zero (p = 95%). Also, the back-calculated concentrations of the calibration standards should be within 15% of the nominal value (20% at LLOQ).

The lower limit of quantification (LLOQ) was established as the lowest OTA concentration inside the linear range that could be quantified with acceptable precision and accuracy (±20%).

5.7.3. Precision and Accuracy

Within- and between-day precision and accuracy of the linearity were studied by analyzing three replicate calibrations standards at low, intermediate, and high concentrations (2.35, 22.83, and 228.33 ng/mL) on 1 day (within days) and on 3 different days (between days). The accuracy was calculated as the relative error (RE%) of back-calculated concentrations with respect to the nominal value. The precision, defined as the closeness of repeated individual measures of the analyte, was expressed as CV% between the different replicates (within days) and runs (between days). Criteria for precision and accuracy were, respectively, CV (%) and RE (%) of less than 15% (20% for LLOQ).

5.7.4. Recovery

Due to the difficulty of obtaining these blank matrices in quantity enough to prepare all the matrix-matched calibrators and QCs needed, linearity has been studied using OTA solutions. For this reason, the recovery of OTA when the extraction procedure was applied to the samples has been assessed. Recovery has been studied for each one of the matrices, fortifying blank samples of each matrix with known OTA concentrations (2.35, 22.83, and 228.33 ng/mL). The spiking process was as follows: 50 µL of the corresponding working OTA solution was poured into an Eppendorf and evaporated under a vacuum at 60 °C. Then, 2.5 µL of ACN acidified with formic acid (0.4%) was added, and after vortexing for 2 min, 50 µL of the blank sample (plasma or tissue homogenates) were added and vortexed for 1 min more. The addition of this small volume of acidified ACN was made to ensure that plasma or homogenates dissolved OTA residue completely following evaporation. The mixture was left in darkness at room temperature to stand for 10 min before starting the procedure of sample treatment as previously described. The OTA recovery value (%) in all the matrices was calculated by dividing the experimental OTA concentration obtained in the spiked samples by the nominal OTA level. The precision of this process was studied by carrying out the recovery experiment for each matrix at three concentration levels, by triplicate, and on 3 different days so precision could be assessed in intermediate conditions. The criteria for precision was CV (%) of less than 15% (20% for LLOQ). The obtained mean recovery values for each one of the matrices were used in the correction of the levels obtained using the calibration curves.

5.7.5. Dilution Integrity

Some plasma samples were expected to have a high OTA concentration, so they were diluted. For this reason, dilution integrity was proven. For that aim, the dilution factor chosen (1/15) was used in a blank plasma sample spiked with an OTA concentration above the upper limit of quantification (ULOQ) (1174.69 ng/mL). Five determinations were made in one day, and the acceptance criteria were accuracy and precision (respectively RE% and CV%) of less than 15%.

5.7.6. Carry-Over Effect

The carry-over effect was investigated by measuring blank samples after calibration standards at the ULOQ five times.

5.7.7. Stability

OTA stability in frozen tissue homogenates was not assessed in this study, as previous data using rat tissues and the same homogenization buffer showed that OTA is stable at least 10 months after being homogenized and frozen [40]. OTA stability in the HPLC autosampler tray was evaluated by analyzing OTA spiked samples of each matrix at LLOQ and at ULOQ just after preparation (0 h) and at 2, 6, and 12 h in the autosampler tray. The stability of OTA in the analysis solution was accepted if RE% of the OTA concentration obtained is between ±15% of the one just after sample preparation. Additionally, an OTA concentration versus time regression analysis was performed; stability was confirmed if the slope was not statistically different from 0 (p = 95%).

5.7.8. Robustness

Finally, robustness was assessed by studying the influence of different batches of the chromatographic column. Linearity on two different Tracer Extrasil ODS2 columns was studied by analyzing calibrators in each one of them. Linearity was studied as presented before in each one of the columns and for both ranges studied (2.35–22.83 ng/mL and 22.83–228.33 ng/mL), and a global calibration curve was obtained in each one of the ranges and columns. After that, the peak areas of OTA obtained in one of the columns for calibrators at the low, medium, and high levels (2.35, 22.83, and 228.33 ng/mL) were extrapolated and quantified with the global calibration curves obtained in the other column.

Robustness was assessed if precision and accuracy for the calibrators were less than 15% in each one of the columns.

5.8. Acceptance Criteria of an Analytical Run

According to the guidelines [21,22], the run was considered acceptable if at least 67% of QCs were within ±15% of the nominal values and 50% or more of QCs per level were within ±15% of their nominal concentrations.

Supplementary Materials: The following supporting information can be downloaded at https://www.mdpi.com/article/10.3390/toxins16050213/s1. Table S1: Raw data of the linearity study of column A; Table S2: Raw data from the precision and accuracy study; Table S3: Raw data from the recovery study; Table S4: Results of the robustness study; Table S5: Global calibration curves of column B.

Author Contributions: Conceptualization, A.V. and E.G.-P.; animal studies, E.B., M.S.-C., M.I. and L.A.-E.; investigation, E.B.; writing—original draft preparation, E.B.; writing—review and editing, A.V., E.G.-P., M.I. and L.A.-E.; supervision, A.V. and E.G.-P.; funding acquisition, A.V. and L.A.-E. All authors have read and agreed to the published version of the manuscript.

Funding: This work has been supported by the Government of Navarra (Project-43, 2019 modality A) and the and European Regional Development Fund (ERDF under Operational Program for Navarra, 2014–2020). E.B. thanks the "Asociación de Amigos de la Universidad de Navarra", Banco Santander, Government of Navarra and the Ministry of Universities of the Spanish Government ("Ayudas para la formación de profesorado Universitario", FPU20/01671) for the predoctoral grants received. M.S.-C. received funding from "la Caixa" Banking Foundation and "Asociación de Amigos de la Universidad de Navarra". L.A.-E. is supported by a Miguel Servet II contract (CPII20/00027) from ISCIII.

Institutional Review Board Statement: The animal study protocol was approved by the Ethics Committee on Animal Experimentation of Universidad de Navarra (CEEA 049-19), and they were conducted according to the National Institute of Health (NIH) Guide for the Care and Use of Laboratory Animals.

Informed Consent Statement: Not applicable.

Data Availability Statement: Data are contained within the article or the Supplementary Materials.

Conflicts of Interest: The authors declare no conflicts of interest.

Abbreviations

ACN	Acetonitrile
CV	Coefficient of variation
EMA	European Medicines Agency
FDA	Food and Drug Administration
HPLC-FLD	High-performance liquid chromatography-fluorescence detection
IARC	International Agency for Research on Cancer
ILO	International Labour Organization
LLOQ	Lower limit of quantification
LOD	Limit of detection
NIH	National Institute of Health
OTA	Ochratoxin A
PD	Parkinson's disease
QC	Quality control
RE	Relative error
TCA	Trichloroacetic acid
TLV	Threshold Limit Value
ULOQ	Upper limit of quantification

References

1. Schrenk, D.; Bodin, L.; Chipman, J.K.; del Mazo, J.; Grasl-Kraupp, B.; Hogstrand, C.; Hoogenboom, L.; Leblanc, J.C.; Nebbia, C.S.; Nielsen, E.; et al. Risk Assessment of Ochratoxin A in Food. *EFSA J.* **2020**, *18*, e06113. [CrossRef] [PubMed]
2. Arce-López, B.; Lizarraga, E.; Vettorazzi, A.; González-Peñas, E. Human Biomonitoring of Mycotoxins in Blood, Plasma and Serum in Recent Years: A Review. *Toxins* **2020**, *12*, 147. [CrossRef] [PubMed]
3. Rached, E.; Hard, G.C.; Blumbach, K.; Weber, K.; Draheim, R.; Lutz, W.K.; Özden, S.; Steger, U.; Dekant, W.; Mally, A. Ochratoxin A: 13-Week Oral Toxicity and Cell Proliferation in Male F344/ N Rats. *Toxicol. Sci.* **2007**, *97*, 288–298. [CrossRef] [PubMed]
4. National Toxicology Program. NTP Toxicology and Carcinogenesis Studies of Ochratoxin A (CAS No. 303-47-9) in F344/N Rats (Gavage Studies). *Natl. Toxicol. Program. Tech. Rep. Ser.* **1989**, *358*, 1–142.
5. Pfohl-Leszkowicz, A.; Pinelli, E.; Bartsch, H.; Mohr, U.; Castegnaro, M. Sex- and Strain-Specific Expression of Cytochrome P450s in Ochratoxin A- Induced Genotoxicity and Carcinogenicity in Rats. *Mol. Carcinog.* **1998**, *23*, 76–85. [CrossRef]
6. IARC (International Agency for Research on Cancer). *Some Naturally Occurring Substances: Food Items and Constituents, Heterocyclic Aromatic Amines and Mycotoxins*; International Agency for Research on Cancer: Lyon, France, 1992. [CrossRef]
7. Sava, V.; Reunova, O.; Velasquez, A.; Harbison, R.; Sánchez-Ramos, J. Acute Neurotoxic Effects of the Fungal Metabolite Ochratoxin-A. *Neurotoxicology* **2006**, *27*, 82–92. [CrossRef] [PubMed]
8. Kumar, P.; Mahato, D.K.; Sharma, B.; Borah, R.; Haque, S.; Mahmud, M.M.C.; Shah, A.K.; Rawal, D.; Bora, H.; Bui, S. Ochratoxins in Food and Feed: Occurrence and Its Impact on Human Health and Management Strategies. *Toxicon* **2020**, *187*, 151–162. [CrossRef] [PubMed]
9. Haq, M.; Gonzalez, N.; Mintz, K.; Jaja-Chimedza, A.; De Jesus, C.L.; Lydon, C.; Welch, A.; Berry, J.P. Teratogenicity of Ochratoxin a and the Degradation Product, Ochratoxin α, in the Zebrafish (Danio Rerio) Embryo Model of Vertebrate Development. *Toxins* **2016**, *8*, 40. [CrossRef]
10. Qi, X.; Yang, X.; Chen, S.; He, X.; Dweep, H.; Guo, M.; Cheng, W.H.; Xu, W.; Luo, Y.; Gretz, N.; et al. Ochratoxin A Induced Early Hepatotoxicity: New Mechanistic Insights from MicroRNA, MRNA and Proteomic Profiling Studies. *Sci. Rep.* **2014**, *4*, 5163. [CrossRef]
11. Sun, Y.; Huang, K.; Long, M.; Yang, S.; Zhang, Y. An Update on Immunotoxicity and Mechanisms of Action of Six Environmental Mycotoxins. *Food Chem. Toxicol.* **2022**, *163*, 112895. [CrossRef] [PubMed]
12. Khatoon, A.; Khan, M.Z.; Khan, A.; Saleemi, M.K.; Javed, I. Amelioration of Ochratoxin A-Induced Immunotoxic Effects by Silymarin and Vitamin e in White Leghorn Cockerels. *J. Immunotoxicol.* **2013**, *10*, 25–31. [CrossRef] [PubMed]
13. Tamaru, M.; Hirata, Y.; Matsutani, T. Neurochemical Effects of Prenatal Treatment with Ochratoxin A on Fetal and Adult Mouse Brain. *Neurochem. Res.* **1988**, *13*, 1139–1147. [CrossRef] [PubMed]
14. Paradells, S.; Rocamonde, B.; Llinares, C.; Herranz-Pérez, V.; Jimenez, M.; Garcia-Verdugo, J.M.; Zipancic, I.; Soria, J.M.; Garcia-Esparza, M.A. Neurotoxic Effects of Ochratoxin A on the Subventricular Zone of Adult Mouse Brain. *J. Appl. Toxicol.* **2015**, *35*, 737–751. [CrossRef] [PubMed]
15. Mateo, E.; Tonino, R.P.B.; Canto, A.; Monroy Noyola, A.; Miranda, M.; Soria, J.M.; Garcia Esparza, M.A. The Neurotoxic Effect of Ochratoxin-A on the Hippocampal Neurogenic Niche of Adult Mouse Brain. *Toxins* **2022**, *14*, 624. [CrossRef] [PubMed]
16. Sava, V.; Reunova, O.; Velasquez, A.; Sanchez-Ramos, J. Can Low Level Exposure to Ochratoxin-A Cause Parkinsonism? *J. Neurol. Sci.* **2006**, *249*, 68–75. [CrossRef] [PubMed]
17. Bhat, P.V.; Anand, T.; Mohan Manu, T.; Khanum, F. Restorative Effect of L-Dopa Treatment against Ochratoxin A Induced Neurotoxicity. *Neurochem. Int.* **2018**, *118*, 252–263. [CrossRef] [PubMed]
18. Izco, M.; Vettorazzi, A.; Forcen, R.; Blesa, J.; de Toro, M.; Alvarez-Herrera, N.; Cooper, J.M.; Gonzalez-Peñas, E.; Lopez de Cerain, A.; Alvarez-Erviti, L. Oral Subchronic Exposure to the Mycotoxin Ochratoxin A Induces Key Pathological Features of Parkinson's Disease in Mice Six Months after the End of the Treatment. *Food Chem. Toxicol.* **2021**, *152*, 112164. [CrossRef]
19. Braak, H.; Del Tredici, K.; Rüb, U.; De Vos, R.A.I.; Jansen Steur, E.N.H.; Braak, E. Staging of Brain Pathology Related to Sporadic Parkinson's Disease. *Neurobiol. Aging* **2003**, *24*, 197–211. [CrossRef]
20. Vettorazzi, A.; González-Peñas, E.; de Cerain, A.L. Ochratoxin A Kinetics: A Review of Analytical Methods and Studies in Rat Model. *Food Chem. Toxicol.* **2014**, *72*, 273–288. [CrossRef]
21. Food and Drug Administration (FDA). *Bionalytical Method Validation*; FDA: Silver Spring, MD, USA, 2018; Volume 1043.
22. EMA Committee for Medicinal Products for Human Use. *Guideline on Bioanalytical Method Validation*; EMEA/CHMP/EWP/192217/2(Rev. 1 Corr. 2**; European Medicines Agency: Amsterdam, The Netherlands, 2011.
23. Szőke, Z.; Babarczi, B.; Mézes, M.; Lakatos, I.; Poór, M.; Fliszár-Nyúl, E.; Oldal, M. Analysis and Comparison of Rapid Methods for the Determination of Ochratoxin a Levels in Organs and Body Fluids Obtained from Exposed Mice. *Toxins* **2022**, *14*, 634. [CrossRef] [PubMed]
24. Fukui, Y.; Hoshino, K.; Kameyama, Y.; Yasui, T.; Toda, C.; Nagano, H. Placental Transfer of Ochratoxin A and Its Cytotoxic Effect on the Mouse Embryonic Brain. *Food Chem. Toxicol.* **1987**, *25*, 17–24. [CrossRef] [PubMed]
25. Wang, J.; Gan, C.; Qi, X.; Lebre, M.C.; Schinkel, A.H. Human Organic Anion Transporting Polypeptide (OATP) 1B3 and Mouse OATP1A/1B Affect Liver Accumulation of Ochratoxin A in Mice. *Toxicol. Appl. Pharmacol.* **2020**, *401*, 115072. [CrossRef] [PubMed]
26. Appelgren, L.-E.; Arora, R.G. Distribution of 14C-Labelled Ochratoxin A in Pregnant Mice. *Food Chem. Toxicol.* **1983**, *21*, 563–568. [CrossRef] [PubMed]
27. Meulenberg, E.P. Immunochemical Methods for Ochratoxin A Detection: A Review. *Toxins* **2012**, *4*, 244–266. [CrossRef] [PubMed]

28. Gentaur. BeadBugTM 3 Position Bead Homogenizer. Available online: https://biodas.org/beadbug-3-position-bead-homogenizer/ (accessed on 1 March 2024).
29. Li, P.; Bartlett, M.G. A Review of Sample Preparation Methods for Quantitation of Small-Molecule Analytes in Brain Tissue by Liquid Chromatography Tandem Mass Spectrometry (LC-MS/MS). *Anal. Methods* **2014**, *6*, 6183–6207. [CrossRef]
30. Mohos, V.; Faisal, Z.; Fliszár-Nyúl, E.; Szente, L.; Poór, M. Testing the Extraction of 12 Mycotoxins from Aqueous Solutions by Insoluble Beta-Cyclodextrin Bead Polymer. *Environ. Sci. Pollut. Res.* **2022**, *29*, 210–221. [CrossRef]
31. Arce-López, B.; Lizarraga, E.; Flores-Flores, M.; Irigoyen, Á.; González-Peñas, E. Development and Validation of a Methodology Based on Captiva EMR-Lipid Clean-up and LC-MS/MS Analysis for the Simultaneous Determination of Mycotoxins in Human Plasma. *Talanta* **2020**, *206*, 120193. [CrossRef]
32. International Labour Organization (United Nations) International Chemical Safety Cards (ICSCs). Available online: https://www.ilo.org/dyn/icsc/showcard.listCards3 (accessed on 12 March 2024).
33. Kane, A.; Creppy, E.E.; Roth, A.; Riischenthaler, R.; Dirheimer, G. Original Investigations Archives of Distribution of the [3HI-Label from Low Doses of Radioactive Ochratoxin A Ingested by Rats, and Evidence for DNA Single-Strand Breaks Caused in Liver and Kidneys. *Arch. Toxicol.* **1986**, *58*, 219–224.
34. Han, Z.; Zhao, Z.; Shi, J.; Liao, Y.; Zhao, Z.; Zhang, D.; Wu, Y.; De Saeger, S.; Wu, A. Combinatorial Approach of LC-MS/MS and LC-TOF-MS for Uncovering in Vivo Kinetics and Biotransformation of Ochratoxin A in Rat. *J. Chromatogr. B* **2013**, *925*, 46–53. [CrossRef]
35. Aoudia, N.; Tangni, E.K.; Larondelle, Y. Distribution of Ochratoxin A in Plasma and Tissues of Rats Fed a Naturally Contaminated Diet Amended with Micronized Wheat Fibres: Effectiveness of Mycotoxin Sequestering Activity. *Food Chem. Toxicol.* **2008**, *46*, 871–878. [CrossRef] [PubMed]
36. Arbillaga, L.; Vettorazzi, A.; Gil, A.G.; van Delft, J.H.; García-Jalón, J.A.; López de Cerain, A. Gene Expression Changes Induced by Ochratoxin A in Renal and Hepatic Tissues of Male F344 Rat after Oral Repeated Administration. *Toxicol. Appl. Pharmacol.* **2008**, *230*, 197–207. [CrossRef] [PubMed]
37. Mally, A.; Völkel, W.; Amberg, A.; Kurz, M.; Wanek, P.; Eder, E.; Hard, G.; Dekant, W. Functional, Biochemical, and Pathological Effects of Repeated Oral Administration of Ochratoxin A to Rats. *Chem. Res. Toxicol.* **2005**, *18*, 1242–1252. [CrossRef] [PubMed]
38. Vettorazzi, A.; De Trocóniz, I.F.; González-Peñas, E.; Arbillaga, L.; Corcuera, L.A.; Gil, A.G.; de Cerain, A.L. Kidney and Liver Distribution of Ochratoxin A in Male and Female F344 Rats. *Food Chem. Toxicol.* **2011**, *49*, 1935–1942. [CrossRef] [PubMed]
39. Wang, H.; Chen, Y.; Zhai, N.; Chen, X.; Gan, F.; Li, H.; Huang, K. Ochratoxin A-Induced Apoptosis of IPEC-J2 Cells through ROS-Mediated Mitochondrial Permeability Transition Pore Opening Pathway. *J. Agric. Food. Chem.* **2017**, *65*, 10630–10637. [CrossRef] [PubMed]
40. Vettorazzi, A.; Gonzalez-Peñas, E.; Arbillaga, L.; Corcuera, L.A.; López de Cerain, A. Simple High-Performance Liquid Chromatography-Fluorescence Detection Method for Plasma, Kidney and Liver of Rat as a Tool for Toxicology Studies. *J. Chromatogr. A* **2008**, *1215*, 100–106. [CrossRef]
41. Valenta, H. Chromatographic Methods for the Determination of Ochratoxin A in Animal and Human Tissues and Fluids. *J. Chromatogr. A* **1998**, *815*, 75–92. [CrossRef]

toxins

MDPI

Article

In Vitro Digestion and Intestinal Absorption of Mycotoxins Due to Exposure from Breakfast Cereals: Implications for Children's Health

Soraia V. M. de Sá, Miguel A. Faria, José O. Fernandes and Sara C. Cunha *

LAQV-REQUIMTE, Laboratory of Bromatology and Hydrology, Faculty of Pharmacy, University of Porto, Rua de Jorge Viterbo Ferreira, 228, 4050-313 Porto, Portugal; nano.soraia@gmail.com (S.V.M.d.S.); mfaria@ff.up.pt (M.A.F.); josefer@ff.up.pt (J.O.F.)

* Correspondence: sara.cunha@ff.up.pt

Abstract: Breakfast cereals play a crucial role in children's diets, providing essential nutrients that are vital for their growth and development. Children are known to be more susceptible than adults to the harmful effects of food contaminants, with mycotoxins being a common concern in cereals. This study specifically investigated aflatoxin B1 (AFB1), enniatin B (ENNB), and sterigmatocystin (STG), three well-characterized mycotoxins found in cereals. The research aimed to address existing knowledge gaps by comprehensively evaluating the bioaccessibility and intestinal absorption of these three mycotoxins, both individually and in combination, when consumed with breakfast cereals and milk. The *in vitro* gastrointestinal method revealed patterns in the bioaccessibility of AFB1, ENNB, and STG. Overall, bioaccessibility increased as the food progressed from the stomach to the intestinal compartment, with the exception of ENNB, whose behavior differed depending on the type of milk. The ranking of overall bioaccessibility in different matrices was as follows: digested cereal > cereal with semi-skimmed milk > cereal with lactose-free milk > cereal with soy beverage. Bioaccessibility percentages varied considerably, ranging from 3.1% to 86.2% for AFB1, 1.5% to 59.3% for STG, and 0.6% to 98.2% for ENNB. Overall, the inclusion of milk in the ingested mixture had a greater impact on bioaccessibility compared to consuming the mycotoxins as a single compound or in combination. During intestinal transport, ENNB and STG exhibited the highest absorption rates when ingested together. This study highlights the importance of investigating the combined ingestion and transport of these mycotoxins to comprehensively assess their absorption and potential toxicity in humans, considering their frequent co-occurrence and the possibility of simultaneous exposure.

Keywords: breakfast cereals; mycotoxin co-occurrence; *in vitro* digestion; LC-MS/MS; children; bioaccessibility; intestinal absorption

Key Contribution: This study goes beyond previous research about bioaccessibility and transport by investigating mycotoxins individually and in combination, considering their co-occurrence in breakfast cereals, and emphasizes the importance of understanding their absorption and potential toxicity in humans, namely, children.

1. Introduction

Mycotoxin toxicity, bioaccessibility, and intestinal absorption form a critical triad for food safety, human health, and agricultural practices. Understanding these interlinked concepts is crucial to assess the impact of mycotoxins on human health and their path to the systemic circulation, where they exert toxic effects.

Produced by various molds, primarily *Aspergillus*, *Penicillium*, and *Fusarium* species, mycotoxins contaminate a wide range of agricultural products [1,2]. Notorious for their potent toxic effects on humans and animals, they can cause acute and chronic illnesses, immunosuppression, cancer, gastrointestinal disorders, and kidney damage [3]. Aflatoxins, a well-recognized group of mycotoxins known for their potent toxicity, include aflatoxin

B1 (AFB1), the most prevalent and carcinogenic member classified as Group 1 by the International Agency for Research on Cancer (IARC) [4]. Aflatoxins are produced by *Aspergillus flavus* and *A. parasiticus*, typically found in soil, decaying vegetation, hay, and grains undergoing microbiological deterioration [5]. The European Food Safety Authority (EFSA) reports a persistent dietary intake of AFB1 in younger age groups, ranging from 0.08 to 1.78 ng/kg body weight per day on the lower end and from 0.58 to 6.95 ng/kg body weight per day on the upper end [3]. This persistent exposure underlines the importance of research on aflatoxin-related health effects in children [3].

Climate change contributes to the emergence of unregulated mycotoxins, such as enniatin B (ENNB) and sterigmatocystin (STG), which require consistent monitoring due to their potential harm [6]. These "emerging mycotoxins" have been detected in various food products like cereals, nuts, and processed foods, including breakfast cereals [7]. As a food highly consumed by children, breakfast cereals raise specific concerns. Children are more vulnerable to food contaminants due to the potential for substantial damage to intestinal enterocytes (cells lining the small intestine) by these toxins [8–10]. Therefore, studying the effects and fate of these compounds within the gastrointestinal tract is of paramount importance.

Concerning these compounds' toxicity, STG serves as a precursor for aflatoxin production and shares structural similarities with it. It is produced by *Aspergillus* species, such as *A. versicolor* and *A. nidulans*, under optimal environmental conditions of 27–29 °C [11]. While STG is associated with low-level acute toxicity, the primary concern lies in its carcinogenic properties, roughly one-tenth of those of aflatoxin B1 [12]. ENNB belongs to a group of cyclic hexadepsipeptides produced by several *Fusarium* species such as *F. avenaceum*, *F. tricinctum*, *F. poae*, *F. sporotrichioides*, and *F. langsethiae* [13,14] and its contamination usually occurs pre-harvest [15]. ENNB exhibits strong cytotoxic effects, inducing cell death via mitochondrial damage [16]. Despite the EFSA asserting that short-term exposure to ENNB does not pose immediate health risks for humans, studies have demonstrated strong cytotoxic activity in different human cell lines, such as intestinal Caco-2 cells [17,18]. This suggests potential health risks with long-term exposure, warranting further investigation.

The quantity of ingested mycotoxin may not reflect the actual amount available for exerting its toxic effects. Only a fraction becomes bioaccessible within the gastrointestinal tract and further bioavailable upon reaching the bloodstream [19]. *In vitro* methodologies, such as simulated digestion models and intestinal absorption cellular models (e.g., Caco-2/HT29MTX co-culture), have gained traction in recent years [20–22]. These models mimic the human digestive and absorption process, allowing for researchers to investigate the influence of the food matrix on molecule bioavailability [19]. Cereal matrices have been extensively studied in this context, demonstrating high bioaccessibility for many toxins [23–25]. However, significant knowledge gaps remain. Although there are studies exploring ENNB bioaccessibility in breakfast cereals [24], and AFB1 bioaccessibility in plant-based milks [26], there is no existing literature on intestinal absorption of any of these compounds in Caco-2/HT29MTX co-cultures, with only one study focusing on the effects of AFB1's metabolite (AFM1) within these co-cultures [27]. Although IARC monographs provide valuable information on AFB1's *in vivo* absorption in humans [28,29], further research is necessary to understand the combined effects of these mycotoxins during intestinal absorption, particularly for ENNB and STG.

Given the limited research on mycotoxin bioaccessibility and intestinal absorption in children [28], particularly concerning combined ingestion, this study aimed to comprehensively assess these processes for AFB1, ENNB, and STG. This will enhance the accuracy of risk evaluation for mycotoxin exposure in this vulnerable population.

This study employed a validated *in vitro* gastrointestinal model mimicking the human digestive system and a well-characterized intestinal absorption cellular model to comprehensively assess the bioavailability of target compounds. We investigated the bioaccessibility of AFB1, ENNB, and STG, both individually and in combination, in artificially contaminated breakfast cereals. We also examined their bioaccessibility in cereals with semi-skimmed milk (including milk alone) and their subsequent isolated and combined absorption in human-derived epithelial intestinal Caco-2/HT-29MTX co-culture cells. These co-culture cells represent the first human intestinal barrier encountered by mycotoxins after ingestion and digestion. Notably, this research appears to be the first to investigate the transport of these mycotoxins within this specific intestinal cell model (Caco-2/HT-29MTX co-culture).

2. Results and Discussion

2.1. Method Performance

Matrix calibration curves for digested matrices were generated to mitigate interferences stemming from matrix components. A linear response, with coefficients of correlation (r) exceeding 0.99, was observed for all mycotoxins in intestinal fraction of digested matrices. In the case of gastric fraction of digested matrices, favorable linear correlations ($r > 0.98$) were achieved only for ENNB (Table S1). Validation of the method for AFB1 and STG in gastric fraction proved challenging due to persistent interferences that the clean-up process failed to eliminate. The limits of detection (LODs) ranged from 0.3–0.9 μg/L to 0.1–0.8 μg/L for gastric and intestinal fractions of digested matrices, respectively, and the limits of quantification (LOQs) ranged from 1.1–3.1 μg/L to 0.4–2.5 μg/L for gastric and intestinal fractions of digested matrices, respectively (Table S1). Recovery percentages surpassed 70% for the four matrices, except for ENNB in breakfast cereal with semi-skimmed milk and in breakfast cereal with soy beverage, AFB1 in breakfast cereal with soy beverage, and STG in breakfast cereal with semi-skimmed lactose-free milk, which exhibited recoveries below 60%, and the coefficient of variation (%CV) values for interday and intraday precision were below 15% and 16%, respectively (Table S2).

Concerning transport assays, a linear response was verified for the three mycotoxins (with $r > 0.98$). The LODs and LOQs were observed to be in the ranges of 0.1 to 0.5 μg/L and 0.2 to 1.5 μg/L, respectively. The coefficient of variation (%CV) values for both interday and intraday precision were found to be below 16% (Table S3).

2.2. Mycotoxins Bioaccessibility

Bioaccessibility values after *in vitro* digestion (gastric and intestinal) of AFB1, ENNB, and STG in breakfast cereals, and breakfast cereals with different milks and soy beverage are presented in Table 1. AFB1 and STG gastric bioaccessibilities were not evaluated, as already described in Section 2.1. Overall, an upward trend was noted throughout the gastrointestinal tract, indicating an increase in mycotoxin bioaccessibility from the gastric to the intestinal phase, except ENNB in the matrices with milk. The overall percentages of bioaccessibility varied within the following range accordingly to the matrix: very low to 98.2% for BC, very low to 47.2% for BCSSM, very low to 21.6% for BCSSMLF, and very low to 26.7% for BCSB. In terms of intestinal bioaccessibility, AFB1 demonstrated values ranging from 3.1% to 86.2%, STG exhibited values between 1.5% and 59.3%, and ENNB showed values spanning from 0.6% to 98.2%.

Concerning the matrix complexity, in general, when the breakfast cereals were consumed mixed with milk, all mycotoxins showed lower bioaccessibilities than when cereals were ingested alone. ENNB and STG presented the lowest percentages of the bioaccessible fractions (less than 2%), namely, when in mixture.

Table 1. Percentage (%) of bioaccessibility of mycotoxins in cereal and cereal with different types of milk and beverages after gastric and intestinal *in vitro* digestion.

Mycotoxin	*In Vitro* Digestion		Bioaccessibility (%)			
			BC	BCSSM	BCSSMLF	BCSB
AFB1	Gastric	Isolated	<LOD	<LOD	<LOD	<LOD
		Mixture	<LOD	<LOD	<LOD	<LOD
	Intestinal	Isolated	57.6 ± 0.1 * ↗	24.2 ± 0.3 * ↗	20.3 ± 1.0 * ↗	3.13 ± 0.17 * ↗
		Mixture	86.2 ± 1.6 * ↗	27.5 ± 10.3 * ↗	18.1 ± 6.8 * ↗	9.06 ± 4.32 * ↗
ENNB	Gastric	Isolated	49.2 ± 1.6 $^{a, b}$	47.2 ± 21.6 ^{b}	18.1 ± 4.2 ^{b}	26.3 ± 6.3 $^{a, b}$
		Mixture	52.4 ± 4.5 $^{a, b}$	40.5 ± 0.3 ^{b}	21.6 ± 4.0 ^{b}	26.7 ± 8.4 $^{a, b}$
	Intestinal	Isolated	91.2 ± 6.4 $^{a, b}$ ↗	18.5 ± 9.5 ^{b}	4.54 ± 1.42 ^{b}	21.9 ± 2.2 ^{b}
		Mixture	98.2 ± 7.3 $^{a, b}$ ↗	10.7 ± 4.2 ^{b}	0.62 ± 0.17 ^{b}	10.8 ± 0.5 ^{b}
STG	Gastric	Isolated	<LOD	<LOD	<LOD	<LOD
		Mixture	<LOD	<LOD	<LOD	<LOD
	Intestinal	Isolated	53.6 ± 1.7 $^{a, b}$ ↗	8.17 ± 0.69 $^{b, *}$ ↗	19.7 ± 1.7 ^{b} ↗	1.45 ± 0.06 $^{a, b, *}$ ↗
		Mixture	59.3 ± 3.7 $^{a, b}$ ↗	9.52 ± 2.80 $^{b, *}$ ↗	8.22 ± 1.49 ^{b} ↗	1.97 ± 0.79 $^{a, b, *}$ ↗

Increase bioaccessibility

Abbreviations: BC—breakfast cereal; BCSSM—breakfast cereal with semi-skimmed milk; BCSSMLF—breakfast cereal with semi-skimmed lactose-free milk; BCSB—breakfast cereal with soy beverage. <LOD—below limit of detection. Data expressed as mean ± standard deviation (n = 6) for samples following normal distribution. The arrows (↗) indicate an increase in bioaccessibility between gastric and intestinal phases. * Significant difference ($p < 0.05$) between isolated and mixed transport of mycotoxins; different letters in the same row indicate differences ($p < 0.05$) betweenransports.

The bioaccessibility of mycotoxins has been reported to be influenced by various factors, including its chemical structure, pH during the digestion process, and the composition of food matrices [25]. The disparities in bioaccessibility (%) presented in Table 1 can also be elucidated by the composition of the digested food matrices as referred to in previous research for both nutrients and toxic compounds [30]. Cereal-based foods typically contain natural compounds, such as adsorbent dietary fibers, which interact with mycotoxins, resulting in a reduction in their bioaccessibility [31]. Interactions with other food components can influence mycotoxin bioaccessibility, with mycotoxins demonstrating the ability to bind primarily with proteins and lipids within the food matrix [24]. Moreover, dietary fibers have already been used as a protective measure against mycotoxicosis by their inclusion in food and feed products, offering a cost-effective method for detoxification [32,33]. Table 3 (Section 4) displays the proximate nutritional composition of each product used in the gastrointestinal digestion process. The bioaccessibility values for ENNB, in the breakfast cereal digested sample (corn with honey), are 91.2% and 98.2% for the mycotoxin in isolated form and in the mixture, respectively. These values surpass those reported in the only study in a similar type of digested matrix (corn flakes) by Prosperini et al. (ranging from 43 to 70%) [24]. These differences can be attributed to the composition of our sample. Although having the same fiber content (3 g/100 g of product), it exhibits lower levels of fat (0.6 g against 0.7 g/100 g of product) and protein (5.5 g against 8.0 g/100 g of product), which enables greater bioaccessibility, given that the levels of fat and proteins are lower, resulting in less retention of mycotoxins in the matrix, making them more bioaccessible.

For the digested breakfast cereal samples with the different types of milk and soy beverage, as can be seen in Table 1, the bioaccessibility percentages of mycotoxins are much lower compared to the digested breakfast cereal sample alone. This behavior could be explained by the fact that the milks and the soy beverage contain a higher fat content (>3 g/100 mL) (Table 3), which leads to lower bioaccessibility. The high fat content can undergo the release of mycotoxins from the matrix, most probably related to the different lipophilic and hydrophilic properties of these molecules [34]. ENNB demonstrates to be the most lipophilic (log $P_{o/w}$ of 3.61), being poorly soluble in water, followed by STG and AFB1 (log $P_{o/w}$ of 2.61 and 2.09, respectively), moderately hydrophilic, which explains the greater retention of these compounds in matrices containing milk. Additionally, the digested breakfast cereal with soy beverage exhibits even lower values compared to the other matrices with milk. One of the reasons for this decrease in bioaccessibility, with the exception of ENNB, is due to the soy beverage's increased fiber content, which is not found in the other types of milk.

2.3. Cell Monolayer Integrity Control

To validate that the transport study adhered to recommended conditions, two distinct measurements were employed to assess monolayer integrity after exposure for 180 min to the detoxified bioaccessible fraction and a mixed standard solution of mycotoxins: (i) the determination of mycotoxin cytotoxicity with the concentrations under investigation and (ii) the measurement of trans-epithelial electrical resistance (TEER). TEER determination furnishes details regarding the consistency of the Caco-2 cell layer on the filter support and the integrity of the tight junctions established between the polarized cells [22], and a decline in that TEER values could indicate an increase in the permeability of the tight junctions due to cell detachment, etc. [35]. TEER values for Caco-2/HT-29MTX monolayers were complying with values reported in the literature [22]. No significant differences ($p < 0.05$) were observed in TEER values, measured before (0 h, ≈1390 Ωcm^2) and after transport experiments (3 h, ≈1730 Ωcm^2), following exposure to mycotoxins, either individually or in a mixture in both matrices.

The detoxified bioaccessible fraction was serially diluted in the transport medium (HBSS) and tested by exposing it to the monolayers for 180 min, as well as a standard mixture of the contaminants in relevant concentrations ENNB (0.31 μM), STG (0.62 μM), and AFB1 (0.64 μM). It is worth noting that these concentrations are within the range found in food [2,36], despite the low concentrations of AFB1 reported in the literature and the stringent regulations governing its maximum content in food [37]. The bioaccessible fraction in a dilution higher than 6× did not impart cell toxicity to the co-cultured cells Caco-2/HT-29MTX; thus, this dilution was selected for further assays. The toxicity of the compounds themselves was also not noticed after 180 min of exposure, thereby not affecting cell viability (Figure 1) as measured by the MTT assay. These preliminary assays assured that the monolayer integrity of differentiated Caco-2/HT-29MTX cells remained uncompromised during transport assays.

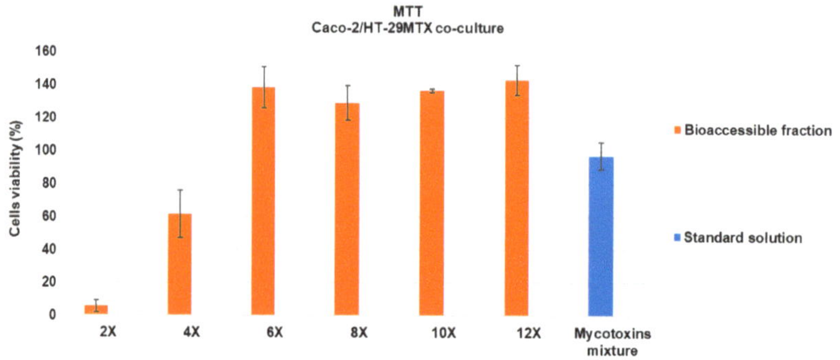

Figure 1. Percentage of cell viability of proliferating Caco-2/HT-29MTX cells, after 180 min of exposure to mycotoxins. Data expressed as mean ± SD of 3 independent experiments (n = 3) against an HBSS control (100% viability). AFB1—aflatoxin B1; ENNB—enniatin B; MIX—mixture of the three mycotoxins; STG—sterigmatocystin. Bioaccessible fraction suffered a heat treatment of 98 °C for 5 min, and was diluted 2, 4, 6, 8, 10, and 12 times. Standard solution: mixture of AFB1 + ENNB + STG of 200 μg/L diluted in HBSS.

2.4. Intestinal In Vitro Absorption Assays

The apical–basolateral transport of individual AFB1, ENNB, STG, and their combined mixture was evaluated across Caco-2/HT-29MTX monolayers. Human cell models, such as the Caco-2/HT-29MTX co-culture model, serve as prominent tools for examining intestinal transport and absorption [22]. This model, involving the co-culture of Caco-2 and HT-29MTX, proves valuable for exploring transport across the intestinal epithelium and studying bacterial adhesion and invasion [22]. The inclusion of the mucin layer by HT-29MTX and the assessment of cell layer permeability are pivotal for such investigations, and the co-culture approach yields results that align more closely with the *in vivo* conditions compared to monocultures [22].

Figure 2 presents the outcomes related to the transport across Caco-2/HT-29MTX monolayers of each mycotoxin individually and in a mixture, expressed as mass percentage transported over time, in both cereal and cereal with semi-skimmed milk matrices. All mycotoxins under study were absorbed; however, varied transport rates were noted based on the mycotoxin type, the presence of another molecule (isolated or mixed), and the type of matrix. In theory, the cumulative fraction of a passively transported drug across Caco-2 cell monolayers should exhibit a linear increase over time when conducted under sink conditions. However, in practical experiments, deviations from this scenario may occur, leading to non-linear transport behavior.

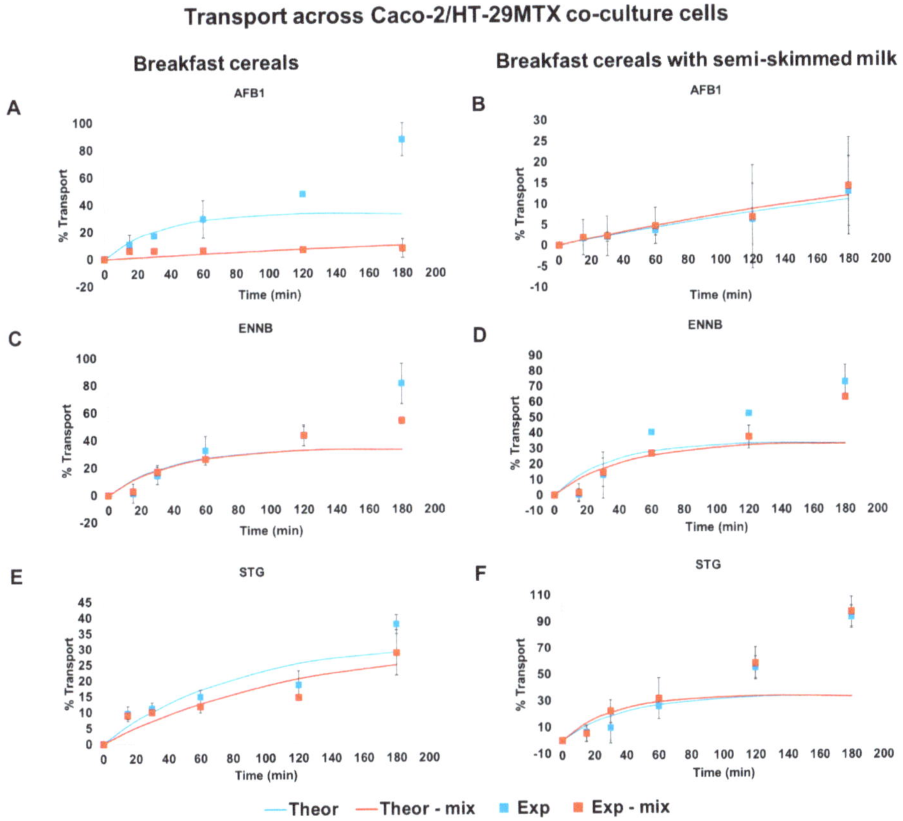

Figure 2. The percentage of AFB1 ((**A**): in breakfast cereals and (**B**): in breakfast cereals with semi-skimmed milk), ENNB ((**C**): in breakfast cereals and (**D**): in breakfast cereals with semi-skimmed milk), and STG ((**E**): in breakfast cereals and (**F**): in breakfast cereals with semi-skimmed milk) transferred to the receiver compartment over 180 min across monolayers of Caco-2/HT-29MTX in the apical → basolateral direction when transported isolated (blue squares and lines) or in mixture (red squares and lines). Data expressed as mean ± SD of 3 independent experiments (n = 3). AFB1—aflatoxin B1; ENNB—Enniatin B; MIX—mixture of the three mycotoxins; STG—sterigmatocystin.

AFB1 exhibited an almost linear transfer to the receiver compartment (transport percentages similar to theoretical values until 60 min) when isolated in the breakfast cereal matrix, resulting in high transport rates of around 87% (Figure 2A). When in a mixture, the transport rate was much lower and in accordance with theoretical values, achieving maximum percentages of around 7%. In contrast, in the presence of semi-skimmed milk, AFB1 showed a linear and slower rate of transport, reaching 13% and 14%, when isolated and in mixture, respectively, being in accordance with theoretical values (Figure 2B). This implies that the addition of milk to the matrix and the combination with other mycotoxins may reduce the intestinal transport of AFB1.

The ENNB transport profile remained consistent across all situations, following the same trend regardless of the matrix or whether ENNB was isolated or in a mixture. ENNB isolated demonstrated a similar trend, whether in breakfast cereal alone (Figure 2C) or with semi-skimmed milk (Figure 2D); the same trend was verified when ENNB was in mixture, achieving maximum percentages of transport around 80 and 72% when isolated and lower ones, around 53 and 62%, when combined. This decrease in transport percentage suggests that combination with other mycotoxins may hinder the intestinal transport of ENNB, as verified already for AFB1.

In the case of STG, the transport rate showed linearity in the breakfast cereal with semi-skimmed milk, rapidly transferred to the receiver compartment whether transported in isolation or in a mixture, resulting in very high transport percentages (92 and 96%, respectively), and only displaying values similar to theoretical transport up to 60 min (Figure 2F). When absorption occurred only in the breakfast cereal matrix (Figure 2E), a slower rate was observed until 120 min, followed by a more rapid transport until 180 min, achieving maximum transport percentages of 38 and 28%, whether isolated or in a mixture, respectively. This suggests that the addition of milk to the matrix may increase the intestinal transport of STG, when the molecule is not in the presence of AFB1 and ENNB.

The absorption of mycotoxins in the intestine is a complex process influenced by several factors, including the type of molecule, the presence of other substances, and the physiological conditions. When mycotoxins are present in a mixture, several mechanisms may contribute to a decrease in their intestinal absorption. As mentioned before, the food matrix can affect their absorption, due to the presence of other food components, such as fats, carbohydrates, proteins, fibers, and various other nutrients, that may influence the solubility and absorption of mycotoxins. Moreover, different mycotoxins may compete for the same transporters or binding sites in the intestine. If these binding sites become saturated with one mycotoxin, it may reduce the absorption of others, leading to a lower overall absorption rate [38].

In addition to detailing the mycotoxin transport over time, we calculated the apparent permeability (P_{app}) to provide a precise estimation of mycotoxin intestinal absorption. As shown in Figure 2, overall, all mycotoxins exhibited rapid initial absorption, followed by a reduction. This suggests that during the assay, sink conditions were not always adequately confirmed, deviating from a linear fit of mycotoxin content [39]. Consequently, permeability values were computed for non-sink conditions using the equation outlined in Section 4.6 [39,40]. According to Tavelin et al., 2002, P_{app} values obtained under this condition align more closely with the actual permeability coefficients of epithelial cell monolayers in vivo. In Table 2, we present the calculated P_{app} values derived from the curves depicting the transport of mycotoxins, as illustrated in Figure 2.

Table 2. Apparent permeabilities ($\times 10^{-6}$ cm/s) of AFB1, ENNB, and STG isolated and in mixture in the apical → basolateral direction. MB (%) shows the mass balance recoveries of transport experiments and the calculated human fraction absorbed (FA %) for each mycotoxin.

			P_{app} ($\times 10^{-6}$ cm/s)	MB (%)	FA (%)
Breakfast cereal	**AFB1**	Isolated	94.4 ± 0.1	156 ± 16	98.8 *
		Mix	3.31 ± 1.10	28.3 ± 2.0	66.1 *
	ENNB	Isolated	83.2 ± 0.0	145 ± 20	98.6
		Mix	78.6 ± 0.0	99.6 ± 2.0	98.5
	STG	Isolated	37.6 ± 0.0	72.6 ± 1.9	96.7
		Mix	25.4 ± 0.0	58.7 ± 3.6	94.9
Breakfast cereal w/semi-skimmed milk	**AFB1**	Isolated	7.35 ± 0.00	29.2 ± 1.4	82.6
		Mix	8.18 ± 0.00	36.7 ± 2.6	84.2
	ENNB	Isolated	91.6 ± 0.0	131 ± 12	98.7
		Mix	72.5 ± 0.0	115 ± 0.3	98.4
	STG	Isolated	81.2 ± 0.2	172 ± 11	98.6
		Mix	100.0 ± 0.2	167 ± 18	98.9

Abbreviations: AFB1—aflatoxin B1; ENNB—enniatin B; STG—sterigmatocystin. Data expressed as mean ± SD (n = 3). * significant difference ($p < 0.05$) between isolated and mixed transport of mycotoxins.

Artursson et al., 2001 suggested that permeability coefficients exceeding 1×10^{-6} cm/s indicate high permeability [41]. Therefore, the P_{app} values derived for mycotoxins imply that AFB1, ENNB, and STG were effectively absorbed in Caco-2/HT-29MTX cells, whether individually or in combination. AFB1 exhibits the highest apparent permeability value in the breakfast cereal matrix when isolated, followed by ENNB and finally STG. However, in the cereals with milk, AFB1 shows the lowest apparent permeability value, with ENNB having the highest. Regarding the permeability values when mycotoxins are in a mixture, in cereal matrix, ENNB demonstrates the highest value, while AFB1 has the lowest result. Overall, the simultaneous transport of mycotoxins led to a significant decrease in the P_{app}, with the exception of STG in breakfast cereal with semi-skimmed milk matrix (Table 2).

Our results indicate differences in the P_{app} of mycotoxins between the two matrices. The presence of milk alongside breakfast cereals demonstrated a notable impact on the absorption kinetics. This finding suggests that the matrix composition plays a crucial role in modulating mycotoxin bioavailability, and the inclusion of milk can be a contributing factor. The observed variations in P_{app} values between the breakfast cereal only and breakfast cereal with milk matrices may be attributed to the complex interplay of components present in milk, as stated above for bioaccessibility differences. Milk contains various bioactive molecules, such as proteins, fats, and carbohydrates, which could potentially interact with mycotoxins, affecting their transport across the intestinal barrier. The specific mechanisms behind these interactions warrant further investigation to fully understand the dynamics of mycotoxin absorption in the presence of milk.

Concerning the mass balance values in both matrices (i.e., the total mycotoxin recovered from the apical and basolateral compartments at the final of the experiment divided by the initial amount in the apical compartment), they varied between 28 and 156% for breakfast cereals and 29 and 172% for breakfast cereals with semi-skimmed milk (Table 2). These values, namely, the lower ones, could be due to some reasons, such as the mycotoxin having been adsorbed to the experimental apparatus (such as the plastic of the well plate, the insert device, or the membrane itself), the compound being held within the cells or in the cell membranes, undergoing metabolic processes, compound precipitation, or degradation during incubation [40,42].

The Caco-2 cell model is frequently employed for predicting the human fraction absorbed (FA%), which is the fraction of a drug absorbed in humans following oral administration [43]. This is achieved by establishing correlations between the apparent permeabilities in the apical-to-basolateral (AB) direction across Caco-2 cell monolayers of molecules and the experimental human fraction absorbed data. The outcome is a sigmoidal relationship between the human fraction absorbed and the logarithm of the apparent permeability (log P_{app}) of molecules [42]. Therefore, by establishing this correlation, the *in vitro* permeability of a compound in Caco-2 cells can serve as a predictive measure for human absorption. Besides being validated by *in vivo* comparison using several drugs, FA% values are deemed more intuitive than P_{app} values when comprehending the intestinal absorption of compounds, providing the added advantage of categorizing permeability into low (0–20%), medium (20–80%), and high (80–100%) absorption. Figure 3 illustrates the FA% of each of the assayed conditions in both matrices under study, showcasing their position on the sigmoidal curve based on their percentage of absorption. AFB1 exhibits the highest FA% when isolated, with a percentage of 98.8% in breakfast cereal but a lower value of 82.6% in breakfast cereal with milk. Following this, ENNB demonstrates FA% values of 98.6% and 98.7% (breakfast cereal and breakfast cereal with milk, respectively), while STG ranges between 96.7% (breakfast cereal) and 98.5% (breakfast cereal with milk). Concerning the transport of mycotoxins in a mixture, STG attains the highest FA% value of 98.9% in breakfast cereal with milk, slightly lower at 94.9% in breakfast cereal alone. ENNB follows with values of 98.5% (breakfast cereal) and 98.4% (breakfast cereal with milk), and AFB1 exhibits percentages of 66.1% and 84.2%, respectively. Consequently, all mycotoxins are categorized as highly absorbed, whether isolated or in a mixture, in both matrices under study. The only exception is AFB1 in the mixture within the breakfast cereal matrix, showing a medium level of absorption (Figure 3). This can lead us to the conclusion that overall, the effect of the milk matrix did not influence the final correspondent predicted FA% in humans, being, overall, all the analyzed molecules highly absorbed in the gastrointestinal tract. On the other side, the presence of other mycotoxins did have an impact in AFB1 absorption, leading to a relevant decrease in absorption from high to medium. Extrapolating P_{app} values to FA% revealed that only substantial differences in P_{app} values significantly impact a compound's *in vivo* absorption.

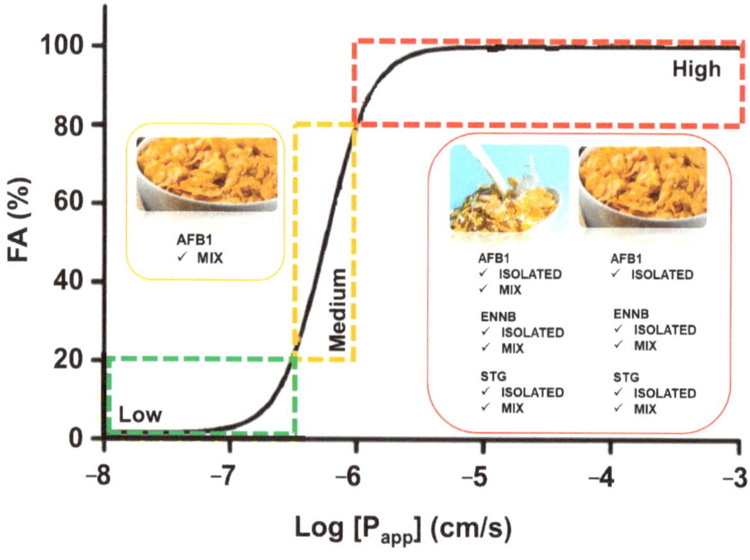

Figure 3. AFB1, ENNB, and STG—isolated and in mixture, in breakfast cereals and breakfast cereals with semi-skimmed milk—positions in the sigmoidal curve according to their FA%—high (red), medium (orange), and low (green) absorption. The sigmoidal curve was built according to (Skolnik et al., 2010 [42]). Data expressed as mean ± SD of 3 independent experiments (n = 3). AFB1—aflatoxin B1; ENNB—enniatin B; and STG—sterigmatocystin.

Transport mechanisms across absorptive epithelia can manifest through various routes, including passive pathways such as transcellular and/or paracellular transport, active transcellular transport mediated by transporters, or transcytosis [38,39]. Mycotoxins are absorbed through the small intestine primarily via passive diffusion, exhibiting a remarkably high absorption rate [44]. The available data (*in vivo* rat whole small intestine) indicate that nearly total absorption of AFB1 can occur within the intestinal tract [45]. Nevertheless, there is a lack of existing literature on the intestinal absorption of AFB1, ENNB, and STG in Caco-2/HT29MTX cell co-cultures and in the presence of the digested food matrix. Only one study has been conducted, focusing on the impact of AFB1 metabolite (AFM1) in these co-cultures [27], and one study *in vitro* using Caco-2 cells for AFB1 [46]; however, numerous studies have been dedicated to utilizing Caco-2 cells for assessing the absorption and toxicity of other crucial mycotoxins like trichothecenes and zearalenone [47–49]. Therefore, this study is highly significant as it aims to contribute with more data, enabling a more precise determination of risk assessment.

When simultaneously considering values of both intestinal bioaccessibility and predicted fraction absorbed, in percentage, data show (Figure 4) that the inclusion of milk imparts an overall decrease in bioaccessibility (all below 50%) and AFB1 appears as the least absorbed (although still in high percentage, as stated above) with the exception of the case in which it is ingested in the cereal matrix and as a single compound. Overall, the effect of the inclusion of milk in the ingested matrix has more impact than substances ingested as a mixture or single mycotoxin. When taken together the bioaccessibility and FA%, which can be correlated with the final bioavailability, it can be verified that the most bioaccessible and absorbed toxin is ENNB (ingested with cereals only) and that the least bioavailable is AFB1 ingested cereals with milk.

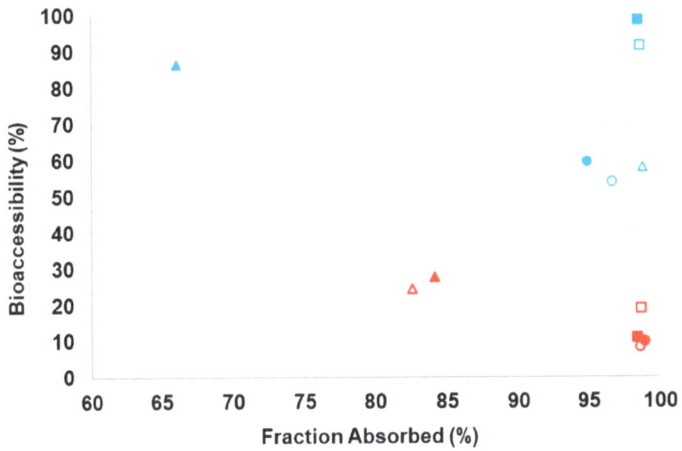

Figure 4. Graphical representation of samples' bioaccessibility and fraction absorbed. AFB1: triangles; ENNB: squares; STG: circles. Solid triangles, squares, and circles are related to mycotoxin mixture, and empty triangles, squares, and circles are related to isolated mycotoxins. **Blue** color indicates values from samples with breakfast cereals only and **red** with both cereals and milk. Data expressed as mean (n = 3).

3. Conclusions

This study emphasizes the importance of examining the combined ingestion and transport of mycotoxins to comprehensively understand their absorption and more accurately assess human exposure. This is particularly crucial considering the frequent co-occurrence and potential for simultaneous exposure to these toxins in breakfast cereals.

The *in vitro* gastrointestinal method revealed varying patterns of bioaccessibility for AFB1, ENNB, and STG in breakfast cereals, including those consumed with different milks and soy beverages. An overall upward trend in mycotoxin bioaccessibility was observed from the gastric to intestinal stages, except for ENNB, which showed variable behavior depending on the milk type. The overall bioaccessibility ranking, from highest to lowest, was digested breakfast cereal, breakfast cereal with semi-skimmed milk, breakfast cereal with lactose-free milk, and breakfast cereal with soy beverage. Bioaccessibility percentages varied widely across all samples and conditions, with AFB1 ranging from 3.1% to 86.2%, STG from 1.5% to 59.3%, and ENNB from 0.6% to 98.2%. Interestingly, ENNB displayed the highest bioaccessibility values, regardless of whether it was present alone or in combination.

Regarding intestinal transport, ENNB and STG exhibited the highest absorption rates when present together in a mixture, irrespective of the matrix. When isolated, their absorption rates were even higher, particularly in the breakfast cereal with milk matrix. Notably, AFB1 demonstrated the highest absorption rate within the isolated breakfast cereal matrix.

This study's findings are significant because they provide novel insights into the bioaccessibility and intestinal absorption of mycotoxins in breakfast cereals. It uniquely considered both bioaccessibility and intestinal transport simultaneously within a real-world food matrix, mimicking realistic consumption scenarios. This information can be used to develop strategies for mitigating mycotoxin exposure risk in children. For instance, parents could be advised to choose cereals with milk, potentially reducing exposure, particularly to the highly toxic and prevalent AFB1.

4. Materials and Methods

4.1. Reagents and Materials

AFB1 (10 mg, >98% purity) was purchased from LGC (Teddington, Middlesex, UK). ENNB (1 mg, ≥95% purity), porcine α-amylase, pepsin, and bile and pancreatin extracts were all purchased from Sigma-Aldrich corp. (St. Louis, MO, USA). STG (1 mg, ≥98% purity) was purchased from BioViotica (Liestal, Switzerland). Acetonitrile (ACN), methanol (MeOH), and acetic and formic acids of high-performance liquid chromatography (HPLC) grade, as well as ammonium acetate (P.A.), were obtained from Merck (Darmstadt, Germany). Anhydrous magnesium sulphate ($MgSO_4$) and bovine serum albumin (BSA) were purchased from Sigma-Aldrich and sodium chloride (NaCl) from VWR (Stříbrná Skalice, Czech Republic); both were calcinated at 500 °C for 5 h before use. Ultrapure water, purified with a "Seral" system (SeralPur Pro 90 CN), was used. MTT (3-(4,5-dimethylthiazol-2-yl)-2,5-diphenyltetrazolium bromide) and dimethyl sulfoxide (DMSO) were purchased from Duchefa Biochemie (Haarlem, The Netherlands). Fetal bovine serum (FBS), 0.25% trypsin solution, minimum essential medium non-essential amino acids (MEM NEAA) 100×, GlutaMAX™ 100×, and penicillin/streptomycin 100× solution (10,000 Units mL^{-1}/10 mg mL^{-1}) were all purchased from Gibco/Life technology corporation (Paisley, UK). High-glucose Dulbecco's modified Eagle's medium (DMEM) and Hanks Balanced Salt Solution (HBSS) were purchased from Biowest (Nuaillé, France). Syringe filters (PES, 0.22 µm and 0.45 µm pore) were purchased from TPP (Zollstarsse, Switzerland). Transwell inserts and plates (PS/PET membrane, 24 mm diameter, 0.4 µm pore size) for the transport assay were obtained from cellQART (Northeim, Germany). Standard stock solutions of each mycotoxin at 10 and 100 mg/L were prepared by diluting them in methanol for LC–MS/MS validation purposes and spiking of samples, respectively. Stock internal solutions of OTA-d5 (500 mg/L) and $^{13}C_{18}$-STG (25 mg/L) were prepared in DMSO and ACN, respectively. All standard solutions were stored at −18 °C.

4.2. Cell Culture

Human-derived intestinal cells Caco-2 and HT-29MTX were kindly provided by the Fisico-Química Molecular group from University of Coimbra and from Faculty of Sciences University of Porto, respectively.

Cells were grown in 75 cm^2 flasks, at 37 °C with 5% CO_2, in complete medium (CM) with the following composition [22]: DMEM with 10% heat inactivate FBS, 1% penicillin/streptomycin, 1% non-essential amino acids (NEAA), and 1% glutamax. Cytotoxicity and permeability assays were performed under passages 73–76 and 74–77 for HT-29MTX and Caco-2 cells, respectively.

4.3. Spiking Samples

Before digestion, 16 µL of a 100 mg/L solution of each mycotoxin (400 µg/L), AFB1, ENNB, and STG, was used to spike 4 g of breakfast cereal homogeneously and left to incubate for 15 min at room temperature. Then, 16.67 mL of the different milks (semi-skimmed, semi-skimmed lactose free, and soy vegetable drink) was added. Each mixture was performed in triplicate (5 g each). The assay with only breakfast cereals (5 g) was spiked with 20 µL of a 100 mg/L solution of each mycotoxin (400 µg/L), homogeneously, and left to incubate during 15 min at room temperature.

The nutritional compositions of the previously mentioned breakfast cereal, milk, and soy beverage samples are presented in Table 3.

Table 3. Nutritional properties of the breakfast cereal, milks, and soy beverage used for *in vitro* digestion (as reported in nutritional labels).

Breakfast Cereal Sample	Fat	Fiber	Protein	Carbohydrates
	(g/100 g)	(g/100 g)	(g/100 g)	(g/100 g)
Breakfast cereal with corn balls with organic honey	0.6	3.0	5.5	82.0
Beverages samples	**(g/100 mL)**	**(g/100 mL)**	**(g/100 mL)**	**(g/100 mL)**
Semi-skimmed milk	1.6	-	3.4	4.9
Semi-skimmed lactose-free milk	1.6	-	3.3	4.9
Soy beverage	1.8	0.5	3.0	2.5

4.4. In Vitro Digestion

The *in vitro* digestion procedure was performed according to the internationally standardized method described by INFOGEST 2.0 [50]. Briefly, 5 g of sample (breakfast cereals and breakfast cereals with milk (semi-skimmed, semi-skimmed lactose-free, and soy vegetable drink)) was mixed with 4 mL of simulated salivary fluid (SSF), 0.5 mL of α-amylase solution at 6.04 U/mg in water, 25 µL 0.3 M $CaCl_2$ solution, and 475 µL water. After 2 min incubation, the mixture was mixed with 8 mL of simulated gastric fluid (SGF), 5 µL 0.3 M $CaCl_2$, 35 µL (in the case of breakfast cereals samples) or 88 µL (in the case of breakfast cereals with milks) of 6 M HCl to adjust to pH = 3, 1.460 mL (in the case of breakfast cereals samples) or 1.407 mL (in the case of breakfast cereals with milks) of water, and 0.5 mL of pepsin (2668.2 U/mg). The gastric mixture was then incubated at 37 °C for 120 min in an orbital shaker–incubator (ES-20, BioSan, Riga, Latvia) with integrated horizontal shaker at 250 rpm. Then, the gastric chime (10 mL) was mixed with 4.25 mL of simulated intestinal fluid (SIF) solution, 20 µL 0.3 M $CaCl_2$, 40 µL (in the case of breakfast cereals samples) or 75 µL (in the case of breakfast cereals with milks) of 1 M NaOH to adjust the pH to 7.0, 1.940 mL (in the case of breakfast cereals samples) or 1.905 mL (in the case of breakfast cereals with milks) of water, 2.5 mL of pancreatin solution (4.7 U/mg based on trypsin activity), and 1.25 mL of bile solution (2.151 mmol/g). All digestions were made in triplicates plus one blank sample used to adjust the pH and used as a matrix for the LC–MS/MS calibration assays as well as for transport assays. After digestion, the samples were immediately centrifuged at 3000× *g* for 5 min to obtain the bioaccessible fraction (supernatant). Supernatants were then frozen at −20 °C until further analysis.

4.5. Cytotoxic Assay

The MTT assay was conducted to assess the impact on cell viability to the tested toxicants prepared in control bioaccessible fractions from *in vitro* digestion after suffering a heat treatment at 98 °C for 5 min then filtered throughout 0.45 and 0.22 μm filters and diluted with HBSS. Proliferating Caco-2 (45,000 cells/mL) and HT-29MTX (5000 cells/mL) were seeded in 96-well plates (TPP; Trasadingen, Switzerland), allowing one week for cell adherence at 37 °C, 5% CO_2, and then exposed to mycotoxins at the concentrations used for transport assay: 0.31 μM (ENNB), 0.62 μM (STG), and 0.64 μM (AFB1) for 3 h. Cells treated with CM alone, and with HBSS, were used as a control on cell viability during the experiments. The medium was removed, after 180 min of exposure, and 100 μL of freshly prepared MTT solution (0.5 mg/mL) was added to each well and the plates placed in the incubator at 37 °C, 5% CO_2 for 20 min, and after that was added to each well 100 μL of DMSO and let the plates rest for another 30 min before absorbance reading at 570 nm using an absorbance plate reader (SPECTROstar Nano, BMG Labtech, Offenburg, Germany). Results were expressed as % of cell viability.

4.6. Transport Assay

Caco-2/HT-29MTX co-culture [22] cells were seeded at a density of 1×10^5 cells/cm^2 (ratio 90:10, Caco-2/HT-29MTX) in 24 mm 6-well Transwell inserts with a pore size of 0.4 μm and a growth area of 4.5 cm^2 (cellQART®, Northeim, Germany—ref.: 9300404). Tests were carried out with 2.5 mL in the basolateral compartment and 1.5 mL in the apical compartment. During the differentiation process, the culture medium was changed every two days, and the cells were used for the transport assay on the 26th day after seeding. For the transport experiments, the culture medium of the Caco-2/HT-29MTX co-culture cells was replaced with HBSS containing 0.5% (*w*/*v*) BSA in the basolateral compartment. BSA was added to minimizing the binding of compounds to the plastic surfaces [40].

Mycotoxins at the initial concentrations of 0.31 μM (ENNB), 0.62 μM (STG), and 0.64 μM (AFB1) (200 μg/L each) each were prepared in control bioaccessible fractions (digested samples without mycotoxins) from *in vitro* digestion after suffering a heat-shock treatment at 98 °C for 5 min, to inhibit the enzymes in order to maintain the cells' viability [51], then filtered through 0.45 and 0.22 μm filters, sequentially, diluted with HBSS 1:8, and introduced in the apical compartments isolated or in mixture. At time points of 15, 30, 60, 120, and 180 min of absorption, 200 μL samples were taken from the basolateral compartments, and the same volume (200 μL) of HBSS was added to maintain the volume. The concentration of mycotoxins used for the absorption assay was selected to allow for the quantification of mycotoxins at all stages of the transport assay. The trans-epithelial electrical resistance (TEER) values were measured at 37 °C using a Millicell ERS-2 Voltohmmeter (Merck Millipore, Darmstadt, Germany) at the beginning and end of the experiment after washing the cell monolayer with HBSS to assess barrier integrity. The resistance was expressed in Ωcm^2, calculated by multiplying the cell monolayer resistance (Ω) by the filter area (cm^2). All transport assays were conducted in triplicate.

The assessment of mycotoxin intestinal permeability involved measuring transport rates across Caco-2/HT-29MTX cell monolayers, as outlined in previous studies [40]. The permeability coefficient (P_{app}) in the apical → basolateral direction was determined using the following equation, specifically designed for experiments conducted under non-sink conditions [39,40]:

$$C_R(t) = \frac{M}{V_D + V_R} + \left(C_{R(0)} - \frac{M}{V_D + V_R}\right) e^{-PappA\left(\frac{1}{VD} + \frac{1}{VR}\right)t}$$

In this context, $C_R(t)$ represents the mycotoxin concentration in the receiver compartment over time, M denotes the total mycotoxin amount within the system, V_D and V_R stand for the volumes of the donor and receiver compartments, respectively. C_{R0} signifies the initial concentration of the mycotoxin in the receiver compartment at the onset of the time interval, A represents the surface area of the filter, and t signifies the elapsed time from the commencement of the interval. The permeability coefficient (P_{app}) was derived through non-linear regression [39]. The mass balance was calculated through the following equation:

$$\text{Mass balance } (\%) = \frac{\left[\left(V_R \times C_{R(final)}\right) + \left(V_D \times C_{D(final)}\right)\right]}{V_{D(0)} \times C_{D(0)}} \times 100$$

C_R and C_D represent concentrations on the receiver (R) and donor (D) sides of the monolayer at the experiment's onset (0) and conclusion (final), with V denoting the respective volumes. The apical-to-basolateral permeability (P_{app}) data obtained from Caco-2/HT-29MTX cells were employed to calculate the human fraction absorbed, FA (%), using the non-linear regression model outlined by Skolnik et al., 2010 [42].

$$\text{FA } (\%) = \frac{100}{1 + e^{((-5.74 - \text{PappX})/0.39)}}$$

In this context, 100 is defined as the sum of the minimum and maximum values of % FA, restricted within the range of 1 to 100%. The log $P_{appA \rightarrow B}$ value at 50% absorption in humans is represented as −5.74, while P_{appX} signifies the log $P_{appA \rightarrow B}$ for mycotoxins in Caco-2/HT-29MTX cells, as determined in the current study. Additionally, 0.39 corresponds to the slope derived from the model fit.

4.7. *Mycotoxins Extraction and Cleanup*

Bioaccessible and Bioavailable Fractions

Two hundred microliters (200 µL) of digested sample obtained as described in Section 2.3 were transferred into a conical microtube and a fixed concentration of OTA-d5 (40 µg/L) was added. Thereafter, 200 µL acidified ACN with 5% formic acid (*v*/*v*) were added along with 70 mg of $MgSO_4$ anhydrous salt and 10 mg of NaCl and the tube was immediately vortexed for 10 s to prevent agglomeration of the salts. The tubes were then centrifuged at 13,000 rpm for 5 min to induce phase separation and mycotoxins partitioning. The organic phase was transferred to a 2 mL vial, evaporated to dryness under a stream of nitrogen, and finally reconstituted in 150 µL of mobile phase B (Section 4.5) and analyzed by LC–MS/MS. Each sample was injected three times.

4.8. *Instrument and Analytical Conditions*

MS/MS analysis was performed on a Quattro Micro triple quadrupole mass spectrometer (Waters, Manchester, UK) interfaced with a high-performance liquid chromatography (HPLC) system Waters Alliance 2695 (Waters, Milford). A Kinetex C18 2.6 µm particle size analytical column (150 × 4.6 mm) with pre-column from Phenomenex (Tecnocroma, Portugal), maintained at 35 °C, was used for chromatographic separation. A gradient elution was performed using a mobile phase (300 µL/min) constituted by a phase A (water/methanol/acetic acid, 94:5:1 (*v*/*v*) and 5 mM ammonium acetate) and a phase B (methanol/water/acetic acid, 98:2 (*v*/*v*) and 5 mM ammonium acetate). The solvent gradient was as follows: 0–7.0 min, 95% A; 7.0–11.0 min, 35% A; 11.0–13.0 min, 25% A; 13.0–15.0 min, 0% A; 15.0–24.0 min 95% A; and 24.0–27.0 min, 95% A. MS/MS acquisition was operated in positive-ion mode with multiple reaction monitoring (MRM), the collision gas was Argon 99.995% (Gasin, Leça da Palmeira, Portugal) with a pressure of 2.9×10^{-3} mbar in the collision cell. Capillary voltages of 3.0 KV were used in the positive ionization mode. Nitrogen was used as desolvation gas and cone gas being the flows of 350 and 60 L/h, respectively. The desolvation temperature was set to 350 °C and the source temperature to 150 °C. Dwell times of 0.1 s/scan were selected. The data were collected using the software program MassLynx 4.1. For each analyte, two transitions were selected for identification, and the corresponding cone voltage and collision energy were optimized for maximum intensity. The optimized MS/MS parameters for the target analytes are listed in Table 4.

Table 4. MS/MS parameters for each mycotoxin under study.

Mycotoxins	Retention Time (min)	MRM Transition (*m/z*)		CV (V)		CE (eV)	
		QIT (*m/z*)	CIT (*m/z*)	QIT	CIT	QIT	CIT
AFB1	15.61	313 > 241.4	313 > 285.3	47	47	36	23
OTA-d5 (IS1)	19.14	409 > 239	409 > 363	32	32	22	22
STG	20.19	325 > 254	325 > 310	35	40	35	25
$^{13}C_{18}$-STG (IS2)	20.19	342.7 > 297.4	342.7 > 326.7	24	24	28	28
ENNB	21.63	663 > 218	663 > 336	60	60	70	70

AFB1—aflatoxin B1; CE—collision energy; CIT—confirmation ion transition; CV—cone voltage; ENNB—enniatin B; OTA-d5—ochratoxin d5; QIT—quantification ion transition; STG—sterigmatocystin; $^{13}C_{18}$-STG—$^{13}C_{18}$ sterigmatocystin.

4.9. Method Validation and Quality Control

Linearity was assessed through matrix-matched calibration curves consisting of six calibration points spanning the range from 25 to 400 µg/L (bioaccessibility assays) and from 12.5 to 200 µg/L (bioavailability assays). Precision was evaluated by examining repeatability (intraday precision) and reproducibility (interday precision) of a spiked sample at three different concentration levels. This assessment was conducted using five replicates (with two injections for each replicate) on each precision day.

The limit of detection (LOD) and limit of quantification (LOQ) were established by repeatedly analyzing chromatographic extracts of sample solutions spiked with decreasing amounts of the analytes until signal-to-noise ratios of 3:1 and 10:1 were achieved, respectively.

4.10. Statistical Analysis

XLSTAT for Windows 11 Pro version 23H2 (Addinsoft, Paris, France) was used for statistical analysis. The normal distribution of variables was assessed through the Shapiro–Wilk test. Mean comparisons were conducted using two-way ANOVA, with a significance level of 5%. Data are presented as mean ± SD from three independent experiments. GraphPad Prism version 9.3.1 for Windows (Graphpad Software, La Jolla, CA, USA) was employed to generate all graphs.

Supplementary Materials: The following supporting information can be downloaded at: https://www.mdpi.com/article/10.3390/toxins16050205/s1, Table S1: Regression equations and coefficients of correlation (r) for the mycotoxins under study in bioaccessible fractions; Table S2: Recoveries (%), repeatability and reproducibility (%CV), limits of detection (LOD), and quantification (LOQ) of the analytical method used to determine the mycotoxins under study in the bioaccessible fractions (gastric and intestinal) after *in vitro* digestion; Table S3: Regression equations and coefficients of correlation (r) for the mycotoxins under study in transport assay.

Author Contributions: S.V.M.d.S.: Writing—Original Draft, Investigation, Data Curation, Formal analysis, Visualization. J.O.F.: Writing—Reviewing and Editing, Resources, Funding acquisition. M.A.F.: Conceptualization, Formal analysis, Writing—Review and Editing, Supervision, Funding acquisition. S.C.C.: Conceptualization, Writing—Review and Editing, Supervision, Funding acquisition. All authors have read and agreed to the published version of the manuscript.

Funding: This work received financial support from FCT/MCTES (UIDB/50006/2020 DOI 10.54499/UIDB/50006/2020) through national funds.

Institutional Review Board Statement: Not applicable.

Informed Consent Statement: Not applicable.

Data Availability Statement: No new data were created or analysed in this study. Data sharing is not applicable to this article.

Acknowledgments: This work received support and help from FCT/MCTES (LA/P/0008/2020 DOI 10.54499/LA/P/0008/2020, and UIDP/50006/2020 DOI 10.54499/UIDP/50006/2020, through national funds. Soraia Sá thanks FCT/MCTES (Fundação para a Ciência e Tecnologia and Ministério da Ciência, Tecnologia e Ensino Superior) for PhD grant ref. 10.54499/SFRH/BD/143928/2019 (https://doi.org/10.54499/SFRH/BD/143928/2019). Sara C. Cunha acknowledges 2022.07841.CEEC IND/CP1724/CT0014- FCT contract. Miguel A. Faria acknowledges FCT the researcher contract (https://doi.org/10.54499/DL57/2016/CP1346/CT0036).

Conflicts of Interest: The authors declare that they have no known competing financial interests or personal relationships that could have appeared to influence the work reported in this paper.

References

1. Diana Di Mavungu, J.; Monbaliu, S.; Scippo, M.-L.; Maghuin-Rogister, G.; Schneider, Y.-J.; Larondelle, Y.; Callebaut, A.; Robbens, J.; Van Peteghem, C.; De Saeger, S. LC-MS/MS multi-analyte method for mycotoxin determination in food supplements. *Food Addit. Contam.* **2009**, *26*, 885–895. [CrossRef] [PubMed]
2. Sá, S.V.d.; Monteiro, C.; Fernandes, J.O.; Pinto, E.; Faria, M.A.; Cunha, S.C. Emerging mycotoxins in infant and children foods: A review. *Crit. Rev. Food Sci. Nutr.* **2021**, *63*, 1707–1721. [CrossRef] [PubMed]
3. EFSA. Risk assessment of aflatoxins in food. *EFSA J.* **2020**, *18*, e06040.
4. International Agency for Research on Cancer, IARC. 2006. Available online: http://www.iarc.fr (accessed on 11 March 2024).
5. HMDB. Human Metabolome Database. Available online: https://hmdb.ca/metabolites (accessed on 11 March 2024).
6. Vaclavikova, M.; Malachova, A.; Veprikova, Z.; Dzuman, Z.; Zachariasova, M.; Hajslova, J. 'Emerging'mycotoxins in cereals processing chains: Changes of enniatins during beer and bread making. *Food Chem.* **2013**, *136*, 750–757. [CrossRef] [PubMed]
7. Mihalache, O.A.; De Boevre, M.; Dellafiora, L.; De Saeger, S.; Moretti, A.; Pinson-Gadais, L.; Ponts, N.; Richard-Forget, F.; Susca, A.; Dall'Asta, C. The Occurrence of Non-Regulated Mycotoxins in Foods: A Systematic Review. *Toxins* **2023**, *15*, 583. [CrossRef] [PubMed]
8. Raiola, A.; Tenore, G.C.; Manyes, L.; Meca, G.; Ritieni, A. Risk analysis of main mycotoxins occurring in food for children: An overview. *Food Chem. Toxicol.* **2015**, *84*, 169–180. [CrossRef] [PubMed]
9. Rebellato, A.P.; dos Santos Caramês, E.T.; Pallone, J.A.L.; de Oliveira Rocha, L. Mycotoxin bioaccessibility in baby food through *in vitro* digestion: An overview focusing on risk assessment. *Curr. Opin. Food Sci.* **2021**, *41*, 107–115. [CrossRef]
10. Raiola, A.; Meca, G.; Mañes, J.; Ritieni, A. Bioaccessibility of deoxynivalenol and its natural co-occurrence with ochratoxin A and aflatoxin B1 in Italian commercial pasta. *Food Chem. Toxicol. Int. J. Publ. Br. Ind. Biol. Res. Assoc.* **2012**, *50*, 280–287. [CrossRef] [PubMed]
11. Battilani, P.; Palumbo, R.; Giorni, P.; Dall'Asta, C.; Dellafiora, L.; Gkrillas, A.; Toscano, P.; Crisci, A.; Brera, C.; De Santis, B. Mycotoxin mixtures in food and feed: Holistic, innovative, flexible risk assessment modelling approach: MYCHIF. *EFSA Support. Publ.* **2020**, *17*, 1757E. [CrossRef]
12. EFSA Panel on Contaminants in the Food Chain (CONTAM). Scientific Opinion on the risk for public and animal health related to the presence of sterigmatocystin in food and feed. *EFSA J.* **2013**, *11*, 3254. [CrossRef]
13. Logrieco, A.; Moretti, A.; Castella, G.; Kostecki, M.; Golinski, P.; Ritieni, A.; Chelkowski, J. Beauvericin Production by Fusarium-Species. *Appl. Environ. Microbiol.* **1998**, *64*, 3084–3088. [CrossRef]
14. Thrane, U. Development in the taxonomy of Fusarium species based on secondary metabolites. In *Fusarium: Paul E. Nelson Memorial Symposium*; APS Press: Eagan, MN, USA, 2001; pp. 29–49.
15. Santini, A.; Raiola, A.; Meca, G.; Ritieni, A. Aflatoxins, ochratoxins, trichotecenes, patulin, fumonisins and beauvericin in finished products for human consumption. *J. Clin. Toxicol* **2015**, *5*, 265–276.
16. Wätjen, W.; Debbab, A.; Hohlfeld, A.; Chovolou, Y.; Kampkötter, A.; Edrada, R.A.; Ebel, R.; Hakiki, A.; Mosaddak, M.; Totzke, F. Enniatins A1, B and B1 from an endophytic strain of Fusarium tricinctum induce apoptotic cell death in H4IIE hepatoma cells accompanied by inhibition of ERK phosphorylation. *Mol. Nutr. Food Res.* **2009**, *53*, 431–440. [CrossRef]
17. Fernández-Blanco, C.; Font, G.; Ruiz, M.J. Interaction effects of enniatin B, deoxinivalenol and alternariol in Caco-2 cells. *Toxicol. Lett.* **2016**, *241*, 38–48. [CrossRef]
18. Prosperini, A.; Juan-García, A.; Font, G.; Ruiz, M.J. Reactive oxygen species involvement in apoptosis and mitochondrial damage in Caco-2 cells induced by enniatins A, A_1, B and B_1. *Toxicol. Lett.* **2013**, *222*, 36–44. [CrossRef]
19. Versantvoort, C.H.; Oomen, A.G.; Van de Kamp, E.; Rompelberg, C.J.; Sips, A.J. Applicability of an *in vitro* digestion model in assessing the bioaccessibility of mycotoxins from food. *Food Chem. Toxicol.* **2005**, *43*, 31–40. [CrossRef]
20. Lucas-González, R.; Viuda-Martos, M.; Pérez-Alvarez, J.A.; Fernández-López, J. In vitro digestion models suitable for foods: Opportunities for new fields of application and challenges. *Food Res. Int.* **2018**, *107*, 423–436. [CrossRef]
21. Fedi, A.; Vitale, C.; Ponschin, G.; Ayehunie, S.; Fato, M.; Scaglione, S. In vitro models replicating the human intestinal epithelium for absorption and metabolism studies: A systematic review. *J. Control. Release* **2021**, *335*, 247–268. [CrossRef]
22. Verhoeckx, K.; Cotter, P.; López-Expósito, I.; Kleiveland, C.; Lea, T.; Mackie, A.; Requena, T.; Swiatecka, D.; Wichers, H. *The Impact of Food Bioactives on Health: In Vitro and Ex Vivo Models*; Verhoeckx, K., Ed.; Springer International Publishing: Berlin/Heidelberg, Germany, 2015.

23. Assunção, R.; Martins, C.; Dupont, D.; Alvito, P. Patulin and ochratoxin A co-occurrence and their bioaccessibility in processed cereal-based foods: A contribution for Portuguese children risk assessment. *Food Chem. Toxicol.* **2016**, *96*, 205–214. [CrossRef]
24. Prosperini, A.; Meca, G.; Font, G.; Ruiz, M.J. Bioaccessibility of enniatins A, A_1, B, and B_1 in different commercial breakfast cereals, cookies, and breads of Spain. *J. Agric. Food Chem.* **2013**, *61*, 456–461. [CrossRef]
25. González-Arias, C.A.; Marín, S.; Sanchis, V.; Ramos, A. Mycotoxin bioaccessibility/absorption assessment using *in vitro* digestion models: A review. *World Mycotoxin J.* **2013**, *6*, 167–184. [CrossRef]
26. Romero-Sánchez, I.; Alonso-Núñez, I.; Gracia Lor, E.; Madrid, Y. Analysis and Evaluation of in Vitro Bioaccessibility of Aflatoxins B1, B2, G1 and G2 in Plant-Based Milks. *SSRN* 2024, *preprint*.
27. Wu, C.; Gao, Y.; Li, S.; Huang, X.; Bao, X.; Wang, J.; Zheng, N. Modulation of intestinal epithelial permeability and mucin mRNA (MUC2, MUC5AC, and MUC5B) expression and protein secretion in Caco-2/HT29-MTX co-cultures exposed to aflatoxin M1, ochratoxin A, and zearalenone individually or collectively. *Toxicol. Lett.* **2019**, *309*, 1–9. [CrossRef]
28. Pereira, C.; Cunha, S.C.; Fernandes, J.O. Mycotoxins of Concern in Children and Infant Cereal Food at European Level: Incidence and Bioaccessibility. *Toxins* **2022**, *14*, 488. [CrossRef] [PubMed]
29. IARC. *Some Traditional Herbal Medicines, Some Mycotoxins, Naphthalene and Styrene*; IARC Monographs on the Evaluation of Carcinogenic Risks to Humans; International Agency for Research on Cancer: Lyon, France, 2002; p. 36.
30. Sobral, M.M.C.; Cunha, S.C.; Faria, M.A.; Martins, Z.E.; Ferreira, I.M. Influence of oven and microwave cooking with the addition of herbs on the exposure to multi-mycotoxins from chicken breast muscle. *Food Chem.* **2019**, *276*, 274–284. [CrossRef] [PubMed]
31. Aoudia, N.; Callu, P.; Grosjean, F.; Larondelle, Y. Effectiveness of mycotoxin sequestration activity of micronized wheat fibres on distribution of ochratoxin A in plasma, liver and kidney of piglets fed a naturally contaminated diet. *Food Chem. Toxicol.* **2009**, *47*, 1485–1489. [CrossRef] [PubMed]
32. Aoudia, N.; Tangni, E.; Larondelle, Y. Distribution of ochratoxin A in plasma and tissues of rats fed a naturally contaminated diet amended with micronized wheat fibres: Effectiveness of mycotoxin sequestering activity. *Food Chem. Toxicol.* **2008**, *46*, 871–878. [CrossRef]
33. Kabak, B.; Dobson, A.D.; Var, I.l. Strategies to prevent mycotoxin contamination of food and animal feed: A review. *Crit. Rev. Food Sci. Nutr.* **2006**, *46*, 593–619. [CrossRef]
34. Feudjio, F.T.; Dornetshuber, R.; Lemmens, M.; Hoffmann, O.; Lemmens-Gruber, R.; Berger, W. Beauvericin and enniatin: Emerging toxins and/or remedies? *World Mycotoxin J.* **2010**, *3*, 415–430. [CrossRef]
35. Hurni, M.A.; Noach, A.; Blom-Roosemalen, M.; de Boer, A.G.; Nagelkerke, J.F.; Breimer, D.D. Permeability enhancement in Caco-2 cell monolayers by sodium salicylate and sodium taurodihydrofusidate: Assessment of effect-reversibility and imaging of transepithelial transport routes by confocal laser scanning microscopy. *J. Pharmacol. Exp. Ther.* **1993**, *267*, 942–950. [PubMed]
36. Kumar, P.; Gupta, A.; Mahato, D.K.; Pandhi, S.; Pandey, A.K.; Kargwal, R.; Mishra, S.; Suhag, R.; Sharma, N.; Saurabh, V. Aflatoxins in Cereals and Cereal-Based Products: Occurrence, Toxicity, Impact on Human Health, and Their Detoxification and Management Strategies. *Toxins* **2022**, *14*, 687. [CrossRef] [PubMed]
37. EC. *Commission Regulation (EC) No 1881_2006 (02006R1881—EN—01.01.2023—033.001)—Setting Maximum Levels for Certain Contaminants in Foodstuffs*; Official Journal of the European Union: Geneva, Switzerland, 2006.
38. Tran, V.N.; Viktorová, J.; Ruml, T. Mycotoxins: Biotransformation and bioavailability assessment using caco-2 cell monolayer. *Toxins* **2020**, *12*, 628. [CrossRef] [PubMed]
39. Tavelin, S.; Gråsjö, J.; Taipalensuu, J.; Ocklind, G.; Artursson, P. Applications of epithelial cell culture in studies of drug transport. *Epithel. Cell Cult. Protoc.* **2002**, 233–272.
40. Hubatsch, I.; Ragnarsson, E.G.; Artursson, P. Determination of drug permeability and prediction of drug absorption in Caco-2 monolayers. *Nat. Protoc.* **2007**, *2*, 2111–2119. [CrossRef] [PubMed]
41. Artursson, P.; Palm, K.; Luthman, K. Caco-2 monolayers in experimental and theoretical predictions of drug transport. *Adv. Drug Deliv. Rev.* **2001**, *46*, 27–43. [CrossRef] [PubMed]
42. Skolnik, S.; Lin, X.; Wang, J.; Chen, X.-H.; He, T.; Zhang, B. Towards prediction of *in vivo* intestinal absorption using a 96-well Caco-2 assay. *J. Pharm. Sci.* **2010**, *99*, 3246–3265. [CrossRef] [PubMed]
43. Tavelin, S.; Taipalensuu, J.; Söderberg, L.; Morrison, R.; Chong, S.; Artursson, P. Prediction of the oral absorption of low-permeability drugs using small intestine-like 2/4/A1 cell monolayers. *Pharm. Res.* **2003**, *20*, 397–405. [CrossRef] [PubMed]
44. Tomková, I.; Ševčíková, Z.; Levkut, M.; Revajová, V.; Čonková, E.; Laciaková, A.; Lenhardt, L. Effect of aflatoxin B 1 on CD3 T cells andalkaline phosphatase in the intestine of mice. *Mycopathologia* **2002**, *154*, 15–19. [CrossRef] [PubMed]
45. Ramos, A.; Hernandez, E. In situ absorption of aflatoxins in rat small intestine. *Mycopathologia* **1996**, *134*, 27–30. [CrossRef] [PubMed]
46. Sobral, M.M.C.; Faria, M.A.; Cunha, S.C.; Miladinovic, B.; Ferreira, I.M. Transport of mycotoxins across human gastric NCI-N87 and intestinal Caco-2 cell models. *Food Chem. Toxicol. Int. J. Publ. Br. Ind. Biol. Res. Assoc.* **2019**, *131*, 110595. [CrossRef] [PubMed]
47. Avantaggiato, G.; Havenaar, R.; Visconti, A. Evaluation of the intestinal absorption of deoxynivalenol and nivalenol by an *in vitro* gastrointestinal model, and the binding efficacy of activated carbon and other adsorbent materials. *Food Chem. Toxicol.* **2004**, *42*, 817–824. [CrossRef] [PubMed]
48. Bony, S.; Carcelen, M.; Olivier, L.; Devaux, A. Genotoxicity assessment of deoxynivalenol in the Caco-2 cell line model using the Comet assay. *Toxicol. Lett.* **2006**, *166*, 67–76. [CrossRef] [PubMed]

49. Wang, J.; Bakker, W.; Zheng, W.; de Haan, L.; Rietjens, I.M.; Bouwmeester, H. Exposure to the mycotoxin deoxynivalenol reduces the transport of conjugated bile acids by intestinal Caco-2 cells. *Arch. Toxicol.* **2022**, *96*, 1473–1482. [CrossRef] [PubMed]
50. Brodkorb, A.; Egger, L.; Alminger, M.; Alvito, P.; Assunção, R.; Ballance, S.; Bohn, T.; Bourlieu-Lacanal, C.; Boutrou, R.; Carrière, F. INFOGEST static *in vitro* simulation of gastrointestinal food digestion. *Nat. Protoc.* **2019**, *14*, 991–1014. [CrossRef] [PubMed]
51. Kondrashina, A.; Arranz, E.; Cilla, A.; Faria, M.A.; Santos-Hernández, M.; Miralles, B.; Hashemi, N.; Rasmussen, M.K.; Young, J.F.; Barberá, R. Coupling *in vitro* food digestion with *in vitro* epithelial absorption; Recommendations for biocompatibility. *Crit. Rev. Food Sci. Nutr.* **2023**, 1–19. [CrossRef] [PubMed]

toxins

MDPI

Article

Discovering the Protective Effects of Quercetin on Aflatoxin B1-Induced Toxicity in Bovine Foetal Hepatocyte-Derived Cells (BFH12)

Marianna Pauletto [1,*,†], Mery Giantin [1,†], Roberta Tolosi [1], Irene Bassan [1,‡], Anisa Bardhi [2], Andrea Barbarossa [2], Ludovica Montanucci [3], Anna Zaghini [2] and Mauro Dacasto [1]

1 Department of Comparative Biomedicine and Food Science, University of Padua, Viale dell'Università 16, I-35020 Legnaro, Italy; mery.giantin@unipd.it (M.G.); roberta.tolosi@unipd.it (R.T.); irene.bassan2@gmail.com (I.B.); mauro.dacasto@unipd.it (M.D.)

2 Department of Veterinary Medical Sciences, Alma Mater Studiorum—University of Bologna, Via Tolara di Sopra 50, Ozzano dell'Emilia, I-40064 Bologna, Italy; anisa.bardhi@unibo.it (A.B.); andrea.barbarossa@unibo.it (A.B.); anna.zaghini@unibo.it (A.Z.)

3 Genomic Medicine Institute, Lerner Research Institute, Cleveland Clinic, 9500 Euclid Avenue, Cleveland, OH 44195, USA; montanl@ccf.org

* Correspondence: marianna.pauletto@unipd.it; Tel.: +39-049-827-2935

† These authors contributed equally to this work.

‡ Current address: Chelab srl—Mérieux NutriSciences Italia, Via Fratta 25, I-31023 Resana, Italy.

Abstract: Aflatoxin B1 (AFB1) induces lipid peroxidation and mortality in bovine foetal hepatocyte-derived cells (BFH12), with underlying transcriptional perturbations associated mainly with cancer, cellular damage, inflammation, bioactivation, and detoxification pathways. In this cell line, curcumin and resveratrol have proven to be effective in mitigating AFB1-induced toxicity. In this paper, we preliminarily assessed the potential anti-AFB1 activity of a natural polyphenol, quercetin (QUE), in BFH12 cells. To this end, we primarily measured QUE cytotoxicity using a WST-1 reagent. Then, we pre-treated the cells with QUE and exposed them to AFB1. The protective role of QUE was evaluated by measuring cytotoxicity, transcriptional changes (RNA-sequencing), lipid peroxidation (malondialdehyde production), and targeted post-transcriptional modifications (NQO1 and CYP3A enzymatic activity). The results demonstrated that QUE, like curcumin and resveratrol, reduced AFB1-induced cytotoxicity and lipid peroxidation and caused larger transcriptional variations than AFB1 alone. Most of the differentially expressed genes were involved in lipid homeostasis, inflammatory and immune processes, and carcinogenesis. As for enzymatic activities, QUE significantly reverted CYP3A variations induced by AFB1, but not those of NQO1. This study provides new knowledge about key molecular mechanisms involved in QUE-mediated protection against AFB1 toxicity and encourages in vivo studies to assess QUE's bioavailability and beneficial effects on aflatoxicosis.

Keywords: aflatoxin B1; quercetin; lipid peroxidation; CYP3A; NQO1; bovine; hepatocyte cell line; transcriptome

Key Contribution: Quercetin is effective in mitigating aflatoxin-mediated toxicity in a bovine foetal hepatocyte-derived cell line. Transcriptional molecular mechanisms underlying quercetin-mediated protection against aflatoxin B1 toxicity have been revealed.

1. Introduction

Mycotoxins are food and feed contaminants produced as secondary metabolites by a few genera of filamentous fungi, eliciting toxic effects in humans and animals. Estimates by the Food and Agricultural Organization (FAO) indicate that ~25% of global food and feed products are contaminated with mycotoxins, with an estimated economic loss of USD 50 million every year [1,2]. However, the FAO estimates appear to fall short of reality [3].

Global warming and climate change, such as an increase in temperature (2°–5°) and CO_2 concentration (twice or thrice the expected value) as well as drought episodes, have been predicted to affect the occurrence of mycotoxins in the years to come [4].

Mycotoxins that would benefit most from climate change are aflatoxins, a class of natural mycotoxins produced by *Aspergillus flavus* and *A. fumigatus* [1,5]. According to their fluorescence, aflatoxins are broadly classified into four classes: B1, B2, G1, and G2 [6]. Aflatoxins usually affect crops from tropical and/or subtropical areas and are commonly found in mouldy corn, soybeans, rice, sorghum, dried figs, spices, hazelnuts, peanuts, and other grain and oil crops [5,7–9]. It is believed that about 4.5 billion people worldwide are at risk of excessive exposure to aflatoxins via contaminated food [7], but until a few years ago, these mycotoxins were not considered a matter of concern in Europe. However, due to global climate change, the presence of fungal strains endemic in tropical or subtropical climate zones has increased in temperate ones. Therefore, the risk of food and feed contamination and, consequently, of human and animal exposure to aflatoxins is expected to become higher than in the past [1,5,10].

Aflatoxin B1 is the primary aflatoxin responsible for food and feed contamination [2], and it is by far the most toxic aflatoxin [2,11]. It exhibits pleiotropic effects in several human and animal tissues; it is immunotoxic, mutagenic, carcinogenic, and teratogenic. The liver is the main target organ of aflatoxin B1. Exposure to aflatoxin B1 has been linked to hepatic carcinogenesis in humans, poultry (e.g., turkey), fish (e.g., trout), and rodents [12–15]. In particular, epidemiological surveys suggest that dietary aflatoxin B1 exposure contributes to the higher incidence of hepatocellular carcinoma (HCC) in Asia and sub-Saharan Africa [16–18]. Additional risk factors are likely to increase the aflatoxin B1-dependent occurrence of HCC, such as hepatitis virus B (HVB) infection [16,19]. Based on these different lines of epidemiological evidence, in 2012, the International Agency for Research on Cancer (IARC) classified aflatoxin B1 and the other aflatoxins, including B1's major metabolite aflatoxin M1, as carcinogenic to humans (Group I) [10,20].

Aflatoxin B1 is not carcinogenic per se; it is a pro-carcinogen, as it undergoes a bioactivation reaction in the human liver, resulting in the formation of the ultimate carcinogen aflatoxin B1-8,9-epoxide and a number of hydroxylated derivatives [10–12]. It is worth mentioning that the cytochromes P450 1A1 and 3A4 (CYP1A1 and CYP3A4, respectively) are phase I drug-metabolizing enzymes playing a critical role in aflatoxin B1 metabolism in human, duck, turkey, and cattle livers [11,21]. Apart from aflatoxin B1 epoxide, the most toxic and carcinogenic aflatoxin derivative is aflatoxin M1. This hydroxylated derivative, together with aflatoxin B1 and variable amounts of aflatoxicol, can be found in the milk and cheese from mycotoxin-exposed dairy cows. Of the farming species, cattle are relatively more resistant to aflatoxins because of the capability of rumen microbiota to convert aflatoxin B1 into less toxic or nontoxic derivatives [22,23]. Nevertheless, dairy milk contamination with aflatoxins remains a serious food safety concern, as it might pose severe health risks to humans consuming dairy food products in general and to susceptible population groups [10,24,25]. To cope with this health concern, approaches focused primarily on limiting aflatoxin B1 absorption and preventing aflatoxin excretion in dairy milk have been adopted, but they are only partially successful [25].

Promising strategies reducing the impact of aflatoxins in cattle farming are those that make use of phytochemical derivatives, such as polyphenols, with strong antioxidant activity [8]. Indeed, besides the aforementioned toxic effects, aflatoxin B1-dependent hepatotoxicity is closely related to the mycotoxin's capability to generate reactive oxygen species (ROS), resulting in cellular oxidative stress and its well-known consequences (e.g., oxidative DNA damage and membrane lipid peroxidation) [9,16,26]. Overall, these botanical extracts can reduce oxidative damage in the body by increasing the activity of antioxidant enzymes and scavenging free radicals [27]. Moreover, natural antioxidants can take advantage of more pathways and molecular targets. Finally, they are less toxic and have fewer side effects [9]. Therefore, some of these plant-based compounds are increasingly used to prevent aflatoxin B1 formation in toxigenic moulds and to detoxify aflatoxin B1-

contaminated food and feed without any deleterious effects on their nutritional value [8,28]. Curcumin, resveratrol, oxidized tea phenolics, and flavonoids such as quercetin are among the most studied anti-aflatoxin polyphenols [8]. To assess the effectiveness and feasibility of these dietary polyphenols in preventing or mitigating aflatoxin B1 contamination of dairy milk, conducting mechanism-based studies in bovine target tissues in vitro and in vivo is a fundamental starting point [9,29].

The aim of the present study was to assess the protective role of quercetin (QUE) in a bovine foetal hepatocyte-derived cell line (BFH12) exposed to aflatoxin B1. To this end, we measured the effects triggered by QUE on aflatoxin B1-induced cytotoxicity and related transcriptional changes (RNA-sequencing). Confirmatory studies on protein expression and biological activity were also performed. This is part of a series of in vitro and in vivo studies aimed at measuring the biological effects of polyphenols on the toxicity of aflatoxin B1 (AFB1) and its derivatives in dairy cows [24,29–32].

2. Results

2.1. *Quercetin's Half-Maximal Inhibitory Concentration and Ability to Counteract AFB1 Cytotoxicity*

Bovine cells were incubated with increasing concentrations of QUE to build up a dose–response curve (Figure S1). The fitting of the data was not excellent (R squared = 0.68), but some sigmoidal trend was observed, thus allowing for the estimation of the QUE half-maximal inhibitory concentration (IC_{50}) at 64 h, which turned out to be 59.97 μM.

Afterwards, we assayed the protective role of QUE against AFB1 cytotoxicity. At the tested sub-cytotoxic concentrations, QUE reduced the AFB1 cytotoxicity in a dose-dependent manner. The decrease was statistically significant ($p \leq 0.01$) with 20 and 30 μM QUE, and the highest QUE concentration reduced the median AFB1 cytotoxicity by 13.39% (from 86.00% to 72.61%; Figure 1).

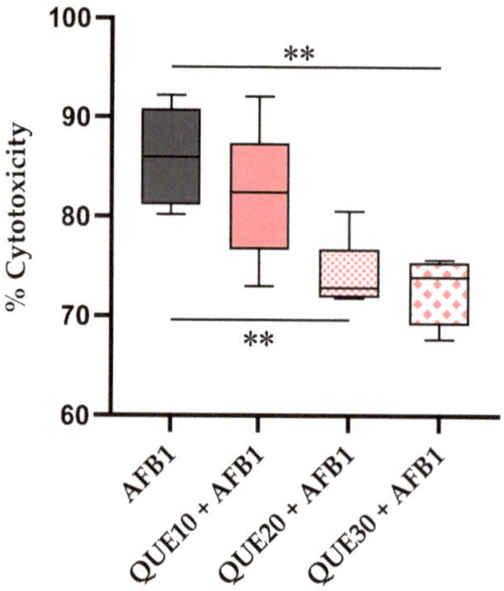

Figure 1. Effect of quercetin on AFB1 cytotoxicity. The box and whiskers plot reports the viability of bovine foetal hepatocyte (BFH12) cells pre-treated (16 h) with increasing QUE concentrations (10, 20, and 30 μM) and exposed (48 h) to a combination of AFB1 3.6 μM and QUE (at the same concentrations above mentioned). **: $p \leq 0.01$ (ANOVA one-way and Dunnett's multicomparisons tests; the mean of each condition was compared with that of AFB1). Graphs were obtained using GraphPad prism software. AFB1 = aflatoxin B1; QUE = quercetin.

2.2. Effect of QUE on AFB1 Biotransformation in BFH12 Cells

Quercetin significantly decreased the amount of aflatoxin M1 (AFM1) in the cellular medium in a dose-dependent manner. In particular, the AFM1 concentration dropped from 57.00 ng/mL (i.e., 0.174 µM) to 0.5 ng/mL (0.002 µM; Figure 2). As for AFL, cells exposed to AFB1 + 10 µM QUE showed significantly higher amounts of this AFB1 derivative, i.e., from 85.5 ng/mL (0.272 µM) to 142.5 ng/mL (0.453 µM). However, at medium and maximum QUE concentrations, the amount of aflatoxicol (AFL) was similar to the amount detected in the presence of AFB1 alone (Figure S2a).

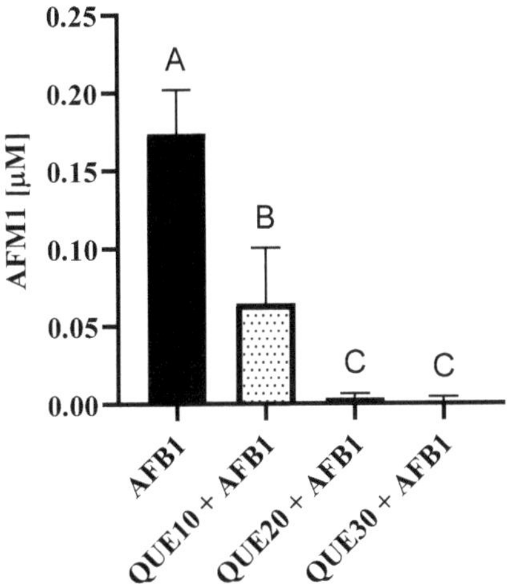

Figure 2. Effect of quercetin on AFB1 biotransformation. Bars represent the amount (µM) of aflatoxin M1 (AFM1) detected in the cellular medium after 48 h of exposure to 3.6 µM AFB1 alone or in combination with increasing QUE concentrations (10, 20, and 30 µM). Data are expressed as mean concentrations ± standard deviations of four independent cell culture experiments. Different letters above error bars indicate significant differences ($p \leq 0.05$) among groups (Tukey's post hoc test). Graphs were obtained using GraphPad prism software.

Detectable amounts of AFM1 and AFL have never been found in cell pellets; however, the incubation with increasing QUE concentrations led to an increase in AFB1 concentration, i.e., from 9.79 ng/mL (0.031 µM) to 47.50 ng/mL (0.152 µM), but these variations were not statistically significant (Figure S2b).

2.3. Effects of QUE, Alone or in the Presence of AFB1, on Selected AFB1 Target Genes (qPCR)

2.3.1. Drug Metabolizing Enzymes (DMEs)

The incubation with increasing amounts of QUE downregulated the expression of glutathione S-transferase A1 (GSTA1) (Figure 3a).

In cells pre-treated with QUE and then exposed to AFB1, substantial transcriptional changes were recorded at the highest QUE concentrations (20 and 30 µM). An overall significant gene upregulation was observed for CYP1A1, CYP1B1, and GSTA1. The most significant variations were observed for the expression of CYP3A28, the bovine orthologue of human CYP3A4 [33]. A significant and dose-dependent fold change decrease was observed with 20 and 30 µM QUE (Figure 3a).

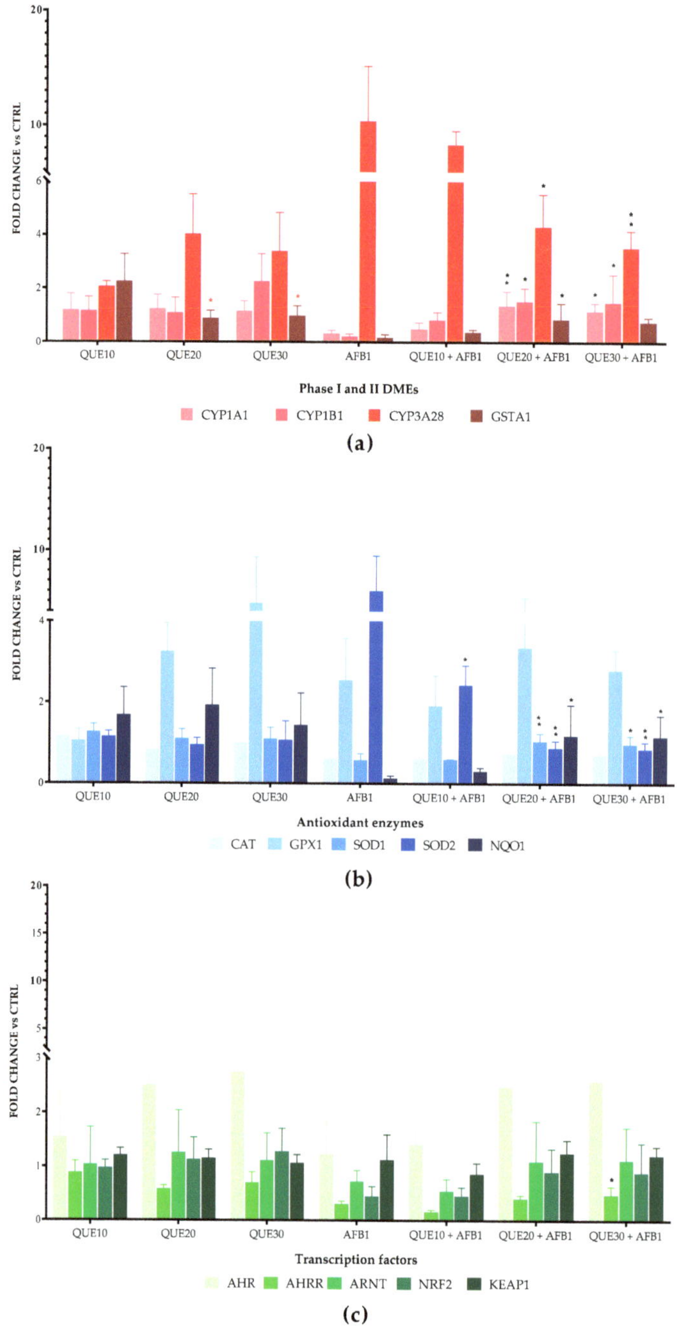

Figure 3. Quercetin-induced changes in mRNA levels of oxidative (phase I) and conjugative (II) drug metabolizing enzymes (DMEs, (**a**)), antioxidant enzymes (AOEs, (**b**)) and related transcription factors (TF, (**c**)). The one-way ANOVA, followed by a Dunnett's multi-comparisons test, was carried out to compare the gene expression level of BFH12 cells exposed to increasing concentrations of QUE (10,

20, and 30 μM; the lowest dose was used as control in the comparisons); significant variations were identified by red asterisks. The same statistical approach was adopted to separately investigate the effects of QUE pre-treatment in the presence of AFB1 3.6 μM: all the co-treatment conditions were compared to the AFB1 condition; significant variations were identified by black asterisks. Data (means ± standard deviations) are expressed as n-fold changes normalized to ΔΔCt mean value of control (i.e., polychlorinated biphenyls 126, PCB126) to whom an arbitrary value of 1 was assigned. *: $p \leq 0.05$; **: $p \leq 0.01$ (ANOVA one-way and Dunnett's multicomparisons test). Graphs were obtained using GraphPad prism software. AHR = aryl hydrocarbon receptor; AHRR = aryl hydrocarbon receptor repressor; ARNT = aryl hydrocarbon receptor nuclear translocator; CAT = catalase; CTRL = control; CYP1A1 = cytochrome P450 1A1; CYP1B1 = cytochrome P450 1B1; CYP3A28 = cytochrome P450 3A28; DMEs = drug metabolizing enzymes; GPX = glutathione peroxidase I; GSTA1 = glutathione S-transferase A1; KEAP = kelch like ECH associated protein 1; NQO1 = NAD(H):quinone oxidoreductase 1; NRF2 = nuclear factor erythroid 2-related factor 2; SOD1 = superoxide dismutase 1; SOD2 = superoxide dismutase 2.

2.3.2. Antioxidant Enzymes (AOEs)

The pre-incubation with the chosen QUE concentrations did not result in statistically significant variations in mRNA levels of the most important AOEs (Figure 3b).

In cells co-exposed to AFB1 and QUE, as compared to cells exposed to AFB1 alone, significant transcriptional changes were observed only with 20 and 30 μM QUE: superoxide dismutase 1 (SOD1) and quinone oxidoreductase 1 (NQO1) genes were upregulated, while SOD2 was significantly and dose-dependently downregulated (Figure 3b).

2.3.3. Transcription Factors (TFs)

Genes coding for DMEs and AOEs recognize a number of TFs involved in their transcriptional mechanisms of gene regulation and play an important role in AFB1 toxicity [34,35]. No statistically significant variations in the mRNA of target TFs were observed in cells pre-incubated with either QUE alone or QUE in combination with AFB1, with the sole exception of the aryl hydrocarbon receptor repressor (AHRR), showing a slight but significant induction in cells pre-treated with 30 μM QUE (Figure 3c).

2.4. *Differential Expression Analysis*

To obtain clear transcriptional results, RNA-sequencing (RNA-seq) investigations were carried out only in cells treated with the highest QUE concentration (30 μM).

A total of 296,495,475 raw reads were sequenced and deposited in GeneBank under BioProject accession PRJNA627332. These reads were subjected to quality control measures, and after trimming and rRNA removal, 24.5 million reads per sample were retained on average. Approximately 99% of the obtained reads were mapped to the *B. taurus* reference genome (Table S1). The MDS plot in Figure S3 shows an unsupervised clustering of the samples. The first dimension (x-axis) clearly separates samples based on the experimental group. Major differences were observed between cells exposed to AFB1 (bottom right) and control cells (bottom left). Both the first and the second dimension (y-axis) separate the transcriptional profiles observed in cells exposed to QUE (alone or in combination with AFB1), thus forming a unique cluster in the plot (top centre). Biological variability within experimental groups is low, as demonstrated by the coherent clusters formed by the replicas.

When comparing the transcriptional profiles of cells exposed to QUE and those of control cells, a total of 1028 differentially expressed genes (DEGs) were found: 487 were upregulated, and 541 were downregulated (Figure S4). The edgeR output of the DE analysis conducted in this study is reported in Table S2. Among the top 10 upregulated genes (Figure 4a), we found QUE transporter solute carrier family 2 member 1 (SLC2A1, also known as GLUT1), with a $\log_2$ fold change (lfc) of 4.72; players in inflammatory processes, cell proliferation, and survival, such as tumour necrosis factor superfamily member 9 (TNFRSF9, lfc = 5.13); high-mobility group at-hook 1 (HMGA1, lfc = 3.34);

secreted phosphoprotein 1 (*SPP1*, also known as osteopontin; lfc = 3.82); and the MAP BZIP transcription factor (MAFF, lfc = 2.37), a gene coding for a small Maf protein and involved in the cellular stress response. Additionally, QUE significantly increased the expression of two enzymes involved in lipid metabolism, i.e., stearoyl-CoA desaturase (SCD, lfc = 3.22) and fatty acid synthase (FASN, lfc = 2.89). Among the top 10 downregulated genes, we found a gene coding for an acute phase protein, pentraxin 3 (PTX3, lfc = −5.42), and two genes implicated in carcinogenic processes, i.e., olfactomedin-like 2B (OLFML2B, lfc = −3.22) and elastin microfibril interfacer 2 (EMILIN2, lfc = −3.33).

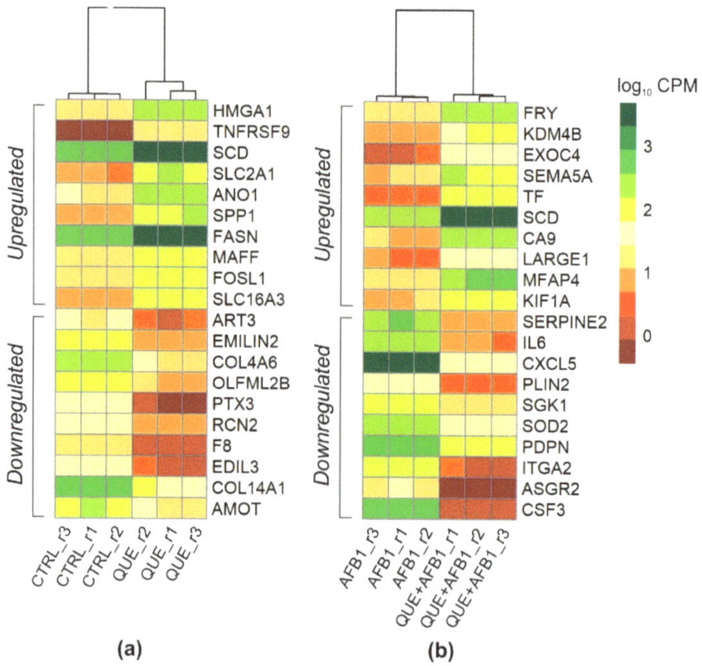

Figure 4. Gene expression levels of differentially expressed genes. Heatmap of top 10 genes up- and downregulated by QUE (vs. CTRL) (**a**) and QUE + AFB1 co-treatment (vs. AFB1) (**b**). Heatmaps were constructed in R environment using the pheatmap package and using the normalized log10CPM (Counts Per Million) as input. Biological replicates are represented by r1, r2, and r3 suffixes.

The enrichment analysis, carried out following a more comprehensive analysis of DEGs, resulted in the significant enrichment of 19 Biological Processes (BPs) and 17 Kyoto Encyclopaedia of Genes and Genomes (KEGGs) (Table S3, Figure S5). Among the significant BPs, there was a "steroid metabolic process" (gene count = 16), represented by genes involved in the metabolism of steroids and other substrates (e.g., fatty acids, prostaglandins, and xenobiotics), such as hydroxysteroid dehydrogenases HSD 3B1 (lfc = 2.54) and HSD17B12 (lfc = 1.95), and key transcription factors in the regulation of cholesterol metabolism, such as SCARB1. Several genes belonging to this enriched BP are involved in regulating cholesterol biosynthesis, including sterol regulatory element-binding transcription factor 2 (SREBF2; lfc = 1.56), HMG-CoA reductase (HMGCR; lfc = 2.62), low-density lipoprotein receptor (LDLR; lfc = 3.24), and 7-dehydrocholesterol reductase (DHCR7; lfc = 2.39).

An additional significantly enriched BP was "response to external stimulus" (gene count = 47). This BP was represented by both up- and downregulated genes. With regard to genes induced by QUE, we highlight prostaglandin-endoperoxide synthase 2 (PTGS2; lfc = 2.82), a member of the Forkhead box (FOXA3; lfc = 2.19), glutathione peroxidase 1 (GPX1; lfc = 1.51), and adrenoceptor beta 2 (ADRB2; lfc = 2.77). With regard to genes downregulated by QUE, we found complement factors C2 (lfc = −2.41) and H (CFH, lfc = −4.78), C-X-C motif chemokine ligands 9, 10, and 11 (lfc = −4.57, −4.15, −5.16, respectively), fatty acid-binding protein 4 (FABP4; lfc = −4.37), and the CD36 molecule (lfc = −2.15).

Like "response to external stimulus", the BP term "response to wounding" (gene count = 13) was also significantly enriched. This BP was represented by some DEGs already listed in the "response to stimulus" BP and some additional genes, such as von Willebrand factor (VWF; lfc = 2.73) and claudin 1 (CLDN1; lfc = 1.88).

Finally, an interesting enriched BP was "inflammatory response" (gene count = 18), with several genes shared with the "response to external stimulus" BP and some additional ones, such as prostaglandin-endoperoxide synthase 1 (PTGS1; lfc = −2.30).

Enriched KEGGs provided further insights into the molecular changes triggered by QUE. Interestingly, some KEGGs related to immunity were significantly enriched, such as "neutrophil extracellular trap formation" (gene count = 26) and "complement and coagulation cascades" (gene count = 10). With regard to complement factors, while C2 and CFH were significantly downregulated, a gene encoding a complement receptor (C3AR1) was upregulated, with an lfc of 2.94.

Notable enriched KEGGs were "PPAR signalling pathway" (gene count = 13) and "drug metabolism—cytochrome P450" (gene count = 9). In the PPAR pathway, significant genes code for fatty acid-binding proteins 3, 5, and 4 (lcf = 3.50, −4.33, −4.37, respectively) and enzymes were involved in the biosynthesis and degradation of cellular lipid fatty acids. The "P450-mediated drug metabolism" KEGG was mostly represented by key antioxidant enzymes and glutathione S-transferases, including GSTA3 (lfc = 8.32), GSTA2 (lfc = 3.01), GSTM1 (lfc= 2.11), GSTK1 (lfc = −1.61), and GSTT4 (lfc = −1.94).

Gene expression profiles of cells co-treated with QUE and AFB1 were compared to those of cells treated with AFB1 alone. This comparison allowed for the identification of 1890 DEGs (Table S2). Most of the significant genes, 1275, were upregulated, while 615 genes were downregulated (Figure S4). The expression levels of the top 10 up- and downregulated genes are reported in a heatmap (Figure 4b). Noteworthily, among the top 10 upregulated DEGs, we found genes related to carcinogenesis, such as lysine demethylase 4B (KDM4B, lfc = 3.41), stearoyl-CoA desaturase (SCD, lfc = 3.75), and carbonic anhydrase 9 (CA9, lfc = 4.60). Top upregulated genes also included FRY microtubule-binding protein (FRY, lfc = 3.24), semaphorin 5A (SEMA5A, lfc = 3.56), and transferrin (TF, lfc = 5.13). Among the top downregulated DEGs, we found genes implicated in inflammatory processes, such as chemokine (C-X-C motif) ligand 5 (CXCL5, lfc = −6.16), interleukin 6 (IL6, lfc = −5.72), colony-stimulating factor 3 (CSF3, lfc = −8.19), and podoplanin (PDPN, lfc = −2.73), which also play an important role in liver fibrosis and cancer. A key role in the liver is also played by another gene, perilipin 2 (PLIN2; lfc = −3.63), a marker of steatosis, which is downregulated here. Like PDPN, serum/glucocorticoid-regulated kinase 1 (SGK1, lfc = −2.86), listed among the top 10 genes downregulated by QUE pre-treatment, is known to play a crucial role in tumourigenesis and cancer progression. A further gene whose expression was strongly reduced by QUE pre-treatment was the key antioxidant enzyme SOD2 (lfc = −2.87), which is consistent with the qPCR data (Figure 3b). As observed in cells exposed to only QUE, microsomal glutathione S-transferase 1 (MGST1), GSTA2, GSTM1, and GSTK1 were significantly regulated, with lfc values of 3.20, 4.96, 3.01, and −1.40, respectively.

The enrichment analysis conducted on the list of 1890 DEGs resulted in the significant enrichment of 6 BPs and 10 KEGGs (Table S3, Figure S6). Overall, some terms modulated by the co-treatment QUE + AFB1 were also significantly affected by QUE alone, e.g., "response to external stimulus", "steroid biosynthesis", and "neutrophil extracellular trap formation", yet the DEGs involved were not completely the same. As far as the DEGs representing the enriched BP "response to external stimulus" (gene count = 73) are concerned, we found many genes playing a role in mediating inflammatory processes, such as prostaglandin I2 synthase (PTGIS, lfc = 3.09), CXCL5 (lfc = −6.16), C-C motif chemokine ligand 2 (CCL2;

lfc = −3.82), CCL20 (lfc = −4.91), and high-mobility group box 1 (HMGB1; lfc = −1.30), in addition to genes regulating apoptosis, such as BCL2 interacting protein 3 like (BNIP3L; lfc = 1.79) and interferon alpha inducible protein 6 (IFI6; lfc = 2.01).

In the analysis of the effects of QUE alone, the pathway of steroid metabolism appeared to be significantly affected by the co-treatment QUE + AFB1, as demonstrated by the enrichment of the BP term "cholesterol biosynthetic process" (gene count = 9) and the KEGG pathway "steroid biosynthesis" (gene count = 11). Specifically, the co-treatment seems to mostly upregulate genes of the cholesterol biosynthesis, rather than steroids in general, such as LPCAT3 (lfc = 1.85), SREBF2 (lfc = 2.49), lanosterol synthase (LSS; lfc = 2.41), DHCR7 (lfc = 3.02), and HMGCR (lfc = 2.86).

Innate and inflammatory host defences were significantly enriched, as demonstrated by the KEGG pathways "rheumatoid arthritis" (gene count = 19), "systemic lupus erythematosus" (gene count = 18), "neutrophil extracellular trap formation" (gene count = 28), and "cytokine-cytokine receptor interaction" (gene count = 31). In these pathways, we found many mediators of inflammation, such as intercellular adhesion molecule-1 (ICAM1; lfc = −2.62), interleukin 1-alpha (IL1A; lfc = −3.19), interleukin 6 signal transducer (IL6ST; lfc = −1.75), interleukin 7 receptor (IL7R; lfc = −2.75), CXCL3 (lfc = −3.89), LIF interleukin 6 family cytokine (LIF; lfc = −2.73), interleukin 1 receptor antagonist (IL1RN; lfc = 2.28), CD40 (lfc = −2.35), colony-stimulating factor 2 (CSF2; lfc = −2.79), and C2 (lfc = −2.73).

A further significantly enriched KEGG was "ABC-transporters" (gene count = 12), with both up- and downregulated genes. Notably, the mRNA expression of ABCG1, ABCA2, and ABCD1 was induced by QUE + AFB1, with lfc values of 2.81, 2.37, and 1.35, while the mRNA expression of ABCB1 and ABCC4 was downregulated, with lfc values of −1.55 and −2.07, respectively.

Finally, gene set enrichment analysis (GSEA) highlighted a suppression of several pathways involved in inflammation in the co-treated cells, compared to cells treated with AFB1 alone (Figure 5a,b, Table S4). Some examples are represented by the KEGG pathways "IL-17 signalling pathway" (normalized enrichment score [NES] = −1.99), "NF-kappa B signalling pathway" (NES = −1.73), and "TNF signalling pathway" (NES = −1.63), as well as the hallmark pathways "TNFA signalling via NFKB" (NES = −1.92), "inflammatory response" (NES = −1.58), and "interferon gamma response" (NES = −1.35). Notably, some pathways related to carcinogenesis were also significantly suppressed in the presence of QUE, such as "MYC targets V1" (NES = −2.33), "MYC targets V2" (NES = −2.07), and "DNA repair" (NES = −1.66).

Although the analysis of the transcriptional effects of AFB1 is beyond the scope of the present study and has already been performed in a previous experiment [32], we compared the DEGs from QUE + AFB1 vs. AFB1 and those from AFB1 vs. control (CTRL) (i.e., 2803 genes). We observed that 60% of the total number of DEGs in the comparison QUE + AFB1 vs. AFB1 (i.e., 1136 out of 1890) exhibited a transcriptional regulation opposite to that in the comparison AFB1 vs. CTRL (Figure 6). Specifically, the transcriptional level of 336 genes was increased by AFB1 when compared to CTRL, but it was decreased by the co-treatment QUE + AFB1 when compared to AFB1 alone. Likewise, the expression of 800 genes decreased in response to AFB1 but increased in the condition QUE + AFB1.

The KEGG functional enrichment of these 1136 genes (Figure 7) revealed some terms identical or similar to those from the functional analyses conducted on the gene expression differences between cells co-treated with QUE + AFB1 and those treated with AFB1 alone (Figures 5 and S4, Tables S3 and S4). In particular, these terms were mainly related to immunity and inflammation (i.e., IL-17 signalling, cytokine-cytokine interactions, TNF signalling, and lipids and atherosclerosis) and calcium signalling.

Finally, the RNA-seq results were compared to those obtained using qPCR (Figure 3) in order to check the consistency between the two methods. The two approaches were mostly in agreement (Figure S7).

Figure 5. Gene Set Enrichment Analysis (QUE + AFB1 vs. AFB1). Gene-concept network of the significantly enriched KEGG pathways (**a**) and Hallmark gene sets (**b**). Node size reflects the gene set size (i.e., number of genes). The dots colour (from blue to red) indicates genes ranking (highly down- or upregulated).

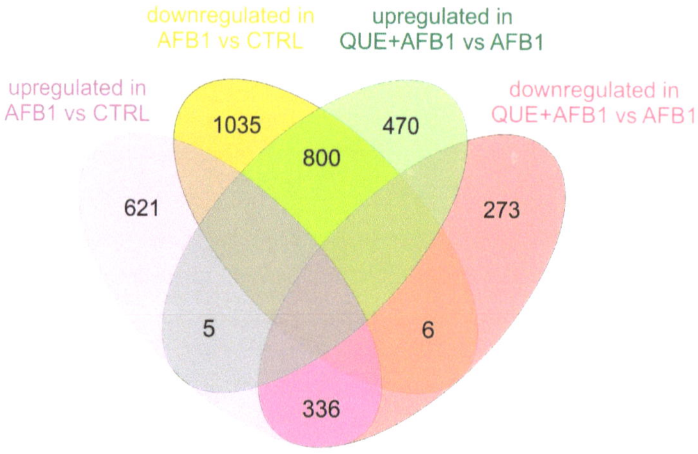

Figure 6. Shared differentially expressed genes. Venn diagram of shared DEGs in the comparison between AFB1 vs. CTRL and QUE + AFB1 vs. AFB1.

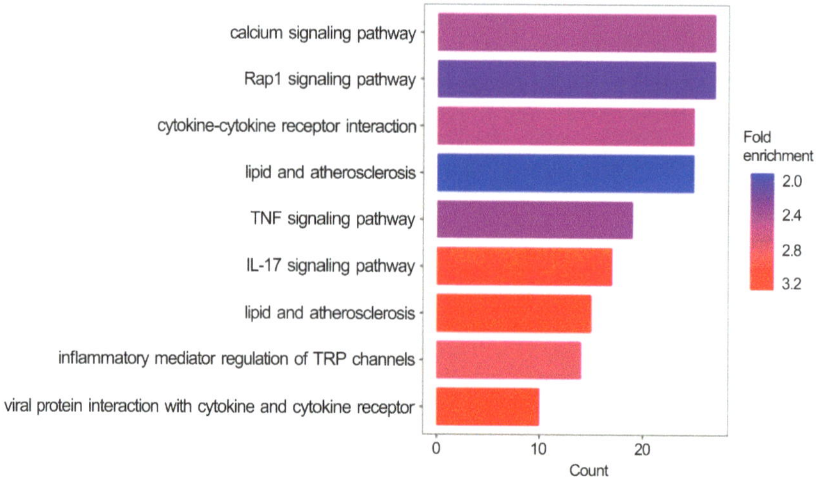

Figure 7. Enrichment analysis of the shared DEGs. Bar plots report the enriched KEGG pathways (FDR $\leq$ 0.05) in the list of shared DEGs in the pair-wise comparisons QUE + AFB1 vs. AFB1 and AFB1 vs. CTRL. Count = number of DEGs in each enriched pathway. The colour gradient reflects the fold enrichment of each term.

2.5. Effect of QUE on Possible AFB1-Dependent Oxidative Stress

The generation of ROS, oxidative stress, and lipid peroxidation are among the mechanisms of AFB1-induced toxicity [2,9,12]. To assess the counteracting effect of QUE on such a sequel of toxic events, we measured the amount of malondialdehyde (MDA), a known marker of lipid peroxidation. AFB1 significantly increased ($p \leq 0.001$) the amount of MDA compared to control cells (Figure 8a). However, QUE did not show a significant increase in this parameter. Importantly, the co-incubation of BFH12 cells with QUE and AFB1 revealed a significant decrease ($p \leq 0.001$) in MDA amount compared to cells exposed to AFB1 alone, thus confirming the flavonoid's potential to counteract AFB1-dependent oxidative damage.

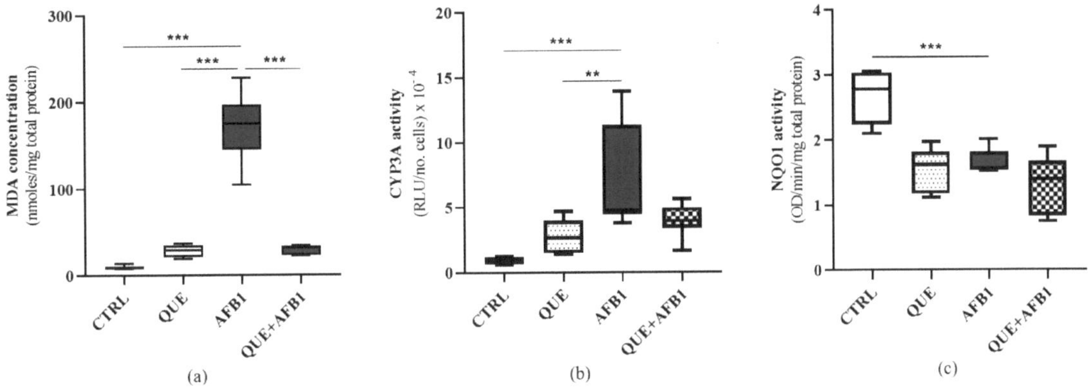

Figure 8. Effect of quercetin on AFB1-dependent oxidative stress, cytochrome P450 3A (CYP3A) and NQO1 activities. The box and whiskers plots report the amount of malondialdehyde (MDA, (**a**)) and the enzyme activity of CYP3A (**b**) and NQO1 (**c**) measured in the cellular medium either from control than in QUE (30 μM), AFB1 (3.6 μM), and QUE + AFB1 (30 μM and 3.6 μM, respectively)-exposed BFH12 cells. The statistical comparisons were established between the median values reported in cells exposed to AFB1 and those observed in all other experimental conditions. **: $p \leq 0.01$; ***: $p \leq 0.001$ (ANOVA one-way and Dunnett's multicomparisons tests). Graphs were obtained using GraphPad prism software. CYP3A = cytochrome P450 3A; MDA = malondialdehyde; OD = optical density; RLU = relative light units.

2.6. Effect of QUE on Target Enzyme Activity (CYP3A and NQO1)

According to the present gene expression results and similarly to previously published results [29,31], we measured the activity of two enzymes known to be involved in AFB1 bioactivation and the antioxidant response(i.e., CYP3A and NQO1, respectively) [11,36].

With regard to CYP3A catalytic activity, cells exposed to AFB1 showed a statistically significant increase in CYP3A activity when compared to that of control and QUE pre-incubated cells ($p \leq 0.001$ and $p \leq 0.01$, respectively). By contrast, the combined exposure to AFB1 and QUE considerably reduced the CYP3A activity, bringing it back to levels slightly above than that of the control (Figure 8b).

An overall decrease in NQO1 activity (and, consequently, in cellular antioxidant response) was observed in BFH12 cells exposed to QUE, AFB1, and their combinations. Nevertheless, such a decrease was statistically significant only in AFB1-exposed cells ($p \leq 0.001$; Figure 8c). This is in opposition to what we observed at the transcriptional level. Indeed, the qPCR revealed a slightly significant increase in NQO1 activity at the highest QUE concentrations (20 and 30 μM + AFB1), compared to AFB1 alone (Figure 3b). Likewise, the RNA-seq analysis (QUE + AFB1 vs. AFB1) identified this gene as upregulated (Table S2).

3. Discussion

For many years, the antioxidant properties of polyphenols were assumed to result merely from their capacity to donate electrons or chelate transition metals. However, recent studies have shown that they might have multiple modes of action, interfering with several cell signalling pathways [37–41]. Although several studies have been conducted on polyphenols' positive effects in human medicine [42,43] and animal farming [44–47], only marginal attention has been paid to the underlying molecular mechanisms, which are not yet fully understood.

In cattle, the in vitro benefits of QUE have been evaluated very recently. A potential application of QUE in treating bovine viral diarrhoea virus infection has been explored in Madin–Darby bovine kidney cells (MDBK) [48]. Likewise, QUE's anti-inflammatory potential has been demonstrated in bovine intestinal epithelial cells (BIECs) [49] and mammary epithelial cells (BMECs) [50]. The above-mentioned studies provided important molecular results to disentangle QUE's mechanism of action in cattle, thus increasing the fundamental knowledge that is necessary to introduce QUE-supplemented feed in farming practice. However, to our knowledge, sequencing technologies querying the entire genome, such as RNA-seq, have never been employed in cattle, neither in vitro nor in vivo.

Therefore, in the present study, we primarily assessed the toxicity and the transcriptional effects of QUE in a selected bovine hepatic cellular model, BFH12. Then, we proved QUE's ability to mitigate AFB1 toxicity in vitro and demonstrated the underlying molecular pathways.

3.1. Cytotoxicity of Quercetin

The effects of QUE have previously been studied in different cell lines, and the published IC_{50} values were consistent with the results obtained in the present study. After 48 h of exposure, QUE's IC_{50} was approximately 50 μM in several in vitro models, such as the human breast cancer cell line MCF-7 [51], the human leukaemia cell lines K562 and CEM [41], and the bovine mammary epithelial cell line BME-UV1 [30]. The IC_{50} of QUE that we observed in this study is consistent with the one estimated after 48 h in the human hepatoma cell line HepG2, i.e., 107 μM [52]. Higher IC_{50} values were found in prostate cancer (PC3) and lung cancer (A549) cell lines, which are clearly less sensitive to QUE [51]. If QUE's IC_{50} is compared with the IC_{50} values obtained in the same cell line for other well-known polyphenolic compounds [29,31], then QUE is less toxic than curcuminoids but as toxic as resveratrol.

The positive impact of this natural flavonoid in cattle was clearly demonstrated by its ability to significantly decrease AFB1-induced cytotoxicity. Notably, compared to resveratrol [31], QUE appeared to be three times less effective in counteracting AFB1-mediated cytotoxicity; this is particularly relevant considering that these two polyphenols have similar IC_{50} values in BFH12 cells. Conversely, in BME-UV1, QUE exhibited a higher protective ability against AFB1 cytotoxicity than resveratrol [30]. We might hypothesize that the efficacy of different polyphenols varies depending on the target tissues, suggesting overall better outcomes when these natural extracts are used in combination, as recently shown in rat models of colon carcinoma [53].

3.2. Biotransformation of AFB1

Aflatoxin M1, one of the most toxic AFB1 derivatives [54], appeared to be inversely correlated with QUE concentration, as previously observed for curcuminoids and resveratrol [29,31]. Most probably, QUE reduces AFM1 production by targeting the major enzymes involved in AFM1 hepatic formation. As an example, CYP3A mRNA expression and catalytic activity were highly enhanced by AFB1 alone, but this increase was reversed via the pre-treatment with QUE, even though at the catalytic activity level, this effect was only close to the threshold of significance. To explain these results, we hypothesize that QUE might modulate the synthesis and/or activity of CYP3A. Nevertheless, to clarify the possible interactions among QUE, AFB1, and CYP3A, target studies are clearly needed. In opposition to what we reported for resveratrol [31], the amount of highly toxic AFL did not increase with flavonoid concentration. It has always been thought that AFL production is not an efficient detoxification reaction because AFL is carcinogenic and a repository for AFB1 [14]. However, AFL production has recently been associated with minor sensitivity to AFB1 by reducing the amount of AFB1 available for bioactivation [55]. Therefore, we can postulate that QUE's failure to shift AFB1 metabolism towards AFL is one of the mechanisms that make this flavonoid not as effective as resveratrol in reducing AFB1-induced toxicity.

3.3. Molecular Effect of QUE Underlying Its Potential Role as an Anti-AFB1

In a previously published study, AFB1 was found to significantly affect the transcriptome of BFH12 cells, eliciting carcinogenesis, cellular damage and apoptosis, inflammation, bioactivation, and detoxification pathways [32]. Here, we demonstrated that a pre-treatment with QUE mitigated AFB1-induced cytotoxicity and profoundly modified the cell line's transcriptional response to AFB1 exposure. The transcriptional modifications induced by QUE, conferring moderate protection against AFB1 toxicity, were mainly related to cholesterol homeostasis, inflammatory processes, carcinogenesis, oxidative stress, and drug transport.

Quercetin administered alone had a considerable impact on the BFH12 cell transcriptome, significantly affecting the expression of more than one thousand genes. Similar results were obtained when treating BFH12 cells with curcumin [29]; conversely, resveratrol induced negligible transcriptional changes in the same cellular model [31]. Notably, of the natural extracts that we have so far tested in BFH12 cells, resveratrol was the most effective in mitigating AFB1-induced cytotoxicity. The present results seem to support the idea that the extent to which polyphenols affect BFH12 cells' transcriptional profiles might not be positively correlated with their efficacy in mitigating AFB1 toxicity. In addition, a particularly interesting fact is that transcriptional profiles of cells exposed to QUE alone were nearly identical to those of cells co-treated with QUE and AFB1. This means that adding the mycotoxin had no substantial effect on cells previously "primed" with the flavonoid and that AFB1 did not produce transcriptional modifications additional to those triggered by QUE. However, further studies are needed to decipher the mechanisms underlying this peculiar transcriptional outcome.

3.3.1. Cholesterol Metabolism

In the present study, we showed that QUE interferes with cholesterol metabolism, as previously demonstrated [56]. Indeed, in BFH12 cells, several genes involved in steroid metabolism were differentially regulated by QUE, such as SCARB1 (scavenger receptor class B type 1), which was also upregulated in this study. Conversely, this transcription factor has been reported to be downregulated by QUE in gut broilers, which in turn decreases triglycerides, total cholesterol, and low-density lipoproteins [57]. This result is also consistent with the observed upregulation of LDLR, responsible for LDL endocytosis and clearance, which has previously been reported to be regulated by QUE in rat models [58]. Other genes involved in steroid metabolism (e.g., FASN, HMGCR, and SCD) were upregulated by QUE, yet previous studies reported opposite results. For instance, several polyphenols, including QUE, have been recognized as FASN inhibitors [59,60]. In hepatocytes, HMGCR is a molecular target of many dietary polyphenols, being inhibited at both mRNA and enzymatic activity levels [61]. In hamsters fed with QUE diets, the mRNA of hepatic HMGCR was inhibited [62]; however, in the same study, HepG2 cells incubated with QUE did not show significant variations in HMGCR mRNA levels. By contrast, QUE was found to increase HMGCR activity in mice fed with a high-fat diet [63]. Likewise, HMGCR and its upstream and downstream genes, SREBF2 and LSS, were induced when cells exposed to QUE + AFB1 were compared to those exposed to AFB1 alone. These conflicting results are likely due to differences in cholesterol metabolism between species [64]. In this cell model, we might interpret this overexpression as a strategy of QUE to alleviate the AFB1-induced oxidative stress, as previously suggested in human lens epithelial cells subjected to UV-B [65].

3.3.2. Inflammatory Processes

The data presented in this study showed that QUE's mechanism of action might also encompass the regulation of the pathway of nuclear factor kappa-light-chain-enhancer of activated B cells (NF-kB), as previously suggested [66]. Notably, in the present study, the expression of this transcription factor, with well-recognized functions in regulating immune and inflammatory responses, was not directly modulated by QUE. Instead, a key gene that is normally inhibited via NF-kB deactivation, PTX3, was greatly downregulated by QUE. The importance of PTX3′s decrease in mediating QUE's anti-inflammatory activity was previously demonstrated in human mesangial cells, in which it blocked the NF-kB signalling pathway and reduced renal damage [67]. Furthermore, suppression of NF-kB signalling is most likely linked to the downregulation of PTGS1 (i.e., COX1), an enzyme with critical roles in the pathophysiological progress of inflammation and cancer. Indeed, inhibitors or silencers of PTGS1 greatly attenuate the inflammatory response by negatively governing the NF-kB signalling pathway [68]. On the other hand, QUE induced the transcription of PTGS2 (i.e., COX2), which is usually activated in response to pro-inflammatory stimuli. In the literature, PTGS2 expression has been reported to be attenuated by QUE [69], so the induction observed here appears to be controversial. QUE's anti-inflammatory properties were also confirmed by the potent downregulation of pro-inflammatory chemokines, in particular the CXCR3 ligands CXCL9, CXCL10, and CXCL11, which are important players in chronic liver diseases [70], inflammation [71], and cancer [72].

QUE's potential to moderate inflammatory processes appeared to be crucial in mitigating AFB1-induced toxicity. Indeed, several genes with pro-inflammatory activity, such as interleukins and related receptors, chemokine ligands, and colony-stimulating factors, were downregulated in QUE + AFB1 vs. AFB1, as previously observed when pre-treating BFH12 with curcumin [29]. Interestingly, genes that mediate toxic pro-inflammatory responses (e.g., IL1A, IL6, and IL7R) are greatly induced by AFB1, both in vivo [73] and in vitro [74], including in BFH12 cells [26,32]. Importantly, the key protein that mediates the decrease in cytokine/chemokine release was, most likely, HMGB1, whose transcription was indeed lowered by QUE pre-treatment. This nuclear protein is involved in various liver injuries leading to inflammation and in regulating specific cell death responses, and QUE has previously been identified as its potential inhibitor [75]. For instance, in normal human hepatocytes, the HMGB1 transcript was lowered by QUE (from 25 to 100 μM), and the level of proteins playing a role in the corresponding signalling pathway was further reduced, while the production of ROS and the pattern of apoptosis were further suppressed [76].

Looking at the downregulated inflammatory mediators (e.g., CCL2, CCL20, CXCL5, and CSF3), we recognized the inhibition of the IL17-driven inflammation, whose hallmark is neutrophil accumulation [77]. Accordingly, the IL17 pathway was a significantly enriched KEGG pathway in the GSEA, together with the TNF and NF-kB pathways, which have several genes in common with the IL17 pathway and are all crucial in inflammatory responses.

Lastly, as far as QUE's ability to hamper inflammatory processes is concerned, the downregulation by QUE of PDPN appeared to be of particular interest. This gene is normally upregulated during inflammation and cancer in different cell types. Some polyphenols, such as epigallocatechin-3-gallate and curcumin, have been reported to suppress PDPN expression in mouse tumours [78,79]. However, to our knowledge, the potential role of QUE as a PDPN inhibitor has never been reported before.

The central role of inflammation was also revealed by comparing genes regulated by AFB1 (compared to the control) and those regulated by QUE + AFB1 (compared to AFB1). Indeed, most of the shared DEGs showing opposite fold changes were related to inflammatory processes. This clearly demonstrated that in BFH12 cells, particularly under our experimental conditions, many benefits of QUE (here demonstrated in terms of reduced AFB1-induced cytotoxicity) are mediated by anti-inflammatory mechanisms.

3.3.3. Oxidative Stress

A further highlight of the present study was QUE's antioxidant potential, which is the reason why natural polyphenols are studied in the first place. Specifically, QUE alone increased the expression of MAFF, a small MAF transcription factor that dimerizes with nuclear factor erythroid 2–related factor 2 (NRF2), therefore activating antioxidant defences and resisting toxicant-induced oxidative stress [80]. It is well known that NRF2 pathway activation in response to cell stress leads to the downstream regulation of cytoprotective genes forming a network of cooperating enzymes involved in drug detoxification reactions (e.g., NQO1, CYPs, GSTs, and UGTs) and elimination (e.g., ABCB1 and BCRP) of pro-oxidants [81,82]. Additional NRF2-induced antioxidant enzymes are those mediating ROS elimination, such as enzymes producing and regenerating glutathione and redox cycling enzymes (e.g., thioredoxin, GPX, SOD, and CAT) [82]. In the present study, of the above-mentioned genes, QUE alone significantly impacted the expression of GPX1, GSTA3, GSTA2, and GSTM1. Although GPXs and GSTs are known targets of natural polyphenols [83], their upregulation, together with that of MAFF, seems not enough to suggest a strong activation of the NRF2 pathway in QUE-treated BFH12 cells. Accordingly, QUE did not increase NQO1 enzyme activity. Noteworthily, when comparing the QUE + AFB1 and AFB1 conditions, QUE's antioxidant activity seems slightly more evident. Indeed, AFB1 greatly increased MDA production in BFH12 cells, but the pre-treatment with QUE was preventive. Accordingly, a previously published study reported that a QUE-supplemented diet significantly reduced the AFB1-induced increase in MDA in mouse brain tissue [84]. Importantly, the increase in MDA production is a general marker of oxidative stress, and the results obtained here demonstrate QUE's antioxidant power in bovine hepatic cells. Regarding gene expression, GSTA2, GSTM1, and MGST1 were significantly upregulated by QUE pre-treatment. Notably, in BFH12, we observed that curcumin too induced the expression of these GSTs [29]. The most interesting regulated gene was NQO1, which was significantly induced by QUE pre-treatment. However, we showed that this transcriptional change was not reflected by an increased NQO1 catalytic activity. Taking these results together, we might speculate that in BFH12 cells, QUE induces a mild NRF2 activation by regulating the protein synthesis and/or activity of this key master regulator, rather than its mRNA transcription. This activation mostly results in an increased expression of antioxidant enzymes, mainly GSTs. Conversely, NQO1 does not seem to play a key role in the QUE-dependent antioxidant response to AFB1, as was demonstrated for curcumin and resveratrol [29,31]. Interestingly, compounds that decrease the inflammatory response by suppressing NF-kB signalling, as demonstrated for QUE, activate the NRF2 pathway [85]. Thus, NRF2 is an anti-inflammatory gene [86], and inhibition of inflammation by NRF2 is associated with the inhibition of the NF-kB pathway and pro-inflammatory cytokine production [87]. Although the molecular events regulating the interaction between NRF2 and inflammatory regulators remain largely unclear, phenolic antioxidants have been demonstrated to trigger these mechanisms [88].

3.3.4. Carcinogenesis

Additional molecular targets of QUE in BFH12 cells were represented by genes involved in cancer progression. QUE alone largely downregulated some genes with a role in promoting cancer (e.g., OLFML2B, EMILIN2, and CD36), which is consistent with the anticancer potential of this natural extract [89]. In particular, QUE has already been shown to inhibit CD36 expression, thus protecting cells against cell proliferation, migration, and invasion [90,91]. Several genes involved in carcinogenesis were also modulated by QUE pre-treatment, as suggested by the GSEA suppression of the "MYC targets V1" and "MYC targets V2" hallmarks. MYC is a regulator of ribosomal biogenesis and protein synthesis [92]; its overexpression is suggested to activate several ribosomal proteins and enhance ribosomal biogenesis [93], which in turn affects cell proliferation. MYC upregulation occurs in up to 75% of cancers [94]. Some natural extracts, e.g., piperine and curcumin, have already been demonstrated to suppress the activity of MYC, resulting in anticancer effects [95,96]. In the present study, QUE did not directly suppress MYC, but it did suppress several genes whose expression is regulated by MYC. Among these were ribosomal proteins (e.g., SNRPD2, SNRPD3, RRP9, and RRP12), cell division cycle 20 (CDC20), heat shock protein family D (hsp60) member 1 (HSPD1), and bystin-like (BYSL). CDC20 has an oncogenic function in tumourigenesis [97], and its expression has been reported to be inhibited in vitro by curcumin and rottlerin, resulting in reduced survival of pancreatic cancer cells [98] and glioma cells [99]. HSPD1 encodes a mitochondrial chaperone promoting cell immortality and proliferation [100]; notably, its inhibition by curcumin has been associated with anti-glioma effects in vitro [101]. BYSL was reported to be crucial in hepatocarcinogenesis, both in vitro and in vivo. Moreover, inhibiting BYSL using RNA interference significantly decreased HCC cell proliferation in vitro, induced cell apoptosis, and partially arrested the cell cycle, leading to the failure of tumour formation [102].

3.3.5. Transporters

AFB1 negatively affected the expression of several ABC transporters involved in the transport of fatty acids (e.g., ABCD1) and lipids (e.g., ABCA2, ABCA7, and ABCG1), whereas it induced the expression of xenobiotic efflux transporters, such as ABCB1 (also known as MDR1) and ABCC4 (MRP4). Intriguingly, the co-treatment with QUE and AFB1 reverted these transcriptional changes. ABCD1 is associated with peroxisomal β-oxidation, and in neuronal N2a cells, the 7-ketocholesterol-induced decrease in its mRNA and protein levels was counteracted by QUE, resveratrol, and apigenin [103]. ABCA2 is a key gene involved in maintaining the homeostasis of sterols, sphingolipids, and cholesterol, mainly in macrophages and neurons, but also in the liver [104]. ABCG1 is a porter of intracellular cholesterol, and ABCA7 is a bridge connecting cholesterol metabolism to the immune system and the body's defence system [105]. Notably, ABCG1 has already been demonstrated in vitro (in macrophages) to be induced by QUE, thus protecting against ox-LDL-induced injury [106]. Looking at the xenobiotic efflux transporters, which are known to interact with several polyphenols [107], the role of ABCB1 in the transport of AFB1 is still unclear. Indeed, contrasting results were reported in vitro and in vivo [108,109], preventing us from interpreting the significance of ABCB1 downregulation observed here in the co-treatment with QUE and AFB1. For instance, curcumin ameliorates the duodenal toxicity of AFB1 in chicken by inducing the mRNA expression of ABCB1 and activating the encoded protein [110]. Likewise, a recent study has demonstrated that in the liver of broilers fed with QUE, ABCB1, mRNA and protein expression were upregulated [111]. On the other hand, QUE has been reported to negatively affect ABCB1 activity in HepG2 cells, thus potentiating the cytotoxic activity of anticancer drugs [112]. With regard to MRP4, it has been shown that QUE is most probably a substrate for this transporter, and its activity can be modulated by flavonoids [107]. To our knowledge, there is no evidence of a possible relationship between AFB1 and MRP4; however, the expression of ABCC4 is upregulated in HCC tissues. Based on this evidence, the ABCC4 downregulation observed in the present study after the co-treatment of QUE and AFB1 might mean that this molecular mechanism contributes to the overall mitigation of AFB1 toxicity in BFH12 cells [113].

4. Conclusions

To the best of our knowledge, QUE's potential benefits in AFB1-exposed bovine cells have never been explored using an integrated approach combining cytotoxicity, RNA-seq, and post-transcriptional assays. The results of the present study, and those obtained in previous experiments on BFH12 cells, allow us to draw the following three conclusions. (1) QUE pre-treatment hampers AFB1-induced cytotoxicity, and this protective mechanism is mediated by the regulation of genes involved in lipid homeostasis, inflammatory and immune processes, and carcinogenesis. (2) In BFH12 cells, QUE possesses antioxidant activity that is mostly mediated by other pathways than previously identified for curcumin and resveratrol (e.g., diaphorase is not a major player). (3) QUE is a promising flavonoid, even though compared to curcuminoids and resveratrol, it exhibits a lower efficacy in mitigating aflatoxin-mediated toxicity in BFH12 cells. This does not mean that using QUE as a feed additive in cattle farming should be abandoned in favour of curcumin or resveratrol. Remarkably, there are plenty of additional factors to be considered when choosing a natural compound as a feed additive, such as the costs of the natural source of the compound (e.g., fruits and vegetables), extraction feasibility and costs, bioavailability, toxicity to tissues and organs, and overall molecular mechanisms (e.g., interference of the compound with physiological proteins, nutrients, or drug transporters), which might significantly impact the overall feasibility of a certain feeding strategy.

In conclusion, targeted molecular studies are envisaged to disentangle the role of specific pathways and genes in AFB1 toxicology and, consequently, to better characterize the protective role of QUE. In the meantime, in vivo studies implementing QUE-supplemented diets are recommended to assess QUE's bioavailability and beneficial effects on aflatoxicosis.

5. Materials and Methods

5.1. Materials and Bovine Cell Line

Cell flasks and plates were purchased from Sarstedt (Verona, Italy). Williams' E Medium, L-alanyl-l-glutamine, penicillin/streptomycin, and foetal bovine serum (FBS) were acquired from Biochrom (Biospa, Milan, Italy). AFB1, QUE ($\geq$95% purity), and dimethyl sulfoxide (DMSO) were obtained from Sigma-Aldrich (St. Louis, MO, USA). PCB126 (99% purity) and AFM1 were purchased from Lab Service Analytica (Bologna, Italy), AFL from DBA Italia (Milano, Italy), and AFB1 13C17 from Orsell (Modena, Italy). All other chemicals used in the study were commercially available and of molecular biology grade. Solvents used for metabolite quantification were all of LC-MS grade.

The bovine SV40 large T-antigen-transduced foetal hepatocyte-derived cell line (BFH12) was kindly provided by Axel Schoeniger (Institute of Biochemistry, University of Leipzig, Germany) and cultured using the experimental procedures detailed in our previous paper [32]. For all the experiments, cells were used from passage 16 to passage 20.

5.2. Cytotoxicity

Cytotoxicity tests were conducted as previously reported [32]. Briefly, four days after seeding (6×10^3 cells/well), BFH12 cells were exposed to increasing concentrations of QUE for a total of 64 h (16 + 48 h; range 5–250 μM). Cell viability was measured using WST-1 Cell Proliferation Reagent (Roche, Basel, Switzerland) and expressed as a percentage relative to the viability of cells exposed to the vehicle only (0.1% DMSO). Experiments were performed in triplicate, and each concentration was tested in sextuplicate.

Afterwards, we assessed the ability of QUE to reduce AFB1-induced cell mortality. Selected sub-cytotoxic concentrations (10, 20, and 30 μM) were defined based on the QUE dose–response curve and the resulting IC_{50} obtained in the present study. In agreement with the methodological approach adopted in our recent studies on AFB1 toxicity in the BFH12 cell line [29,31,32], we pre-treated cells with an aryl hydrocarbon receptor (AHR) agonist, i.e., the most potent dioxin-like PCB (PCB126). The rationale for this choice was the hypothesis that the metabolic competence of BFH12 cells could be lower than that of adult cells. Hence, PCB126 pre-treatment aimed at increasing cell responsiveness to AFB1.

5.3. Incubation of Cells for Gene Expression Analysis

To assess the effects of AFB1 on the BFH12 transcriptional profile, cells were cultured in 6-well plates at a density of 5×10^4 cells/well. Four days after seeding, monolayers were pre-treated with 1 nM PCB126 for 24 h. Then, cells were exposed to QUE (10, 20, or 30 μM), 3.6 μM AFB1, or their combination (Figure 9). Cells pre-treated with PCB126 and exposed to the vehicle (0.1% DMSO) were used as control. A total of four independent cell culture experiments were set up. All the details about the experimental procedures, including RNA isolation, RNA concentration, and quality assessment, are available in our previous papers [29,31,32].

5.4. Quantitative Real-Time PCR (qPCR)

Targeted qPCR analyses were preliminarily carried out to assess the effects of sub-cytotoxic QUE concentrations (i.e., 10, 20, and 30 μM). Target genes were those known to be somehow involved in AFB1 mechanistic toxicology (i.e., genes contributing to AFB1 biotransformation and/or involved in the antioxidant response) [32]. Four biological replicates (i.e., independent cell culture experiments) were performed. In summary, eight experimental conditions were taken into consideration: CTRL, QUE (10, 20, and 30 μM), AFB1, and QUE (10, 20, and 30 μM) + AFB1. The procedures of reverse transcription and qPCR amplification, primer sequences, and the principles of qPCR data analysis are described elsewhere [32].

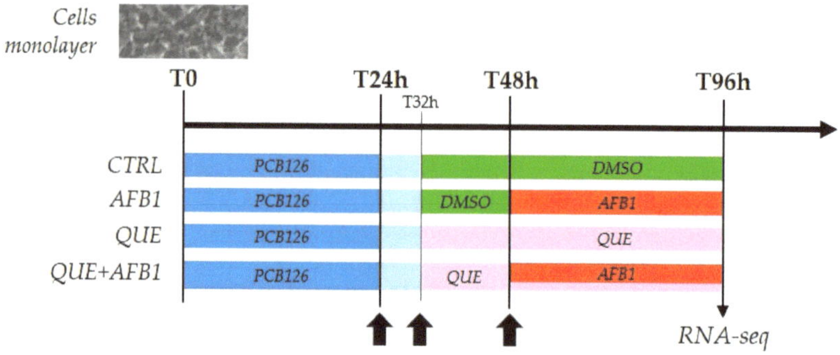

Figure 9. Cell treatments. Scheme reporting the cell treatments performed in this study and the resulting experimental groups assayed with the RNA-sequencing approach. DMSO = dimethyl sulfoxide. Arrows indicate complete medium changes (fresh medium and treatment solutions).

5.5. Preparation of Libraries and RNA-seq

In addition to the CTRL and AFB1 experimental conditions, only cells exposed to the highest QUE concentration (i.e., 30 μM), either alone or in combination with AFB1, were subjected to RNA-seq. The QUE concentration of 30 μM was selected because it was the one provoking the most significant transcriptional variations using qPCR (Section 5.4). Three independent biological replicates per condition (i.e., independent cell culture experiments) were assessed. A total of 12 tagged RNA-seq libraries were prepared and sequenced following a 50-bp single-end strategy in an Illumina Hi-Seq 4000 instrument (Fasteris SA, Geneva, Switzerland). These libraries were prepared as detailed elsewhere [32], using Agilent's SureSelect Strand Specific RNA Library Preparation Kit (Agilent Technologies, Santa Clara, CA, USA) and following the manufacturer's instructions. Notably, the CTRL and AFB1 libraries were previously analysed in a stand-alone study assessing the transcriptional effects of PCB126 and AFB1 on BFH12 cells [32]. In the present study, these libraries were analysed again in the context of a larger dataset including new data (i.e., cells treated with QUE and QUE + AFB1).

5.6. Analysis of RNA-seq Data

Differential expression (DE) was conducted using edgeR [114] and grouping samples according to the treatment (i.e., CTRL, QUE, AFB1, and QUE + AFB1). Pairwise analyses were performed to assess the transcriptional changes induced by QUE, either alone (QUE vs. CTRL) or in combination with AFB1 (QUE + AFB1 vs. AFB1). Common and tagwise dispersions were estimated (*estimateDisp*), a linear model was fitted (*glmQLFit*), and the DEGs were determined using the function *glmTreat* with the following thresholds of significance: false discovery rate (FDR) ≤ 0.05 and $\log_2$ fold change (lfc) ≥ 1.

A functional interpretation of significant DEGs was obtained through GO and KEGG over-representation analysis (ORA) implemented in the R environment, using functions included in the ClusterProfiler package (i.e., *enrichGO, enrichKEGG*) [115]. Ensemble gene identifiers were used to establish two different gene lists (i.e., significantly up- and downregulated genes) and a "background" (i.e., all the expressed genes). Dot plots and gene-concept networks were also constructed via specific functions available in the ClusterProfiler package. Redundancy of gene ontology terms was removed using the *simplify* function (similarity cutoff = 0.5). Dot plots display the most significant enriched terms ($p \leq$

0.05), while gene-concept networks highlight which genes were involved in the significant GO terms.

In order to analyse the transcriptional differences triggered by the cotreatment QUE + AFB1 (compared to AFB1 alone), a pre-ranked GSEA [116] was also performed. This statistical approach is helpful in determining whether gene sets that are defined a priori show statistically significant enrichment at either end of the ranking. A statistically significant enrichment value (Benjamini–Hochberg adjusted $p \leq 0.05$) indicates that the biological activity (e.g., the biomolecular pathway), characterized by the gene set, is correlated with the supplied ranking. The ranked input was prepared as follows. A score ($-\log 10 \times p \times \text{sign(lfc)}$) was calculated from the edgeR differential expression results. A gene differentially expressed at a significant level (a low p close to 0) will be assigned a high score. The sign of lfc indicates whether the gene has an expression that is higher in the QUE + AFB1 condition (i.e., if lfc > 0, then the score will have a positive sign) or lower in the QUE + AFB1 condition (i.e., if lfc < 0, then the score will have a negative sign). In this way, all the expressed genes were ranked from top upregulated to top downregulated. The analysis was carried out using the *gseKEGG* and *GSEA* functions provided by the ClusterProfiler package [115]. While the first function was run for the enrichment of the KEGG pathways, the second one was run using the hallmark gene sets of the Molecular Signatures Database (MSigDB).

5.7. Analytical Investigations

Medium and cell samples were collected at the end of the experiment. Total AFB1, AFM1, and AFL were measured using LC-MS/MS in all the experimental conditions according to [32]. The LC-MS/MS system consisted of a Waters Acquity UPLC binary pump coupled to a Quattro Premier XE triple quadrupole mass spectrometer (Waters, Milford, MA, USA). Waters Acquity BEH C18 (50 × 2.1 mm, 1.7 μm) reversed-phase columns were used for the chromatographic separation. The instrument was equipped with an electrospray ionization source (ESI) operating in positive mode at a capillary voltage of 3.75 kV, with source and desolvation temperatures of 120 and 350 °C, respectively. The desolvation gas flow was 650 L/h, and the cone gas flow was 100 L/h. For each analyte, the following specific transitions were monitored: 313.1 > 284.7 m/z (CV 53 V; CE 25 eV) for AFB1, 329.1 > 272.7 m/z (CV 42 V; CE 26 eV) for AFM1, 297.1 > 268.4 m/z (CV 48 V; CE 20 eV) for AFL, and 330.3 > 300.6 m/z (CV 50 V; CE 22 eV) for the internal standard 13C17-AFB1.

5.8. Oxidative Stress

The oxidative damage induced by AFB1 and the possible counteracting effect of QUE were assessed by measuring the amount of MDA, a marker of lipid peroxidation. The experimental groups were the same as those used for the RNA-seq investigations. The MDA production was measured in six independent biological replicates (i.e., independent cell culture experiments) using the ab233471 lipid peroxidation colourimetric assay kit (Abcam, Prodotti Gianni S.p.A., Milan, Italy) according to [29].

5.9. Enzymatic Activity

As post-transcriptional confirmatory assays, CYP3A and NQO1 enzyme activity were measured. CYP3A and NQO1 are two key enzymes involved in AFB1 bioactivation and the antioxidant response, respectively. The former enzyme activity was measured using the P450-Glo^{TM} CYP3A4 assay, with luciferin-IPA as a substrate (Promega Corporation, Madison, WI, USA), and following the manufacturer's instructions. For both enzymes, further protocol details are reported elsewhere [29]. As for NQO1, its catalytic activity was assessed using the ab184867 NQO1 activity assay kit (Abcam, Prodotti Gianni S.p.A., Milan, Italy).

5.10. Statistical Analysis

Dose–response curves were obtained using the GraphPad Prism software (version 8.0.2, San Diego, CA, USA), plotting a nonlinear regression [log(inhibitor) vs. normalized response, variable slope]. The IC_{50} and the goodness of fit (R squared) were obtained via the software.

The statistical analyses of cytotoxicity, qPCR data, and MDA content, as well as of NQO1 and CYP3A enzyme activities, were performed using a one-way ANOVA followed by Dunnett's multiple comparisons test, with the level of significance set at $p \leq 0.05$. The same statistical approach was adopted for the analytical data, but Tukey's multiple comparisons test was used instead of Dunnett's test.

Supplementary Materials: The following supporting information can be downloaded at https://www.mdpi.com/article/10.3390/toxins15090555/s1, Figure S1. Cytotoxicity; Figure S2. Effects of QUE in AFB1 biotransformation; Figure S3: MDS plot; Figure S4: MD plots; Figure S5: Over-representation analysis QUE vs. CTRL; Figure S6: Over-representation analysis: QUE+AFB1 vs. AFB1; Figure S7: Concordance between qPCR and RNA-seq results; Table S1: Sequencing and mapping results; Table S2: Differential expression analysis; Table S3: Enrichment analysis; Table S4: Over-representation analysis (GSEA).

Author Contributions: Conceptualization, M.P., M.G., A.Z. and M.D.; Data curation, M.P.; Formal analysis, M.P., M.G., A.B. (Andrea Barbarossa) and L.M.; Funding acquisition, M.P., A.Z. and M.D.; Investigation, M.P., M.G., R.T., I.B. and A.B. (Anisa Bardhi); Methodology, M.P., M.G., I.B. and A.B. (Andrea Barbarossa); Supervision, M.D.; Writing—original draft, M.P.; Writing—review and editing, M.G., A.B. (Andrea Barbarossa), A.Z., L.M. and M.D. All authors have read and agreed to the published version of the manuscript.

Funding: This research was supported by (i) a grant to Mauro Dacasto from the Italian Ministry of Education, Universities and Research (MIUR), Project of national interest (PRIN) 2015, "Effects of curcuminoids on the toxicity and mammary excretion of Aflatoxins (B1 and B2) and their metabolites in dairy cows", (ii) a grant to Marianna Pauletto from the University of Padova ("DOR2019—prot. DOR1917008").

Institutional Review Board Statement: Not applicable.

Informed Consent Statement: Not applicable.

Data Availability Statement: Raw Illumina sequencing data have been deposited in GenBank (SRA) under the BioProject accession PRJNA627332.

Acknowledgments: Authors are grateful to Axel Schoeniger (Institute of Biochemistry, University of Leipzig, Germany) who provided the BFH12 cell line.

Conflicts of Interest: The authors declare no conflict of interest.

References

1. Leggieri, M.C.; Toscano, P.; Battilani, P. Predicted Aflatoxin B1 Increase in Europe Due to Climate Change: Actions and Reactions at Global Level. *Toxins* **2021**, *13*, 292. [CrossRef] [PubMed]
2. Li, C.; Liu, X.; Wu, J.; Ji, X.; Xu, Q. Research progress in toxicological effects and mechanism of aflatoxin B $_1$ toxin. *PeerJ* **2022**, *10*, e13850. [CrossRef] [PubMed]
3. Eskola, M.; Kos, G.; Elliott, C.T.; Hajšlová, J.; Mayar, S.; Krska, R. Worldwide contamination of food-crops with mycotoxins: Validity of the widely cited 'FAO estimate' of 25%. *Crit. Rev. Food Sci. Nutr.* **2020**, *60*, 2773–2789. [CrossRef] [PubMed]
4. Medina, Á.; González-Jartín, J.M.; Sainz, M.J. Impact of global warming on mycotoxins. *Curr. Opin. Food Sci.* **2017**, *18*, 76–81. [CrossRef]
5. Battilani, P.; Toscano, P.; Van der Fels-Klerx, H.J.; Moretti, A.; Camardo Leggieri, M.; Brera, C.; Rortais, A.; Goumperis, T.; Robinson, T. Aflatoxin B1 contamination in maize in Europe increases due to climate change. *Sci. Rep.* **2016**, *6*, 24328. [CrossRef] [PubMed]
6. Dalvi, R.R. An overview of aflatoxicosis of poultry: Its characteristics, prevention and reduction. *Vet. Res. Commun.* **1986**, *10*, 429–443. [CrossRef]
7. Pickova, D.; Ostry, V.; Toman, J.; Malir, F. Aflatoxins: History, Significant Milestones, Recent Data on Their Toxicity and Ways to Mitigation. *Toxins* **2021**, *13*, 399. [CrossRef]

8. Rasouli, H.; Nayeri, F.D.; Khodarahmi, R. May phytophenolics alleviate aflatoxins-induced health challenges? A holistic insight on current landscape and future prospects. *Front. Nutr.* **2022**, *9*, 981984. [CrossRef]
9. Wang, X.; Wang, T.; Nepovimova, E.; Long, M.; Wu, W.; Kuca, K. Progress on the detoxification of aflatoxin B1 using natural anti-oxidants. *Food Chem. Toxicol.* **2022**, *169*, 113417. [CrossRef]
10. Rushing, B.R.; Selim, M.I. Aflatoxin B1: A review on metabolism, toxicity, occurrence in food, occupational exposure, and detoxification methods. *Food Chem. Toxicol.* **2019**, *124*, 81–100. [CrossRef]
11. Deng, J.; Zhao, L.; Zhang, N.-Y.; Karrow, N.A.; Krumm, C.S.; Qi, D.-S.; Sun, L.-H. Aflatoxin B1 metabolism: Regulation by phase I and II metabolizing enzymes and chemoprotective agents. *Mutat. Res. Rev. Mutat. Res.* **2018**, *778*, 79–89. [CrossRef]
12. Dai, C.; Tian, E.; Hao, Z.; Tang, S.; Wang, Z.; Sharma, G.; Jiang, H.; Shen, J. Aflatoxin B1 Toxicity and Protective Effects of Curcumin: Molecular Mechanisms and Clinical Implications. *Antioxidants* **2022**, *11*, 2031. [CrossRef] [PubMed]
13. Rawal, S.; Kim, J.E.; Coulombe, R. Aflatoxin B1 in poultry: Toxicology, metabolism and prevention. *Res. Vet. Sci.* **2010**, *89*, 325–331. [CrossRef]
14. Wang, L.; Huang, Q.; Wu, J.; Wu, W.; Jiang, J.; Yan, H.; Huang, J.; Sun, Y.; Deng, Y. The metabolism and biotransformation of AFB1: Key enzymes and pathways. *Biochem. Pharmacol.* **2022**, *199*, 115005. [CrossRef] [PubMed]
15. Williams, D.E. The rainbow trout liver cancer model: Response to environmental chemicals and studies on promotion and chemoprevention. *Comp. Biochem. Physiol. Part C Toxicol. Pharmacol.* **2012**, *155*, 121–127. [CrossRef]
16. Benkerroum, N. Chronic and Acute Toxicities of Aflatoxins: Mechanisms of Action. *Int. J. Environ. Res. Public Health* **2020**, *17*, 423. [CrossRef] [PubMed]
17. Melaram, R. Environmental Risk Factors Implicated in Liver Disease: A Mini-Review. *Front. Public Health* **2021**, *9*, 683719. [CrossRef]
18. Wu, H.-C.; Santella, R. The Role of Aflatoxins in Hepatocellular Carcinoma. *Hepat. Mon.* **2012**, *12*, e7238. [CrossRef] [PubMed]
19. Han, C.; Yu, T.; Qin, W.; Liao, X.; Huang, J.; Liu, Z.; Yu, L.; Liu, X.; Chen, Z.; Yang, C.; et al. Genome-wide association study of the TP53 R249S mutation in hepatocellular carcinoma with aflatoxin B1 exposure and infection with hepatitis B virus. *J. Gastrointest. Oncol.* **2020**, *11*, 1333–1349. [CrossRef]
20. IARC. *A Review of Human Carcinogens*; Centre International de Recherche sur le Cancer, Ed.; IARC Monographs on the Evaluation of Carcinogenic Risks to Humans; International Agency for Research on Cancer: Lyon, France, 2012; ISBN 978-92-832-1329-1.
21. Kuilman, M.E.M.; Maas, R.F.M.; Fink-Gremmels, J. Cytochrome P450-mediated metabolism and cytotoxicity of aflatoxin B1 in bovine hepatocytes. *Toxicol. Vitr.* **2000**, *14*, 321–327. [CrossRef]
22. Fink-Gremmels, J. Mycotoxins in cattle feeds and carry-over to dairy milk: A review. *Food Addit. Contam. Part A* **2008**, *25*, 172–180. [CrossRef]
23. Intanoo, M.; Kongkeitkajorn, M.B.; Pattarajinda, V.; Bernard, J.K.; Callaway, T.R.; Suriyasathaporn, W.; Phasuk, Y. Isolation and screening of aflatoxin-detoxifying yeast and bacteria from ruminal fluids to reduce aflatoxin B_1 contamination in dairy cattle feed. *J. Appl. Microbiol.* **2018**, *125*, 1603–1613. [CrossRef]
24. Girolami, F.; Barbarossa, A.; Badino, P.; Ghadiri, S.; Cavallini, D.; Zaghini, A.; Nebbia, C. Effects of Turmeric Powder on Aflatoxin M1 and Aflatoxicol Excretion in Milk from Dairy Cows Exposed to Aflatoxin B1 at the EU Maximum Tolerable Levels. *Toxins* **2022**, *14*, 430. [CrossRef]
25. Min, L.; Fink-Gremmels, J.; Li, D.; Tong, X.; Tang, J.; Nan, X.; Yu, Z.; Chen, W.; Wang, G. An overview of aflatoxin B1 biotransformation and aflatoxin M1 secretion in lactating dairy cows. *Anim. Nutr.* **2021**, *7*, 42–48. [CrossRef] [PubMed]
26. Iori, S.; Pauletto, M.; Bassan, I.; Bonsembiante, F.; Gelain, M.E.; Bardhi, A.; Barbarossa, A.; Zaghini, A.; Dacasto, M.; Giantin, M. Deepening the Whole Transcriptomics of Bovine Liver Cells Exposed to AFB1: A Spotlight on Toll-like Receptor 2. *Toxins* **2022**, *14*, 504. [CrossRef]
27. Neha, K.; Haider, M.R.; Pathak, A.; Yar, M.S. Medicinal prospects of antioxidants: A review. *Eur. J. Med. Chem.* **2019**, *178*, 687–704. [CrossRef] [PubMed]
28. Makhuvele, R.; Naidu, K.; Gbashi, S.; Thipe, V.C.; Adebo, O.A.; Njobeh, P.B. The use of plant extracts and their phytochemicals for control of toxigenic fungi and mycotoxins. *Heliyon* **2020**, *6*, e05291. [CrossRef]
29. Pauletto, M.; Giantin, M.; Tolosi, R.; Bassan, I.; Barbarossa, A.; Zaghini, A.; Dacasto, M. Curcumin Mitigates AFB1-Induced Hepatic Toxicity by Triggering Cattle Antioxidant and Anti-inflammatory Pathways: A Whole Transcriptomic In Vitro Study. *Antioxidants* **2020**, *9*, 1059. [CrossRef]
30. Ghadiri, S.; Spalenza, V.; Dellafiora, L.; Badino, P.; Barbarossa, A.; Dall'Asta, C.; Nebbia, C.; Girolami, F. Modulation of aflatoxin B1 cytotoxicity and aflatoxin M1 synthesis by natural antioxidants in a bovine mammary epithelial cell line. *Toxicol. In Vitro* **2019**, *57*, 174–183. [CrossRef]
31. Pauletto, M.; Giantin, M.; Tolosi, R.; Bassan, I.; Barbarossa, A.; Zaghini, A.; Dacasto, M. Discovering the Protective Effects of Resveratrol on Aflatoxin B1-Induced Toxicity: A Whole Transcriptomic Study in a Bovine Hepatocyte Cell Line. *Antioxidants* **2021**, *10*, 1225. [CrossRef]
32. Pauletto, M.; Tolosi, R.; Giantin, M.; Guerra, G.; Barbarossa, A.; Zaghini, A.; Dacasto, M. Insights into Aflatoxin B1 Toxicity in Cattle: An In Vitro Whole-Transcriptomic Approach. *Toxins* **2020**, *12*, 429. [CrossRef]
33. Zancanella, V.; Giantin, M.; Dacasto, M. Absolute quantification and modulation of cytochrome P450 3A isoforms in cattle liver. *Vet. J.* **2014**, *202*, 106–111. [CrossRef]

34. Arenas-Huertero, F.; Zaragoza-Ojeda, M.; Sánchez-Alarcón, J.; Milić, M.; Klarić, M.Š.; Montiel-González, J.M.; Valencia-Quintana, R. Involvement of Ahr Pathway in Toxicity of Aflatoxins and Other Mycotoxins. *Front. Microbiol.* **2019**, *10*, 2347. [CrossRef] [PubMed]
35. Zhou, Y.; Jin, Y.; Yu, H.; Shan, A.; Shen, J.; Zhou, C.; Zhao, Y.; Fang, H.; Wang, X.; Wang, J.; et al. Resveratrol inhibits aflatoxin B1-induced oxidative stress and apoptosis in bovine mammary epithelial cells and is involved the Nrf2 signaling pathway. *Toxicon* **2019**, *164*, 10–15. [CrossRef] [PubMed]
36. Sang, R.; Ge, B.; Li, H.; Zhou, H.; Yan, K.; Wang, W.; Cui, Q.; Zhang, X. Taraxasterol alleviates aflatoxin B1-induced liver damage in broiler chickens via regulation of oxidative stress, apoptosis and autophagy. *Ecotoxicol. Environ. Saf.* **2023**, *251*, 114546. [CrossRef]
37. Atrahimovich, D.; Samson, A.O.; Barsheshet, Y.; Vaya, J.; Khatib, S.; Reuveni, E. Genome-wide localization of the polyphenol quercetin in human monocytes. *BMC Genom.* **2019**, *20*, 606. [CrossRef] [PubMed]
38. Huang, Z.; Fang, F.; Wang, J.; Wong, C.-W. Structural activity relationship of flavonoids with estrogen-related receptor gamma. *FEBS Lett.* **2010**, *584*, 22–26. [CrossRef]
39. Rudrapal, M.; Khairnar, S.J.; Khan, J.; Dukhyil, A.B.; Ansari, M.A.; Alomary, M.N.; Alshabrmi, F.M.; Palai, S.; Deb, P.K.; Devi, R. Dietary Polyphenols and Their Role in Oxidative Stress-Induced Human Diseases: Insights Into Protective Effects, Antioxidant Potentials and Mechanism(s) of Action. *Front. Pharmacol.* **2022**, *13*, 806470. [CrossRef]
40. Spencer, J.P.E. Beyond antioxidants: The cellular and molecular interactions of flavonoids and how these underpin their actions on the brain. *Proc. Nutr. Soc.* **2010**, *69*, 244–260. [CrossRef] [PubMed]
41. Srivastava, S.; Somasagara, R.R.; Hegde, M.; Nishana, M.; Tadi, S.K.; Srivastava, M.; Choudhary, B.; Raghavan, S.C. Quercetin, a Natural Flavonoid Interacts with DNA, Arrests Cell Cycle and Causes Tumor Regression by Activating Mitochondrial Pathway of Apoptosis. *Sci. Rep.* **2016**, *6*, 24049. [CrossRef]
42. Cory, H.; Passarelli, S.; Szeto, J.; Tamez, M.; Mattei, J. The Role of Polyphenols in Human Health and Food Systems: A Mini-Review. *Front. Nutr.* **2018**, *5*, 87. [CrossRef]
43. Fraga, C.G.; Croft, K.D.; Kennedy, D.O.; Tomás-Barberán, F.A. The effects of polyphenols and other bioactives on human health. *Food Funct.* **2019**, *10*, 514–528. [CrossRef]
44. Abdel-Moneim, A.E.; Shehata, A.M.; Alzahrani, S.O.; Shafi, M.E.; Mesalam, N.M.; Taha, A.E.; Swelum, A.A.; Arif, M.; Fayyaz, M.; Abd El-Hack, M.E. The role of polyphenols in poultry nutrition. *J. Anim. Physiol. Anim. Nutr.* **2020**, *104*, 1851–1866. [CrossRef]
45. Bešlo, D.; Došlić, G.; Agić, D.; Rastija, V.; Šperanda, M.; Gantner, V.; Lučić, B. Polyphenols in Ruminant Nutrition and Their Effects on Reproduction. *Antioxidants* **2022**, *11*, 970. [CrossRef] [PubMed]
46. Chen, J.; Huang, Z.; Cao, X.; Zou, T.; You, J.; Guan, W. Plant-derived polyphenols in sow nutrition: An update. *Anim. Nutr.* **2023**, *12*, 96–107. [CrossRef] [PubMed]
47. Formato, M.; Cimmino, G.; Brahmi-Chendouh, N.; Piccolella, S.; Pacifico, S. Polyphenols for Livestock Feed: Sustainable Perspectives for Animal Husbandry? *Molecules* **2022**, *27*, 7752. [CrossRef] [PubMed]
48. Chen, N.; Liu, Y.; Bai, T.; Chen, J.; Zhao, Z.; Li, J.; Shao, B.; Zhang, Z.; Zhou, Y.; Wang, X.; et al. Quercetin Inhibits Hsp70 Blocking of Bovine Viral Diarrhea Virus Infection and Replication in the Early Stage of Virus Infection. *Viruses* **2022**, *14*, 2365. [CrossRef]
49. Gong, X.; Huang, Y.; Ma, Q.; Jiang, M.; Zhan, K.; Zhao, G. Quercetin Alleviates Lipopolysaccharide-Induced Cell Damage and Inflammation via Regulation of the TLR4/NF-κB Pathway in Bovine Intestinal Epithelial Cells. *Curr. Issues Mol. Biol.* **2022**, *44*, 5234–5246. [CrossRef] [PubMed]
50. Jiang, M.; Lv, Z.; Huang, Y.; Cheng, Z.; Meng, Z.; Yang, T.; Yan, Q.; Lin, M.; Zhan, K.; Zhao, G. Quercetin Alleviates Lipopolysaccharide-Induced Inflammatory Response in Bovine Mammary Epithelial Cells by Suppressing TLR4/NF-κB Signaling Pathway. *Front. Vet. Sci.* **2022**, *9*, 915726. [CrossRef]
51. Karimian, A.; Majidinia, M.; Moliani, A.; Alemi, F.; Asemi, Z.; Yousefi, B.; Fazlollahpour naghibi, A. The modulatory effects of two bioflavonoids, quercetin and thymoquinone on the expression levels of DNA damage and repair genes in human breast, lung and prostate cancer cell lines. *Pathol. Res. Pract.* **2022**, *240*, 154143. [CrossRef]
52. Abdu, S.; Juaid, N.; Amin, A.; Moulay, M.; Miled, N. Effects of Sorafenib and Quercetin Alone or in Combination in Treating Hepatocellular Carcinoma: In Vitro and In Vivo Approaches. *Molecules* **2022**, *27*, 8082. [CrossRef] [PubMed]
53. Tezerji, S.; Abdolazimi, H.; Fallah, A.; Talaei, B. The effect of resveratrol and quercetin intervention on azoxymethane-induced colon cancer in Rats model. *Clin. Nutr. Open Sci.* **2022**, *45*, 91–102. [CrossRef]
54. El Khoury, A.; Atoui, A.; Yaghi, J. Analysis of aflatoxin M1 in milk and yogurt and AFM1 reduction by lactic acid bacteria used in Lebanese industry. *Food Control* **2011**, *22*, 1695–1699. [CrossRef]
55. Murcia, H.W.; Diaz, G.J. In vitro hepatic aflatoxicol production is related to a higher resistance to aflatoxin B1 in poultry. *Sci. Rep.* **2020**, *10*, 5508. [CrossRef] [PubMed]
56. Zhang, M.; Xie, Z.; Gao, W.; Pu, L.; Wei, J.; Guo, C. Quercetin regulates hepatic cholesterol metabolism by promoting cholesterol-to-bile acid conversion and cholesterol efflux in rats. *Nutr. Res.* **2016**, *36*, 271–279. [CrossRef]
57. Wang, M.; Mao, Y.; Wang, B.; Wang, S.; Lu, H.; Ying, L.; Li, Y. Quercetin Improving Lipid Metabolism by Regulating Lipid Metabolism Pathway of Ileum Mucosa in Broilers. *Oxidative Med. Cell. Longev.* **2020**, *2020*, 8686248. [CrossRef]
58. Li, W.; Yang, C.; Mei, X.; Huang, R.; Zhang, S.; Tang, Y.; Dong, Q.; Zhou, C. Effect of the polyphenol-rich extract from *Allium cepa* on hyperlipidemic sprague-dawley rats. *J. Food Biochem.* **2021**, *45*, e13565. [CrossRef]

59. Jiang, H.Z.; Quan, X.F.; Tian, W.X.; Hu, J.M.; Wang, P.C.; Huang, S.Z.; Cheng, Z.Q.; Liang, W.J.; Zhou, J.; Ma, X.F.; et al. Fatty acid synthase inhibitors of phenolic constituents isolated from Garcinia mangostana. *Bioorganic Med. Chem. Lett.* **2010**, *20*, 6045–6047. [CrossRef]
60. Tian, W.-X. Inhibition of Fatty Acid Synthase by Polyphenols. *Curr. Med. Chem.* **2006**, *13*, 967–977. [CrossRef]
61. Sun, P.; Zhao, L.; Zhang, N.; Zhou, J.; Zhang, L.; Wu, W.; Ji, B.; Zhou, F. Bioactivity of Dietary Polyphenols: The Role in LDL-C Lowering. *Foods* **2021**, *10*, 2666. [CrossRef]
62. Liang, N.; Li, Y.-M.; He, Z.; Hao, W.; Zhao, Y.; Liu, J.; Zhu, H.; Kwek, E.; Ma, K.-Y.; He, W.-S.; et al. Rutin and Quercetin Decrease Cholesterol in HepG2 Cells but Not Plasma Cholesterol in Hamsters by Oral Administration. *Molecules* **2021**, *26*, 3766. [CrossRef] [PubMed]
63. Lu, J.; Wu, D.; Zheng, Y.; Hu, B.; Zhang, Z.; Shan, Q.; Zheng, Z.; Liu, C.; Wang, Y. Quercetin activates AMP-activated protein kinase by reducing PP2C expression protecting old mouse brain against high cholesterol-induced neurotoxicity: Quercetin attenuates neurodegeneration. *J. Pathol.* **2010**, *222*, 199–212. [CrossRef] [PubMed]
64. Straniero, S.; Laskar, A.; Savva, C.; Härdfeldt, J.; Angelin, B.; Rudling, M. Murine bile acids explain species differences in the regulation of bile acid and cholesterol metabolism. *Atherosclerosis* **2021**, *331*, e128. [CrossRef]
65. Hua, H.; Yang, T.; Huang, L.; Chen, R.; Li, M.; Zou, Z.; Wang, N.; Yang, D.; Liu, Y. Protective Effects of Lanosterol Synthase Up-Regulation in UV-B-Induced Oxidative Stress. *Front. Pharmacol.* **2019**, *10*, 947. [CrossRef]
66. González-Gallego, J.; García-Mediavilla, M.V.; Sánchez-Campos, S.; Tuñón, M.J. Fruit polyphenols, immunity and inflammation. *Br. J. Nutr.* **2010**, *104*, S15–S27. [CrossRef]
67. Liu, Y.; Yu, C.; Ji, K.; Wang, X.; Li, X.; Xie, H.; Wang, Y.; Huang, Y.; Qi, D.; Fan, H. Quercetin reduces TNF-α-induced mesangial cell proliferation and inhibits PTX3 production: Involvement of NF-κB signaling pathway. *Phytother. Res.* **2019**, *33*, 2401–2408. [CrossRef]
68. Wang, Y.; Liu, Y.; Zhang, M.; Lv, L.; Zhang, X.; Zhang, P.; Zhou, Y. Inhibition of PTGS1 promotes osteogenic differentiation of adipose-derived stem cells by suppressing NF-kB signaling. *Stem Cell Res. Ther.* **2019**, *10*, 57. [CrossRef]
69. Xiao, X.; Shi, D.; Liu, L.; Wang, J.; Xie, X.; Kang, T.; Deng, W. Quercetin Suppresses Cyclooxygenase-2 Expression and Angiogenesis through Inactivation of P300 Signaling. *PLoS ONE* **2011**, *6*, e22934. [CrossRef]
70. Tacke, F.; Zimmermann, H.W.; Berres, M.-L.; Trautwein, C.; Wasmuth, H.E. Serum chemokine receptor CXCR3 ligands are associated with progression, organ dysfunction and complications of chronic liver diseases: CXCR3 chemokines in liver diseases. *Liver Int.* **2011**, *31*, 840–849. [CrossRef]
71. Kameda, M.; Otsuka, M.; Chiba, H.; Kuronuma, K.; Hasegawa, T.; Takahashi, H.; Takahashi, H. CXCL9, CXCL10, and CXCL11; biomarkers of pulmonary inflammation associated with autoimmunity in patients with collagen vascular diseases–associated interstitial lung disease and interstitial pneumonia with autoimmune features. *PLoS ONE* **2020**, *15*, e0241719. [CrossRef]
72. Tokunaga, R.; Zhang, W.; Naseem, M.; Puccini, A.; Berger, M.D.; Soni, S.; McSkane, M.; Baba, H.; Lenz, H.-J. CXCL9, CXCL10, CXCL11/CXCR3 axis for immune activation—A target for novel cancer therapy. *Cancer Treat. Rev.* **2018**, *63*, 40–47. [CrossRef] [PubMed]
73. Qin, H.; Li, H.; Zhou, X.; Peng, C.; Tan, H.; Wang, M. Effect of superoxide and inflammatory factor on aflatoxin B1 triggered hepatocellular carcinoma. *Am. J. Transl. Res.* **2016**, *8*, 4003–4008. [PubMed]
74. Mehrzad, J.; Malvandi, A.M.; Alipour, M.; Hosseinkhani, S. Environmentally relevant level of aflatoxin B 1 elicits toxic pro-inflammatory response in murine CNS-derived cells. *Toxicol. Lett.* **2017**, *279*, 96–106. [CrossRef] [PubMed]
75. Musumeci, D.; Roviello, G.N.; Montesarchio, D. An overview on HMGB1 inhibitors as potential therapeutic agents in HMGB1-related pathologies. *Pharmacol. Ther.* **2014**, *141*, 347–357. [CrossRef]
76. Fang, P.; Liang, J.; Jiang, X.; Fang, X.; Wu, M.; Wei, X.; Yang, W.; Hou, W.; Zhang, Q. Quercetin Attenuates d-GaLN-Induced L02 Cell Damage by Suppressing Oxidative Stress and Mitochondrial Apoptosis via Inhibition of HMGB1. *Front. Pharmacol.* **2020**, *11*, 608. [CrossRef]
77. Monin, L.; Gaffen, S.L. Interleukin 17 Family Cytokines: Signaling Mechanisms, Biological Activities, and Therapeutic Implications. *Cold Spring Harb. Perspect. Biol.* **2018**, *10*, a028522. [CrossRef]
78. Mineva, N.D.; Paulson, K.E.; Naber, S.P.; Yee, A.S.; Sonenshein, G.E. Epigallocatechin-3-Gallate Inhibits Stem-Like Inflammatory Breast Cancer Cells. *PLoS ONE* **2013**, *8*, e73464. [CrossRef]
79. Da, W.; Zhu, J.; Wang, L.; Sun, Q. Curcumin suppresses lymphatic vessel density in an in vivo human gastric cancer model. *Tumor Biol.* **2015**, *36*, 5215–5223. [CrossRef]
80. Sykiotis, G.P.; Bohmann, D. Keap1/Nrf2 Signaling Regulates Oxidative Stress Tolerance and Lifespan in Drosophila. *Dev. Cell* **2008**, *14*, 76–85. [CrossRef]
81. Hayes, J.D.; Dinkova-Kostova, A.T. The Nrf2 regulatory network provides an interface between redox and intermediary metabolism. *Trends Biochem. Sci.* **2014**, *39*, 199–218. [CrossRef]
82. He, F.; Ru, X.; Wen, T. NRF2, a Transcription Factor for Stress Response and Beyond. *Int. J. Mol. Sci.* **2020**, *21*, 4777. [CrossRef]
83. Zhou, Y.; Jiang, Z.; Lu, H.; Xu, Z.; Tong, R.; Shi, J.; Jia, G. Recent Advances of Natural Polyphenols Activators for Keap1-Nrf2 Signaling Pathway. *Chem. Biodivers.* **2019**, *16*, e1900400. [CrossRef]
84. Gugliandolo, E.; Peritore, A.F.; D'Amico, R.; Licata, P.; Crupi, R. Evaluation of Neuroprotective Effects of Quercetin against Aflatoxin B1-Intoxicated Mice. *Animals* **2020**, *10*, 898. [CrossRef] [PubMed]

85. Bellezza, I.; Giambanco, I.; Minelli, A.; Donato, R. Nrf2-Keap1 signaling in oxidative and reductive stress. *Biochim. Biophys. Acta Mol. Cell Res.* **2018**, *1865*, 721–733. [CrossRef]
86. Ma, Q. Role of Nrf2 in Oxidative Stress and Toxicity. *Annu. Rev. Pharmacol. Toxicol.* **2013**, *53*, 401–426. [CrossRef] [PubMed]
87. Li, W.; Khor, T.O.; Xu, C.; Shen, G.; Jeong, W.-S.; Yu, S.; Kong, A.-N. Activation of Nrf2-antioxidant signaling attenuates NFκB-inflammatory response and elicits apoptosis. *Biochem. Pharmacol.* **2008**, *76*, 1485–1489. [CrossRef]
88. Ma, Q.; Kinneer, K.; Ye, J.; Chen, B.J. Inhibition of Nuclear Factor κB by Phenolic Antioxidants: Interplay between Antioxidant Signaling and Inflammatory Cytokine Expression. *Mol. Pharmacol.* **2003**, *64*, 211–219. [CrossRef] [PubMed]
89. Rauf, A.; Imran, M.; Khan, I.A.; ur-Rehman, M.-; Gilani, S.A.; Mehmood, Z.; Mubarak, M.S. Anticancer potential of quercetin: A comprehensive review: Quercetin as an anticancer agent. *Phytother. Res.* **2018**, *32*, 2109–2130. [CrossRef] [PubMed]
90. Chen, L.; Xia, J.-S.; Wu, J.-H.; Chen, Y.-G.; Qiu, C.-J. Quercetin suppresses cell survival and invasion in oral squamous cell carcinoma via the miR-1254/CD36 cascade in vitro. *Hum. Exp. Toxicol.* **2021**, *40*, 1413–1421. [CrossRef] [PubMed]
91. Jia, Q.; Cao, H.; Shen, D.; Li, S.; Yan, L.; Chen, C.; Xing, S.; Dou, F. Quercetin protects against atherosclerosis by regulating the expression of PCSK9, CD36, PPARγ, LXRα and ABCA1. *Int. J. Mol. Med.* **2019**, *44*, 893–902. [CrossRef]
92. van Riggelen, J.; Yetil, A.; Felsher, D.W. MYC as a regulator of ribosome biogenesis and protein synthesis. *Nat. Rev. Cancer* **2010**, *10*, 301–309. [CrossRef]
93. Guo, Q.M.; Malek, R.L.; Kim, S.; Chiao, C.; He, M.; Ruffy, M.; Sanka, K.; Lee, N.H.; Dang, C.V.; Liu, E.T. Identification of c-myc responsive genes using rat cDNA microarray. *Cancer Res.* **2000**, *60*, 5922–5928. [PubMed]
94. Dang, C.V.; O'Donnell, K.A.; Juopperi, T. The great MYC escape in tumorigenesis. *Cancer Cell* **2005**, *8*, 177–178. [CrossRef] [PubMed]
95. Hosseini, S.A.; Zand, H.; Cheraghpour, M. The Influence of Curcumin on the Downregulation of MYC, Insulin and IGF-1 Receptors: A Possible Mechanism Underlying the Anti-Growth and Anti-Migration in Chemoresistant Colorectal Cancer Cells. *Medicina* **2019**, *55*, 90. [CrossRef]
96. Pandya, N.; Kumar, A. Piperine analogs arrest c-myc gene leading to downregulation of transcription for targeting cancer. *Sci. Rep.* **2021**, *11*, 22909. [CrossRef]
97. Wang, L.; Zhang, J.; Wan, L.; Zhou, X.; Wang, Z.; Wei, W. Targeting Cdc20 as a novel cancer therapeutic strategy. *Pharmacol. Ther.* **2015**, *151*, 141–151. [CrossRef]
98. Zhang, Y.; Xue, Y.; Li, H.; Qiu, D.; Wang, Z.; Tan, S. Inhibition of Cell Survival by Curcumin Is Associated with Downregulation of Cell Division Cycle 20 (Cdc20) in Pancreatic Cancer Cells. *Nutrients* **2017**, *9*, 109. [CrossRef]
99. Wang, L.; Hou, Y.; Yin, X.; Su, J.; Zhao, Z.; Ye, X.; Zhou, X.; Zhou, L.; Wang, Z. Rottlerin inhibits cell growth and invasion via down-regulation of Cdc20 in glioma cells. *Oncotarget* **2016**, *7*, 69770–69782. [CrossRef]
100. Tang, Y.; Zhou, Y.; Fan, S.; Wen, Q. The multiple roles and therapeutic potential of HSP60 in cancer. *Biochem. Pharmacol.* **2022**, *201*, 115096. [CrossRef]
101. Bi, F.; Wang, J.; Zheng, X.; Xiao, J.; Zhi, C.; Gu, J.; Zhang, Y.; Li, J.; Miao, Z.; Wang, Y.; et al. HSP60 participates in the anti-glioma effects of curcumin. *Exp. Ther. Med.* **2021**, *21*, 204. [CrossRef] [PubMed]
102. Wang, H.; Xiao, W.; Zhou, Q.; Chen, Y.; Yang, S.; Sheng, J.; Yin, Y.; Fan, J.; Zhou, J. Bystin-like protein is upregulated in hepatocellular carcinoma and required for nucleologenesis in cancer cell proliferation. *Cell Res.* **2009**, *19*, 1150–1164. [CrossRef] [PubMed]
103. Yammine, A.; Zarrouk, A.; Nury, T.; Vejux, A.; Latruffe, N.; Vervandier-Fasseur, D.; Samadi, M.; Mackrill, J.J.; Greige-Gerges, H.; Auezova, L.; et al. Prevention by Dietary Polyphenols (Resveratrol, Quercetin, Apigenin) Against 7-Ketocholesterol-Induced Oxiapoptophagy in Neuronal N2a Cells: Potential Interest for the Treatment of Neurodegenerative and Age-Related Diseases. *Cells* **2020**, *9*, 2346. [CrossRef] [PubMed]
104. Davis, W.; Tew, K.D. ATP-binding cassette transporter-2 (ABCA2) as a therapeutic target. *Biochem. Pharmacol.* **2018**, *151*, 188–200. [CrossRef] [PubMed]
105. Ye, Z.; Lu, Y.; Wu, T. The impact of ATP-binding cassette transporters on metabolic diseases. *Nutr Metab* **2020**, *17*, 61. [CrossRef] [PubMed]
106. Li, S.; Cao, H.; Shen, D.; Jia, Q.; Chen, C.; Xing, S. Quercetin protects against ox-LDL-induced injury via regulation of ABCAl, LXR-α and PCSK9 in RAW264.7 macrophages. *Mol. Med. Rep.* **2018**, *18*, 799–806. [CrossRef] [PubMed]
107. Wu, C.-P.; Calcagno, A.M.; Hladky, S.B.; Ambudkar, S.V.; Barrand, M.A. Modulatory effects of plant phenols on human multidrug-resistance proteins 1, 4 and 5 (ABCC1, 4 and 5). *FEBS J.* **2005**, *272*, 4725–4740. [CrossRef]
108. Lorico, A.; Nesland, J.; Emilsen, E.; Fodstad, O.; Rappa, G. Role of the multidrug resistance protein 1 gene in the carcinogenicity of aflatoxin B1: Investigations using mrp1-null mice. *Toxicology* **2002**, *171*, 201–205. [CrossRef]
109. Loe, D.W.; Stewart, R.K.; Massey, T.E.; Deeley, R.G.; Cole, S.P.C. ATP-Dependent Transport of Aflatoxin B_1 and Its Glutathione Conjugates by the Product of the Multidrug Resistance Protein (*MRP*) Gene. *Mol. Pharmacol.* **1997**, *51*, 1034–1041. [CrossRef] [PubMed]
110. Cheng, P.; Ishfaq, M.; Yu, H.; Yang, Y.; Li, S.; Li, X.; Fazlani, S.A.; Guo, W.; Zhang, X. Curcumin ameliorates duodenal toxicity of AFB1 in chicken through inducing P-glycoprotein and downregulating cytochrome P450 enzymes. *Poult. Sci.* **2020**, *99*, 7035–7045. [CrossRef]
111. Bhutto, Z.A.; He, F.; Zloh, M.; Yang, J.; Huang, J.; Guo, T.; Wang, L. Use of quercetin in animal feed: Effects on the P-gp expression and pharmacokinetics of orally administrated enrofloxacin in chicken. *Sci. Rep.* **2018**, *8*, 4400. [CrossRef]

112. Hassan, S.; Peluso, J.; Chalhoub, S.; Idoux Gillet, Y.; Benkirane-Jessel, N.; Rochel, N.; Fuhrmann, G.; Ubeaud-Sequier, G. Quercetin potentializes the respective cytotoxic activity of gemcitabine or doxorubicin on 3D culture of AsPC-1 or HepG2 cells, through the inhibition of HIF-1α and MDR1. *PLoS ONE* **2020**, *15*, e0240676. [CrossRef] [PubMed]
113. Zhou, X.; Huang, J.; Li, T.; Liu, J.; Wei, Z.; Lan, C.; Zhu, G.; Liao, X.; Ye, X.; Peng, T. Clinical Significance and Potential Mechanisms of ATP Binding Cassette Subfamily C Genes in Hepatocellular Carcinoma. *Front. Genet.* **2022**, *13*, 805961. [CrossRef] [PubMed]
114. Robinson, M.D.; McCarthy, D.J.; Smyth, G.K. edgeR: A Bioconductor package for differential expression analysis of digital gene expression data. *Bioinformatics* **2010**, *26*, 139–140. [CrossRef] [PubMed]
115. Wu, T.; Hu, E.; Xu, S.; Chen, M.; Guo, P.; Dai, Z.; Feng, T.; Zhou, L.; Tang, W.; Zhan, L.; et al. clusterprofiler 4.0: A universal enrichment tool for interpreting omics data. *Innovation* **2021**, *2*, 100141. [CrossRef] [PubMed]
116. Subramanian, A.; Tamayo, P.; Mootha, V.K.; Mukherjee, S.; Ebert, B.L.; Gillette, M.A.; Paulovich, A.; Pomeroy, S.L.; Golub, T.R.; Lander, E.S.; et al. Gene set enrichment analysis: A knowledge-based approach for interpreting genome-wide expression profiles. *Proc. Natl. Acad. Sci. USA* **2005**, *102*, 15545–15550. [CrossRef]

toxins

MDPI

Article

The Impact of Storage Temperature and Time on Ergot Alkaloid Concentrations

Jensen E. Cherewyk [1,*], Taylor J. Grusie-Ogilvie [2], Sarah E. Parker [3], Barry R. Blakley [1] and Ahmad N. Al-Dissi [4]

1 Department of Veterinary Biomedical Sciences, Western College of Veterinary Medicine, University of Saskatchewan, Saskatoon, SK S7N 5B4, Canada; brb237@mail.usask.ca
2 Prairie Diagnostic Services (PDS), Saskatoon, SK S7N 5B4, Canada; taylor.ogilvie@pds.usask.ca
3 Centre for Applied Epidemiology, Large Animal Clinical Sciences, Western College of Veterinary Medicine, University of Saskatchewan, Saskatoon, SK S7N 5B4, Canada; sarah.parker@usask.ca
4 Department of Veterinary Pathology, Western College of Veterinary Medicine, University of Saskatchewan, Saskatoon, SK S7N 5B4, Canada; ahmad.aldissi@usask.ca
* Correspondence: jensen.cherewyk@usask.ca

Abstract: Ergot sclerotia produce toxic secondary metabolites, ergot alkaloids, that infect cereal crops and grasses. Ergot alkaloids have two isomeric configurations: the C-8-*R*-isomer (*R*-epimer), and the C-8-*S*-isomer (*S*-epimer). Ergot contaminated matrices, such as cereal grains or grasses, may be stored for extended periods at various temperatures before being analyzed, utilized, or consumed. This study assessed the concentration of six common ergot alkaloids in both configurations found in naturally contaminated wheat over time (one, two, and four months) at different temperatures (room temperature, +4 °C, and −20 °C) using ultra-high-performance liquid chromatography–tandem mass spectrometry. The data indicate that the total ergot concentration within a natural contaminated sample varies over time at room temperature, +4 °C, and −20 °C. The total ergot concentration increased until month two, and decreased at month four, independent of temperature ($p < 0.05$). The total *R*-epimer concentration appeared to be less stable over time than the total *S*-epimer concentration. The changes in the total *R* and total *S*-epimer concentrations may have been caused by changes in the ergocristine and ergocristinine concentrations, respectively. Time and temperature should be considered when storing potentially contaminated matrices in a laboratory or practical agriculture situations. Quantification of ergot contaminated matrices should occur prior to their use to ensure the most reliable estimates of the concentration of ergot.

Keywords: contamination; grain; epimer

Key Contribution: This study demonstrates for the first time the effect of extended storage time and multiple temperatures on the *R*-epimers and *S*-epimers of the six common ergot alkaloids. To ensure accurate ergot concentrations, ergot contaminated products should be analyzed prior to use if stored under various conditions.

1. Introduction

Ergot sclerotia contaminate cereal crops or grasses. If the infected crops or forage are harvested, food or feed quality may be impacted. Within recent years, samples contaminated with ergot sclerotia have increased in western Canada [1,2]. This increase in the incidence of ergot-contaminated samples may be associated with changing climate [3], or agricultural practices [1].

Ergot sclerotia contain toxic secondary metabolites known as ergot alkaloids [1]. Ergot alkaloids are divided into three structural classes: clavines, lysergic acid amides, and ergopeptines [4]. Each class exists in two configurations, the C-8-*R*-isomer and the C-8-*S*-isomer. Each configuration is known as the *R*-epimer and *S*-epimer, respectively, with either a -ine (*R*), or -inine (*S*) suffix. There are six common ergot alkaloids worldwide [5,6]. The six common *R* and six common *S*-epimers are, ergocornine/ergocorninine, ergocristine/ergocristinine, ergocryptine/ergocryptinine, ergometrine/ergometrinine, ergosine/ergosinine, and ergotamine/ergotaminine.

Ergot infected grains or grasses can be stored for months before they are used for food or feed products. The stability of ergot alkaloids in natural ergot contaminated samples has been studied [1,7–10]. The concentration of an individual ergot alkaloid, ergovaline, in infected tall fescue has been assessed over time at multiple temperatures [7–10]. Temperature and time appeared to decrease the concentration of ergovaline. In contrast, total ergot alkaloid concentration in rye was reported to be stable over days [11]. However, several studies demonstrated that total ergot alkaloid concentration was affected by longer time periods of months and years [1,8,10,12]. Study variability may be associated with differences in the time/temperature investigated. Discrepancies in the ergot alkaloids or epimers quantified may also contribute to the variation. Previous studies have not assessed the effects of time and temperature on the *R* and *S*-epimers separately.

The *R* and *S*-epimers of ergot alkaloids have demonstrated varied stability in solvents and raw material with altered experimental factors [13–16]. The *R* and *S*-epimers of ergot alkaloids can epimerize from one configuration to another [4]. Epimerization and degradation of ergot alkaloids can occur at high temperatures, greater than room temperature [13,16]. The effects of temperature lower than room temperature have been studied to a lesser extent. Certain *R* and *S*-epimers of ergot alkaloids are found at high concentrations worldwide [3,5,17–19]. All common ergot alkaloids should be analyzed in contaminated samples due to the potential concentration differences.

The concentration of ergot alkaloids in food and feed are determined for the health and safety of humans and animals. Safety standards on ergot alkaloids for human food [18,20] and animal feed [21] consumption have been established. The standards for ergot alkaloid concentration include only the *R*-epimer of ergot alkaloids and do not include the *S*-epimer in Canada at the present time. The European Union (EU) Commission Recommendation for ergot contaminated foodstuff includes both the *R* and *S*-epimers of ergot alkaloids in the regulatory standard [20]. The quantification of both *R* and *S*-epimers, under various storage conditions, should be conducted to prevent underestimation of the total ergot alkaloid concentration and provide an accurate estimate of exposure [4,17,18]. Recent studies have demonstrated that *S*-epimers may be bioactive and have toxic effects [22–24], further supporting the need for quantification of each configuration.

Samples suspected to be contaminated with ergot can be sent to laboratories to assess the concentration of ergot alkaloids [25]. Time and temperature may affect the concentration of ergot alkaloids in the submitted samples [7]. Various temperature conditions have been reported during sample storage after harvest or sample preparation, prior to analysis [1,8–10,26], such as ambient temperature, +5 °C, −4 °C, −20 °C, and −29 °C. However, the length of time prior to analysis of the stored samples is seldomly reported. Some studies did not report the storage temperature of natural ergot contaminated samples prior to or after analysis [27–30].

In an agricultural setting, storage factors for crops consist of temperature, moisture, pests including fungi [31], and time [10]. Stored crops with ergot contamination may have an economic impact on producers through crop quality [2] or livestock consumption [32]. Harvested grain may be downgraded to animal feed if it is contaminated with ergot [12], and livestock may show a reduction in milk yield and weight if ergot contaminated feed is consumed [33]. The effects of storage factors on the concentration of ergot alkaloids could have practical implications if the stored crop quality is impacted.

Assessing the effects of various storage temperatures over an extended time on ergot alkaloid concentrations is limited. Studies with extended storage periods only assessed one temperature, and a study assessing multiple temperatures did not assess the effects of an extended time period at those temperatures. Furthermore, the quantification and investigation of the effect of long-term storage and temperature on multiple *R* and *S* epimers separately has not been assessed to the authors knowledge. It is important to quantify and assess the S-epimers of ergot alkaloids because of the recent recommended guideline changes to include the S-epimers and the potential bioactivity of the S-epimers. Understanding the R and S-epimer concentration stability in natural contaminated grain over time at multiple temperatures may help ascertain ideal storage conditions and determine the storage factors affecting concentration stability. The objective of this research is to assess the effects of storage time and temperature on the six R and six S-epimers of the common ergot alkaloids in natural ergot-contaminated wheat.

2. Results

2.1. Total Ergot Concentration

A small pilot study using the ground ergot contaminated samples was conducted with three groups, prior to drying, freeze drying, and heat drying. The percent change after the drying treatments were minimal: 5% and 7% for the freeze and heat drying, respectively. The concentrations of ergot epimers were assessed in all three groups and the drying treatment did not have an affect on the ergot epimer concentrations. Therefore, moisture content would not impact the results of the present study.

Time and temperature influenced the total ergot concentration. There was a significant interaction between the effect of time and temperature on the mean total ergot concentration (MTEC) ($p < 0.05$). At room temperature, +4 °C, and −20 °C storage there was a significant effect of time on the MTEC ($p < 0.05$) (Figure 1). At room temperature, the MTEC did not significantly change after one month of storage, but significantly increased by 15% after two months of storage ($p = 0.037$) and decreased by 19% after four months of storage ($p < 0.001$), compared to the initial analysis. At +4 °C, the MTEC significantly increased over time at one and two months ($p < 0.001$) but returned to the initial concentration after four months. At −20 °C there was a 25% increase at month two ($p < 0.001$) and a 9% decrease at month four ($p = 0.006$), compared to the initial analysis. The specific ergot alkaloids that are influencing the total ergot concentration can be found below.

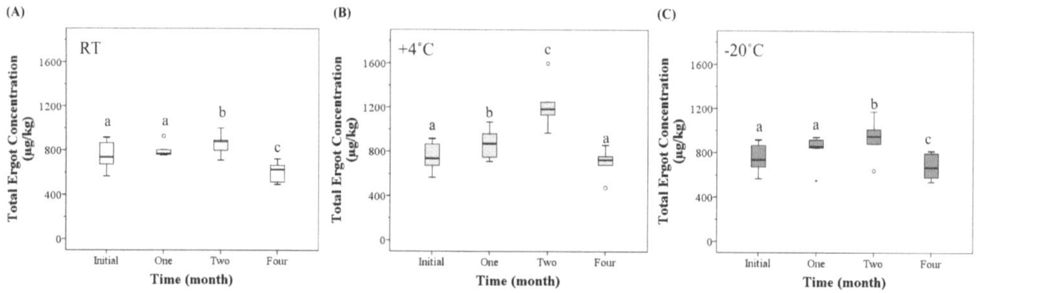

Figure 1. The total (six R and six S-epimers) ergot concentration (µg/kg) in natural ergot-contaminated hard red spring wheat over time (month) at (**A**) room temperature (RT), (**B**) +4 °C, and (**C**) −20 °C, analyzed utilizing high-performance liquid chromatography–tandem mass spectrometry. [Box-plot: whiskers are defined at the minimum and maximum values, top of box is defined as the 75th percentile, bottom of box is 25th percentile, and middle line is defined at the median. The ° are defined as outliers (>1.5 × interquartile range) and the * are defined as an extreme outlier (>3 × interquartile range)]. All outliers were included in the statistical analysis. Different lowercase letters represent statistical differences between each time period at each temperature ($p < 0.05$, generalized estimating equation, pairwise comparison with sequential Sidak correction, $n = 6$/temperature and time.

2.2. Total R and S-epimer Concentration

There was a time effect on the mean total *R* and mean total *S*-epimer concentrations (MTRC and MTSC, respectively) at each storage temperature ($p < 0.05$). From the initial analysis, the MTRC significantly increased by 19–87% at month one and two at room temperature ($p = 0.01$, $p < 0.001$), +4 °C ($p < 0.001$, $p < 0.001$) and −20 °C ($p = 0.007$, $p < 0.001$). At month four, the MTRC was not significantly different from the initial analysis at each temperature ($p > 0.05$). For the MTSC, at room temperature there was a 10% and 41% decrease in the MTSC from the initial analysis compared to month one ($p = 0.006$) and month four ($p < 0.001$), respectively. At 4 °C, there was a 30% MTSC increase from the initial analysis compared to month two ($p < 0.001$), and a 28% decrease compared to month four ($p < 0.001$). At −20 °C, the only difference from the initial analysis was a 24% decrease at month four ($p < 0.001$) (Figure 2).

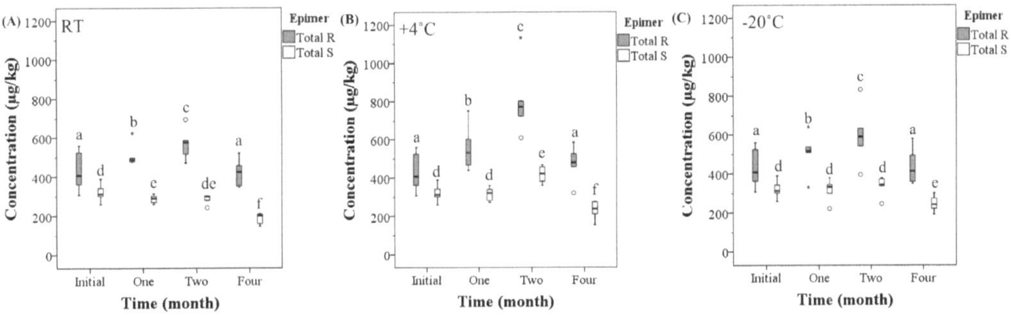

Figure 2. The concentration (µg/kg) of the total *R*-epimers and total *S*-epimers in natural ergot-contaminated hard red spring wheat over time (month) at (**A**) room temperature (RT), (**B**) +4 °C, and (**C**) −20 °C, analyzed utilizing high-performance liquid chromatography–tandem mass spectrometry. [Box-plot: whiskers are defined at the minimum and maximum values, top of box is defined as the 75th percentile, bottom of box is 25th percentile, and middle line is defined at the median. The ° are defined as outliers (>1.5 × interquartile range) and the * are defined as an extreme outlier (>3 × interquartile range)]. All outliers were included in the statistical analysis. Different lowercase letters represent statistical differences between each time period at each temperature for either the total *R*-epimers (a–c) or total *S*-epimers (d–f) ($p < 0.05$, generalized estimating equation, pairwise comparison with sequential Sidak correction, $n = 6$/temperature and time for each epimer.

2.3. Temperature Effects on Total, Total R, and Total S-epimer Concentration

The effect of storage temperature on the concentration of mean total ergot, mean total *R*-epimers, and mean total *S*-epimers at each time period was analyzed (Table 1). The MTEC was not different between temperature groups at month one ($p < 0.05$). At month two, +4 °C and −20 °C had significantly higher MTEC than room temperature ($p < 0.001$ and $p = 0.035$, respectively). The same results occurred at month four. The MTRC was not significantly different between temperature groups at each time period ($p > 0.05$). With the exception of month two, the +4 °C group was significantly higher than the room temperature and −20 °C groups ($p < 0.001$). The MTSC between temperature groups at each time period demonstrated significant differences ($p < 0.05$). At month two and four, the MTSC was significantly higher at the +4 °C and −20 °C temperature groups ($p < 0.001$), compared to the room temperature group.

Table 1. Effects of temperature at separate time periods on the mean concentration of total ergot, total *R*-epimers, and total *S*-epimers in natural ergot-contaminated wheat analyzed with high-performance liquid chromatography tandem mass spectrometry [1].

Time (month)	Temperature (°C)	Mean Total Ergot Alkaloid Concentration	Mean Total *R*-Epimer Concentration	Mean Total *S*-Epimer Concentration
Initial	Initial	749 ± 131	429 ± 99	320 ± 44
1	Room	797 ± 65 [a]	509 ± 56 [a]	288 ± 21 [a]
	+4	870 ± 132 [a]	554 ± 114 [a]	316 ± 34 [b]
	−20	827 ± 140 [a]	509± 100 [a]	318 ± 54 [ab]
2	Room	859 ± 97 [a]	570 ± 76 [a]	289 ± 25 [a]
	+4	1219 ± 210 [b]	802 ± 177 [b]	417 ± 41 [b]
	−20	934 ± 175 [c]	597 ± 142 [a]	337 ± 48 [c]
4	Room	609 ± 89 [a]	421 ± 66 [a]	188 ± 27 [a]
	+4	704 ± 127 [b]	474 ± 88 [a]	230 ± 46 [b]
	−20	679 ± 113 [b]	435 ± 88 [a]	245 ± 39 [c]

[1] Concentration (µg/kg) values are mean ± standard deviation ($n = 6$). Different letters represent significant differences between temperature groups at each time period. ($p < 0.05$, GEE, Pairwise Comparison, Sequential Sidak correction, $n = 6$).

2.4. Individual Ergot Epimer Concentration

The concentrations of individual epimers analyzed at each temperature group over time were assessed. Ergocristine and ergocristinine had the highest mean concentration compared to the other epimers with an average of 34% and 17%, respectively, of the total ergot concentration (Table S1). Since ergocristine and ergocristinine constituted the greatest percentage of the total ergot concentration, the effects on their concentrations were analyzed. Ergocristine appeared to have influenced the concentration of the total ergot alkaloids to a greater extent, especially at +4 °C in month two, compared to the other ergot alkaloids analyzed. The effects of storage temperature and time on the *R* and *S*-epimers of all other analyzed ergot alkaloids are in the supplemental material (Figures S1–S5). Following the analysis, it was determined that storage time and temperature had an effect on the mean concentrations of ergocristine and ergocristinine (Figure 3). The mean ergocristine concentration increased at month one and two, compared to the initial analysis, independent of temperature ($p \leq 0.003$), and returned to the initial concentration at month four. The mean ergocristinine concentration decreased at room temperature and +4 °C over time ($p \leq 0.005$), with the exception at +4 °C in month two. At −20 °C, the concentration did not change from the initial analysis until a decrease at month four (p < 0001).

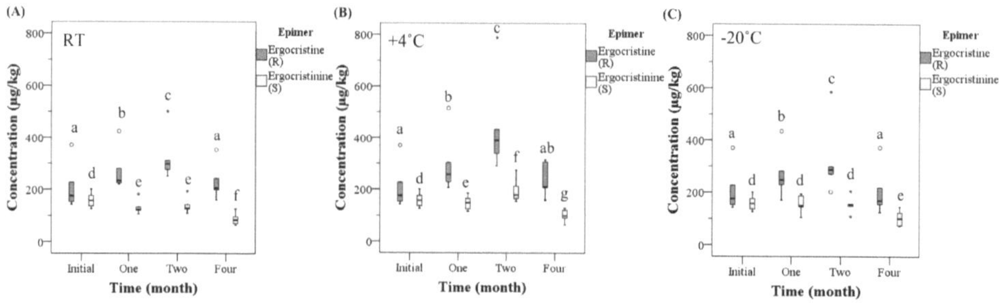

Figure 3. The concentrations (µg/kg) of ergocristine and ergocristinine in natural ergot-contaminated hard red spring wheat over time (month) at (**A**) room temperature (RT), (**B**) +4 °C, and (**C**) −20 °C, analyzed utilizing high-performance liquid chromatography–tandem mass spectrometry. [Box-plot: whiskers are defined at the minimum and maximum values, top of box is defined as the 75th percentile, bottom of box is 25th percentile and middle line is defined at the median. The ° are defined as outliers (>1.5 × interquartile range) and the * are defined as an extreme outlier (>3 × interquartile range)]. All outliers were included in the statistical analysis. Different lowercase letters represent statistical differences between each time period at each temperature for ergocristine (a–c) or ergocristinine (d–g) ($p < 0.05$, generalized estimating equation, pairwise comparison with sequential Sidak correction, $n = 6$/temperature and time for each epimer.

3. Discussion

Studies have found time or temperature to have an affect on the concentration of ergot. Specifically, ergot alkaloid concentrations in ergot contaminated matrices have increased [8,9], decreased [10,34], or have shown variation [1] after long term storage. In the present study, the total ergot concentration increased until month two and decreased at month four at all the temperatures evaluated. The total ergot concentration is not stable under certain storage conditions. Previous studies assessing the total ergot concentration have not assessed the effects of time and temperature on the *R* and *S*-epimers separately. One study assessed individual epimer concentrations of harvested and shipped samples; however, the temperature of the stored shipped samples was not provided and not all six common ergot alkaloids were assessed [1]. Similar increases and decreases of the total ergot epimers over time were observed.

The concentration of total ergot alkaloids after storage may vary due to the analysis of not only the *R*-epimers but also the *S*-epimers of ergot alkaloids. An increase in the total *S*-epimer concentration after long term storage has been suggested [8,12], potentially associated with epimerization of the *R*-epimer to the *S*-epimer [1,12,35–38]. However, the quantification and assessment of the *S*-epimers separately after storage for an extended time at different temperatures was not assessed. The present study demonstrated an increase in the total *R*-epimer concentration until month two, whereas the *S*-epimer concentration mostly decreased or remained stable at all the temperatures. The results observed in the current study do not suggest epimerization of the *R*-epimer to the *S*-epimer. The results may suggest slight back epimerization of the *S*-epimer to the *R*-epimer over time. The *S*-epimer concentrations may be more stable over time compared to the *R*-epimers at various temperatures.

The concentration of specific ergot epimers vary with fungal strain, region, geographical location, year, and host grain type [2,19,39]. Similar to the present study, ergocristine and ergocristinine were found in high concentrations in Canadian grain [1,18,40]. The ergocristine/-inine concentration has previously been reported to be unstable in raw material exposed to different temperatures [13,16]. In the current study, ergocristine concentration varied over time at each temperature, whereas ergocristinine concentration was more stable, especially at −20 °C. Overall, the impact of time and temperature on the ergocristine and ergocristinine concentrations is similar to the pattern observed with the total *R* and total *S*-epimer concentrations. All six common *R* and *S*-epimers of ergot alkaloids should be quantified in ergot contaminated samples since they have different stabilities under the assessed storage conditions.

The rationale for the instability of the concentrations of the total, total *R*, total *S*, and certain individual epimers is unknown. The reason for increase or decrease in the concentrations is out of the scope of the present study. A hypothesis for a decrease in concentrations may be related to the degradation of ergot alkaloids, which has been previously observed [8,35–38]. The degradation may be associated with the formation of ergot derivatives which can be catalyzed by oxidation, reduction, hydrolysis, and under alkaline or acidic conditions [37]. Microbial activity has also demonstrated ergot alkaloid degradation [8]. In the present study, samples were concealed in airtight tubes, which could produce an anaerobic environment and result in degradation. Other factors such as bacteria or other fungi in the samples may also contribute to the degradation of the ergot alkaloids, which have been noted previously [4]. The samples in the present study were covered in plastic, therefore light degradation should not have been a reason for degradation. The rationale for the increase in concentration is less known. In the studies that also observed an increase in total ergot concentration, this was attributed it to a decrease in ergovaline, in which the ergovaline degradation products would be included in the enzyme-linked immunosorbent assay (ELISA) method for quantification [8,9]. In the present study, an ELISA was not used, therefore, the same rationale cannot be applied. Another study observed a significant increase in total ergovaline after 28 days at −20 °C for particular samples, however, no rationale or discussion was provided [7]. A hypothesis for an increase in concentration in the present study could be that the environmental conditions of the samples, as mentioned above, would increase the extraction efficiency of the specific epimer. However, this hypothesis is purely speculative and would need further assessment and data to support it. The specific configurations of the *R* and *S*-epimers may be more susceptible to degradation, epimerization, or extraction. The variability in the concentrations is likely due to multiple factors.

Ergot contaminated samples are sent routinely to laboratories to determine the concentration of ergot alkaloids [25]. The EU Commission Recommendation on sampling and analysis [41], which has been used to evaluate ergot-contaminated samples [5], reports that storage temperature should not alter the contaminant composition, as such, it does not recommend a specific temperature. A laboratory storage temperature of −20 °C has been suggested so that the concentration of ergot alkaloids does not decrease [7], and this has been implemented [3,42]. However, the suggested storage temperature was only examined for a one-month period. The present study supports the concentration stability of total ergot at −20 °C for one month. However, the total ergot concentration may change in the following months. Since ergot concentrations are not stable under certain storage conditions, time and temperature should be considered when storing natural ergot-contaminated samples in laboratories. Under field conditions, certain storage conditions, such as −20 °C, may not be recommended.

Management and environmental factors during storage in a field setting affect the presence and production of mycotoxins [31]. Ergot contamination of cereal crops and feed intended for selling or livestock consumption is an issue [2,3]. Ergot contaminated matrices may be retained for months before they are utilized [2]. The present study suggests that ergot alkaloids are not stable over time at various temperatures. Storage time and temperature should be considered when storing potential ergot contaminated crops. Changes in the concentration of ergot alkaloids due to storage conditions may affect the quality of the grain or feed. Ergot contaminated samples should be analyzed prior to use to ensure the most reliable estimate of risk assessment.

4. Conclusions

Storage time and temperature affect the concentration of total ergot concentration in natural contaminated grain. Total *R* and *S*-epimer concentrations vary at specific temperatures and time periods, and the concentration changes differ from one another. Ergocristine and ergocristinine may influence the change in total *R* and total *S*-epimer concentrations when exposed to the various time and temperatures. The current study does not explicitly demonstrate ideal storage conditions for ergot contaminated samples intended for laboratory or agriculture practices. The rationale for the instability of the ergot alkaloid concentrations is most likely due to multiple abiotic and biotic factors. Long-term storage and various temperatures affect the concentration of ergot alkaloids; therefore, stored grain should be analyzed prior to use to ensure the best ergot alkaloid concentration estimates. Both the *R* and *S*-epimers of the six common ergot alkaloids should be quantified based on varying concentration changes after storage. Storage time and temperature, associated with ergot alkaloid contamination, should be considered for samples intended for consumption.

5. Materials and Methods

5.1. Sample Preparation

Hard red spring wheat contaminated with ergot (n = 6) was obtained from the Canadian Feed Research Centre (North Battleford, SK, Canada). The hard red spring wheat were highly contaminated with ergot sclerotia and were therefore processed as described in detail previously [43]. In summary, the ergot contaminated wheat was diluted with clean wheat to obtain similar starting concentrations among the samples and to achieve concentrations that are observed under practical agricultural situations. The prepared ergot contaminated samples were ground (UDY Cyclone Sample Mill, Model #3010-060, 1 mm mesh, Fort Collins, CO, USA) and hand mixed to ensure a homogenized sample. To ensure a homogenous mixture, six replicates from each of the six samples were analyzed for the concentration of ergot alkaloids. The standard deviation of the six replicates from each of the six prepared samples was relatively low [43]. The starting mean concentration of total ergot alkaloids from the six prepared samples was 841 ± 92 μg/kg. From each prepared ergot contaminated sample, 5 g were weighed(Sartorius BP2100 balance, Elk Grove, CA, USA) and placed directly within a 50 mL plastic trace metal free centrifuge tube and capped.

5.2. Pilot Study

A small pilot study was conducted using the samples to assess both the moisture content of the samples and the ergot alkaloid concentrations following freeze and heat drying. This pilot study was used to determine if moisture would have an effect on the results of the present study. Two hard red spring wheat samples, with two replicates each, were weighed (Mettler Toledo NewClassic ML204 balance, Mississauga, ON, Canada) into 50 mL centrifuge tubes and capped. The weight for each sample and replicate was approximately 5 g. This was repeated three times for the samples to be assessed, prior to drying, after freeze drying, and after heat drying. For freeze drying, the centrifuge tubes containing the samples had the caps replaced with Kimwipes and were secured with elastic bands. The samples were placed into a freeze dryer (VirTis Genesis, SP Industries, Warminster, PA, USA) at −20 °C, under vacuum (10–20 m Torr), for 24 h. For heat drying, samples within the centrifuge tubes with the caps removed were placed into adrying oven (Isotemp, Fisher Scientific, Waltham, MA, USA) at +80 °C for 24 h. All samples, including the samples not exposed to drying, were re-weighed into centrifuge tubes and capped for the total ergot alkaloid concentration analysis, which is described below.

5.3. Quantification

Concentrations of ergot alkaloids, ergocornine, ergocristine, ergocryptine, ergometrine, ergosine, and ergotamine (*R*-epimers), along with the corresponding -inine epimers (*S*-epimers) (Romer Labs, Tulln, Lower Austria, Austria), were measured in all the prepared ergot contaminated samples using ultra-high-performance liquid chromatography–tandem mass spectrometry (UHPLC-MS/MS) (ThermoFisher Scientific, Waltham, MA, USA) as described previously [43]. Briefly, samples were extracted with acetonitrile:water (80:20) and spun (Benchmixer Multi-tube vortex, Sayreville, NJ, USA). The supernatant was filtered and placed into amber vials along with the internal standard, deuterated lysergic acid diethylamide (Sigma Aldrich, Oakville, ON, Canada). Once the samples were dried down using nitrogen gas (Multivap nitrogen evaporator, Organomation, Berlin, MA, USA) and reconstituted in methanol:water (50:50), they were ready for analysis. A C18 column was used to separate the epimers. A triple quadrupole with electrospray ionization in positive mode and selective reaction monitoring were used for the mass spectrometry analysis. The method of analysis was validated and described in detail in Cherewyk et al., 2022 [43]. The inter-day precision of the method was <24% relative standard deviation for all quantified analytes.

5.4. Experimental Design and Statistical Analysis

An initial quantification (time zero) of the prepared ergot contaminated samples (n = 6), with two replicates to assess sampling variability, was conducted to serve as the reference control. The initial reference control sample concentrations were within the standard deviation of the starting mean concentration of the prepared samples. Sub-samples, from each of the prepared ergot contaminated samples, with two replicates were placed at stable room temperature (+22 °C), +4 °C, and −20 °C on the day of the initial quantification. All sub samples used in this study were taken from the starting homogenous prepared ergot contaminated samples. At one, two, and four months after the initial analysis, sub-samples were removed from their respective temperature groups. The samples were analyzed for the concentration of the six *R* and six *S*-epimers as described above. The measured values for the two replicates of each sub sample were averaged. There were six independent samples repeatedly measured for each temperature and time group (n = 6/temperature/time). All prepared ergot contaminated samples were covered with black plastic bags to minimize exposure to light.

All statistics were conducted utilizing SPSS 23 (IBM SPSS Statistics for Windows, version 23, IBM Corp., Armonk, NY, USA). A statistical analysis was undertaken using generalized estimating equations (GEEs) with an identity link function and an unstructured correlation matrix. The GEE was used to account for the repeated measures of samples over time and multiple temperatures. The analysis was conducted on the mean total ergot concentration, mean total *R*-epimer concentration, mean total *S*-epimer concentration, and the mean concentration of each individual epimer. Total ergot concentration is defined as all quantified epimers. Total *R*-epimer concentration and total *S*-epimer concentration are the sum of the concentration of the six *R* and six *S*-epimers, respectively. Differences were considered significant at $p < 0.05$. In the presence of a significant interaction between the effects of time and temperature on the mean concentration, GEE was preformed to analyze the effect of time at each temperature. Differences between temperature groups at each time period were also assessed. Multiple pairwise comparisons with a sequential Sidak correction were used to assess differences.

Supplementary Materials: The following supporting information can be downloaded at: https://www.mdpi.com/article/10.3390/toxins15080497/s1. The supplemental materials file includes Figures S1–S5, the effects of time and temperature on the concentration of the other analyzed individual *R* and *S*-epimers; and Table S1, the percentage of each individual epimer compared to the total epimer concentration for each time and temperature group.

Author Contributions: Conceptualization, J.E.C. and B.R.B.; methodology, J.E.C., T.J.G.-O. and S.E.P.; formal analysis, J.E.C., T.J.G.-O.; investigation, J.E.C.; resources, J.E.C. and T.J.G.-O.; data curation, J.E.C.; writing—original draft preparation, J.E.C.; writing—review and editing, J.E.C., T.J.G.-O., S.E.P., B.R.B. and A.N.A.-D.; supervision, B.R.B. and A.N.A.-D.; funding acquisition, J.E.C., B.R.B. and A.N.A.-D. All authors have read and agreed to the published version of the manuscript.

Funding: This research was funded by the Toxicology Devolved Scholarship, Western College of Veterinary Medicine (WCVM) internal ergot grant, the government of Saskatchewan - Ministry of Agriculture - Agriculture Development Fund (ADF) grant number: 20180361, and the Natural Sciences and Engineering Research Council of Canada (Canada Graduate Scholarships–Doctoral program).

Institutional Review Board Statement: Not applicable.

Informed Consent Statement: Not applicable.

Data Availability Statement: The data presented in this study are available through the corresponding author.

Acknowledgments: The authors would like to thank the kind contributions from Prairie Diagnostic Services (PDS), Canadian Feed Research Centre, and Cherewyk Farms.

Conflicts of Interest: The authors declare no conflict of interest.

References

1. Tittlemier, S.A.; Drul, D.; Roscoe, M.; Mckendry, T. Occurrence of Ergot and Ergot Alkaloids in Western Canadian Wheat and Other Cereals. *J. Agric. Food Chem.* **2015**, *63*, 6644–6650. [CrossRef] [PubMed]
2. Walkowiak, S.; Taylor, D.; Fu, X.; Drul, D.; Pleskach, K.; Tittlemier, S.A. Ergot in Canadian Cereals-Relevance, Occurrence, and Current Status. *Can. J. Plant Pathol.* **2022**, *44*, 793–805. [CrossRef]
3. Poapolathep, S.; Klangkaew, N.; Zhang, Z.; Giorgi, M.; Logrieco, A.F.; Poapolathep, A. Simultaneous Determination of Ergot Alkaloids in Swine and Dairy Feeds Using Ultra High-Performance Liquid Chromatography-Tandem Mass Spectrometry. *Toxins* **2021**, *13*, 724. [CrossRef] [PubMed]
4. Agriopoulou, S. Ergot Alkaloids Mycotoxins in Cereals and Cereal-Derived Food Products: Characteristics, Toxicity, Prevalence, and Control Strategies. *Agronomy* **2021**, *11*, 931. [CrossRef]

5. Babič, J.; Tavčar-Kalcher, G.; Celar, F.A.; Kos, K.; Červek, M.; Jakovac-Strajn, B. Ergot and Ergot Alkaloids in Cereal Grains Intended for Animal Feeding Collected in Slovenia: Occurrence, Pattern and Correlations. *Toxins* **2020**, *12*, 730. [CrossRef] [PubMed]
6. Chung, S.W.C. A Critical Review of Analytical Methods for Ergot Alkaloids in Cereals and Feed and in Particular Suitability of Method Performance for Regulatory Monitoring and Epimer-Specific Quantification. *Food Addit. Contam. A* **2021**, *38*, 997–1012. [CrossRef] [PubMed]
7. Lea, K.; Smith, L.; Gaskill, C.; Coleman, R.; Smith, S.R. Ergovaline Stability in Tall Fescue Based on Sample Handling and Storage Methods. *Front. Chem.* **2014**, *2*, 76. [CrossRef] [PubMed]
8. Roberts, C.A.; Davis, D.K.; Looper, M.L.; Kallenbach, R.L.; Rottinghaus, G.E.; Hill, N.S. Ergot Alkaloid Concentrations in High-and Low-Moisture Tall Fescue Silage. *Crop. Sci.* **2014**, *54*, 1887–1892. [CrossRef]
9. Roberts, C.A.; Kallenbach, R.L.; Rottinghaus, G.E.; Hill, N.S. Ergovaline and Ergot Alkaloid Concentrations Change in Conserved Tall Fescue. *Forage Grazinglands* **2011**, *9*, 1–9. [CrossRef]
10. Roberts, C.A.; Kallenbach, R.L.; Hill, N.S.; Rottinghaus, G.E.; Evans, T.J. Ergot Alkaloid Concentrations in Tall Fescue Hay during Production and Storage. *Crop. Sci.* **2009**, *49*, 1496–1502. [CrossRef]
11. Tkachenko, A.; Benson, K.; Mostrom, M.; Guag, J.; Reimschuessel, R.; Webb, B. Extensive Evaluation via Blinded Testing of an UHPLC-MS/MS Method for Quantitation of Ten Ergot Alkaloids in Rye and Wheat Grains. *J. AOAC Int.* **2021**, *104*, 546–554. [CrossRef] [PubMed]
12. Coufal-Majewski, S.; Stanford, K.; McAllister, T.; Blakley, B.; McKinnon, J.; Chaves, A.V.; Wang, Y. Impacts of Cereal Ergot in Food Animal Production. *Front. Vet. Sci.* **2016**, *3*, 15. [CrossRef] [PubMed]
13. Young, J.C.; Chen, Z.J.; Marquardt, R.R. Reduction in Alkaloid Content of Ergot Sclerotia by Chemical and Physical Treatment. *J. Agric. Food Chem.* **1983**, *31*, 413–415. [CrossRef] [PubMed]
14. Hafner, M.; Sulyok, M.; Schuhmacher, R.; Crews, C.; Krska, R. Stability and Epimerisation Behaviour of Ergot Alkaloids in Various Solvents. *World Mycotoxin J.* **2008**, *1*, 67–78. [CrossRef]
15. Crews, C. Analysis of Ergot Alkaloids. *Toxins* **2015**, *7*, 2024–2050. [CrossRef]
16. Schummer, C.; Zandonella, I.; van Nieuwenhuyse, A.; Moris, G. Epimerization of Ergot Alkaloids in Feed. *Heliyon* **2020**, *6*, e04336. [CrossRef]
17. Carbonell-Rozas, L.; Mahdjoubi, C.K.; Arroyo-Manzanares, N.; García-Campaña, A.M.; Gámiz-Gracia, L. Occurrence of Ergot Alkaloids in Barley and Wheat from Algeria. *Toxins* **2021**, *13*, 316. [CrossRef]
18. Tittlemier, S.A.; Drul, D.; Roscoe, M.; Turnock, D.; Taylor, D.; Fu, B.X. Fate of Ergot Alkaloids during Laboratory Scale Durum Processing and Pasta Production. *Toxins* **2019**, *11*, 195. [CrossRef]
19. Kodisch, A.; Oberforster, M.; Raditschnig, A.; Rodemann, B.; Tratwal, A.; Danielewicz, J.; Korbas, M.; Schmiedchen, B.; Eifler, J.; Gordillo, A.; et al. Covariation of Ergot Severity and Alkaloid Content Measured by HPLC and One ELISA Method in Inoculated Winter Rye across Three Isolates and Three European Countries. *Toxins* **2020**, *12*, 676. [CrossRef]
20. European Commission. Commission Regulation (EU) 2021/1399 of 24 August 2021 Amending Regulation (EC) No 1881/2006 as Regards Maximum Levels of Ergot Sclerotia and Ergot Alkaloids in Certain Foodstuffs. Official Journal of the European Union 2021. Available online: https://eur-lex.europa.eu/legal-content/EN/TXT/?uri=CELEX%3A32021R1399 (accessed on 1 October 2022).
21. CFIA RG-8 Regulatory Guidance: Contaminants in Feed (Formerly RG-1, Chapter 7). Available online: https://www.inspection.gc.ca/animalhealth/%0Alivestock-feeds/regulatory-guidance/rg-8/eng/%0A1347383943203/1347384015909?chap=0 (accessed on 13 December 2022).
22. Mulac, D.; Hüwel, S.; Galla, H.J.; Humpf, H.U. Permeability of Ergot Alkaloids across the Blood-Brain Barrier in Vitro and Influence on the Barrier Integrity. *Mol. Nutr. Food Res.* **2012**, *56*, 475–485. [CrossRef]
23. Cherewyk, J.E.; Parker, S.E.; Blakley, B.R.; Al-Dissi, A.N. Assessment of the Vasoactive Effects of the (S)-Epimers of Ergot Alkaloids in Vitro. *J. Anim. Sci.* **2020**, *98*, skaa203. [CrossRef] [PubMed]
24. Cherewyk, J.E.; Parker, S.E.; Blakley, B.R.; Al-Dissi, A.N. Sustained Vascular Contractile Response Induced by an R-and S-Epimer of the Ergot Alkaloid Ergocristine and Attenuation by a Noncompetitive Antagonist. *J. Anim. Sci.* **2022**, *100*, skac235. [CrossRef] [PubMed]
25. Ensley, S.M.; Radke, S.L. Mycotoxins in Grains and Feeds. *Dis. Swine* **2019**, *11*, 1055–1071.
26. Schummer, C.; Brune, L.; Moris, G. Development of a UHPLC-FLD Method for the Analysis of Ergot Alkaloids and Application to Different Types of Cereals from Luxembourg. *Mycotoxin Res.* **2018**, *34*, 279–287. [CrossRef] [PubMed]
27. Krska, R.; Stubbings, G.; MacArthur, R.; Crews, C. Simultaneous Determination of Six Major Ergot Alkaloids and Their Epimers in Cereals and Foodstuffs by LC-MS-MS. *Anal. Bioanal. Chem.* **2008**, *391*, 563–576. [CrossRef]
28. Diana Di Mavungu, J.; Malysheva, S.V.; Sanders, M.; Larionova, D.; Robbens, J.; Dubruel, P.; Van Peteghem, C.; De Saeger, S. Development and Validation of a New LC-MS/MS Method for the Simultaneous Determination of Six Major Ergot Alkaloids and Their Corresponding Epimers. Application to Some Food and Feed Commodities. *Food Chem.* **2012**, *135*, 292–303. [CrossRef]
29. Guo, Q.; Shao, B.; Du, Z.; Zhang, J. Simultaneous Determination of 25 Ergot Alkaloids in Cereal Samples by Ultraperformance Liquid Chromatography–Tandem Mass Spectrometry. *J. Agric. Food Chem.* **2016**, *64*, 7033–7039. [CrossRef] [PubMed]

30. Arroyo-Manzanares, N.; De Ruyck, K.; Uka, V.; Gámiz-Gracia, L.; García-Campaña, A.M.; De Saeger, S.; Diana Di Mavungu, J. In-House Validation of a Rapid and Efficient Procedure for Simultaneous Determination of Ergot Alkaloids and Other Mycotoxins in Wheat and Maize. *Anal. Bioanal. Chem.* **2018**, *410*, 5567–5581. [CrossRef]
31. van der Fels-Klerx, H.J.; Liu, C.; Focker, M.; Montero-Castro, I.; Rossi, V.; Manstretta, V.; Magan, N.; Krska, R. Decision Support System for Integrated Management of Mycotoxins in Feed and Food Supply Chains. *World Mycotoxin J.* **2022**, *15*, 119–133. [CrossRef]
32. Craig, A.M.; Klotz, J.L.; Duringer, J.M. Cases of Ergotism in Livestock and Associated Ergot Alkaloid Concentrations in Feed. *Front. Chem.* **2015**, *3*, 8. [CrossRef]
33. Koester, L.R.; Poole, D.H.; Serão, N.V.L.; Schmitz-Esser Id, S. Beef Cattle That Respond Differently to Fescue Toxicosis Have Distinct Gastrointestinal Tract Microbiota. *PLoS ONE* **2020**, *15*, e0229192. [CrossRef] [PubMed]
34. Caradus, J.R.; Card, S.D.; Finch, S.C.; Hume, D.E.; Johnson, L.J.; Mace, W.J.; Popay, A.J. Ergot Alkaloids in New Zealand Pastures and Their Impact. *New Zeal. J. Agric. Res.* **2020**, *65*, 1–41. [CrossRef]
35. Krska, R.; Crews, C. Significance, Chemistry and Determination of Ergot Alkaloids: A Review. *Food Addit. Contam. Part A Chem. Anal. Control. Expo Risk Assess* **2008**, *25*, 722–731. [CrossRef] [PubMed]
36. Smith, D.J.; Shappell, N.W. Technical Note: Epimerization of Ergopeptine Alkaloids in Organic and Aqueous Solvents. *J. Anim. Sci.* **2002**, *80*, 1616–1622. [CrossRef] [PubMed]
37. Komarova, E.L.; Tolkachev, O.N. The Chemistry of Peptide Ergot Alkaloids. Part 1. Classification and Chemistry of Ergot Peptides. *Pharm. Chem. J.* **2001**, *35*, 504–513. [CrossRef]
38. Merkel, S.; Dib, B.; Maul, R.; Köppen, R.; Koch, M.; Nehls, I. Degradation and Epimerization of Ergot Alkaloids after Baking and in Vitro Digestion. *Anal. Bioanal. Chem.* **2012**, *404*, 2489–2497. [CrossRef]
39. Grusie, T.; Cowan, V.; Singh, J.; McKinnon, J.; Blakley, B. Proportions of Predominant Ergot Alkaloids (Claviceps Purpurea) Detected in Western Canadian Grains from 2014 to 2016. *World Mycotoxin J.* **2018**, *11*, 259–264. [CrossRef]
40. Coufal-Majewski, S.; Stanford, K.; McAllister, T.; Wang, Y.; Blakley, B.; McKinnon, J.; Swift, M.L.; Chaves, A.V. Effects of Continuously Feeding Diets Containing Cereal Ergot Alkaloids on Nutrient Digestibility, Alkaloid Recovery in Feces, and Performance Traits of Ram Lambs. *Toxins* **2017**, *9*, 405. [CrossRef]
41. European Commission. Commission Recommendation (EU) No 691/2013 of 19 July 2013 Amending Regulation (EC) No 152/2009 as Regards Methods of Sampling and Analysis. Official Journal of the European Union 2013. Available online: https://eur-lex.europa.eu/legal-content/EN/TXT/?uri=celex%3A32013R0691 (accessed on 1 October 2022).
42. Drakopoulos, D.; Sulyok, M.; Krska, R.; Logrieco, A.F.; Vogelgsang, S. Raised Concerns about the Safety of Barley Grains and Straw: A Swiss Survey Reveals a High Diversity of Mycotoxins and Other Fungal Metabolites. *Food Control.* **2021**, *125*, 107919. [CrossRef]
43. Cherewyk, J.; Grusie-Ogilvie, T.; Blakley, B.; Al-Dissi, A. Validation of a New Sensitive Method for the Detection and Quantification of R and S-Epimers of Ergot Alkaloids in Canadian Spring Wheat Utilizing Deuterated Lysergic Acid Diethylamide as an Internal Standard. *Toxins* **2022**, *14*, 22. [CrossRef]

toxins

MDPI

Article

Proteomic Analysis of the Murine Liver Response to Oral Exposure to Aflatoxin B1 and Ochratoxin A: The Protective Role to Bioactive Compounds

Silvia Trombetti [1,†], Alessandra Cimbalo [2,†], Michela Grosso [1], Pilar Vila-Donat [2,*], Jordi Mañes [2] and Lara Manyes [2]

1 Department of Molecular Medicine and Medical Biotechnology, University of Naples Federico II, Via Pansini 5, 80131 Naples, Italy; silvia.trombetti@unina.it (S.T.); michela.grosso@unina.it (M.G.)

2 Biotech Agrifood, Faculty of Pharmacy and Food Sciences, Universitat de València, Avda. Vicent Andrés Estellés s/n, 46100 Burjassot, Spain; alessandra.cimbalo@uv.es (A.C.); jordi.manes@uv.es (J.M.); lara.manyes@uv.es (L.M.)

* Correspondence: pilar.vila@uv.es

† These authors contributed equally to this work.

Abstract: Aflatoxin B1 (AFB1) and Ochratoxin A (OTA) are considered the most important mycotoxins in terms of food safety. The aim of this study was to evaluate the hepatotoxicity of AFB1 and OTA exposure in Wistar rats and to assess the beneficial effect of fermented whey (FW) and pumpkin (P) as functional ingredients through a proteomic approach. For the experimental procedures, rats were fed AFB1 and OTA individually or in combination, with the addition of FW or a FW-P mixture during 28 days. For proteomics analysis, peptides were separated using a LC-MS/MS-QTOF system and differentially expressed proteins (DEPs) were statistically filtered ($p < 0.05$) distinguishing males from females. Gene ontology visualization allowed the identification of proteins involved in important biological processes such as the response to xenobiotic stimuli and liver development. Likewise, KEGG pathway analysis reported the metabolic routes as the most affected, followed by carbon metabolism and biosynthesis of amino acids. Overall, the results highlighted a strong downregulation of DEPs in the presence of AFB1 and OTA individually but not with the mixture of both, suggesting a synergistic effect. However, FW and P have helped in the mitigation of processes triggered by mycotoxins.

Keywords: proteomics; mycotoxin; bioactive compounds; in vivo; LC-MS/MS-QTOF

Key Contribution: The inclusion of bioactive ingredients (fermented whey and pumpkin) mitigated liver damage in rats caused by 28 days of oral exposure to feed produced with AFB1- and/or OTA-contaminated flours.

1. Introduction

Aflatoxin B1 (AFB1) and Ochratoxin A (OTA) are two major mycotoxins which contaminate a wide range of food commodities, especially cereals and derived products, representing a serious concern for human and animal health [1]. These toxic compounds are secondary metabolites produced by filamentous fungi that belong primarily to *Aspergillus* and *Penicillium* species and grow on crops under conditions of improper storage and humidity. Within the two, AFB1 is the most potent hepatotoxic and carcinogenic member of the aflatoxin family whereas OTA is a nephrotoxic and immunosuppressive compound. Due to their severe toxicity, AFB1 and OTA are classified by the International

Agency for Research on Cancer (IARC) as a Group 1 carcinogen (carcinogenic to human) and Group 2B carcinogen (possibly carcinogenic to humans), respectively [2].

Over the years, it has been demonstrated that these toxins are widespread in various species of cereals such as wheat, maize, barley, and rice and can occur during pre-harvest, post-harvest, or storage stages. In fact, diverse climatic factors including high humidity and temperature markedly increase mold growth, leading to its contamination. Therefore, the consumption of contaminated cereals and their derivatives, such as flour, breakfast cereals, and processed food products, represents a significant health hazard, since these toxins are stable even under cooking and processing conditions [3]. With regard to their toxicological effects on humans, it has been demonstrated that long-term exposure to AFB1 and OTA can lead to several health disorders, among which is the onset of liver damage and cancer [4,5].

Accordingly, the liver plays a central role in AFB1 and OTA metabolism. In fact, it has been demonstrated that AFB1 is primarily bioactivated by hepatic microsomal phase I cytochrome P450 enzymes, which are able to convert it into its electrophilic reactive epoxide form (AFBO). Consequently, this metabolite form adducts to DNA and proteins, causing mutations and promoting liver carcinogenesis. However, AFBO can also be metabolized by phase II detoxifying enzymes, leading to its degradation and elimination [6]. Similarly, OTA is biotransformed in the liver by phase I and II enzymes, but nonetheless, it is not the sole organ to metabolize this toxin [7].

Given the broad presence of AFB1 and OTA in food commodities and especially the difficulty of their elimination, the research has focused on the possibility of employing substances capable of modifying their metabolism and reducing their bioaccumulation. Lactic acid bacteria (LAB), for instance, are able to increase the quality of food matrices by producing a rapid fermentation and synthesizing a wide range of beneficial molecules [8]. Among them, organic acids reduce the pH of the substrates, preventing growth of undesirable microorganisms such as mycotoxigenic fungi [9]. In addition to probiotics, plant-based foods like pumpkin (P) (*Cucurbita* spp.) are rich in antioxidants, making them effective in combating oxidative stress. In fact, they contain high levels of bioactive compounds such as carotenoids, vitamin C, and phenolic compounds, which contribute to its strong antioxidant capacity [10]. Moreover, these compounds help reduce chronic inflammation, a factor in diseases like cancer and cardiovascular conditions, making pumpkin a valuable dietary component for mitigating the harmful effects of environmental toxins. In this study, fermented whey (FW) and P as functional ingredients were used either individually or in combination to replicate a realistic scenario in the Mediterranean diet. Moreover, the intake of a single functional compound is implausible as natural foods always contain numerous bioactive compounds [11].

From the perspective of the food industry, the production of 1 kg of cheese generates about 9 L of whey, almost half of which is disposed of as waste. This disposal, often untreated, poses significant environmental problems [12]. Considering that whey offers a promising solution to counter the harmful effects of mycotoxins, harnessing the bioactive components of fermented whey not only solves whey disposal problems but also provides a sustainable way to mitigate the associated risks, turning an environmental liability into a valuable resource [13]. Moreover, several studies have focused on its hepatoprotective effects against acute or chronic toxicity induced by xenobiotics [14–17].

It is also important to emphasize that the use of proteomics has proven to be a valuable tool for deepening the understanding of the mechanisms of action that cause hepatotoxicity, since it enables the identification and quantification of specific proteins associated with toxic responses and protective pathways, which are derived from FW and P interventions.

Furthermore, it is a key element for identifying important biomarkers related to various liver diseases and even cancer [18,19].

In light of this, the aim of the present study was to investigate the advantageous role of goat milk FW and P as functional ingredients in safeguarding the sub-chronic hepatotoxic effects of AFB1 and OTA in male and female rats through a proteomics approach.

2. Results and Discussion

2.1. Identification and Quantification of Proteins

Gel-free shotgun proteomics analysis of rat liver was initiated by identifying peptides features through Spectrum Mill MS Proteomics Workbench Package Rev BI.07.09 (Agilent Technologies, Santa Clara, CA, USA). Thereafter, the proteins with different abundances between groups were statistically filtered by Mass Profiler Professional 15.0 version software (Agilent Technologies, Santa Clara, CA, USA) through an unpaired t-test ($p < 0.05$) distinguishing males from females of each experimental group. More specifically, each group exposed to single or combined mycotoxins was compared with its counterpart supplemented with functional ingredients, once with FW and once with FW + P, in order to identify the DEPs involved.

In male rats exposed to mycotoxins (Figure 1A), 95 proteins were differentially expressed in the AFB1 group compared to the male control group, 67 with OTA versus the control group, and 81 with the combination (AFB1 + OTA vs. control). In females (Figure 1B), more DEPs were observed for each comparison: 134 were identified with AFB1, 101 with OTA, and 140 with the combination.

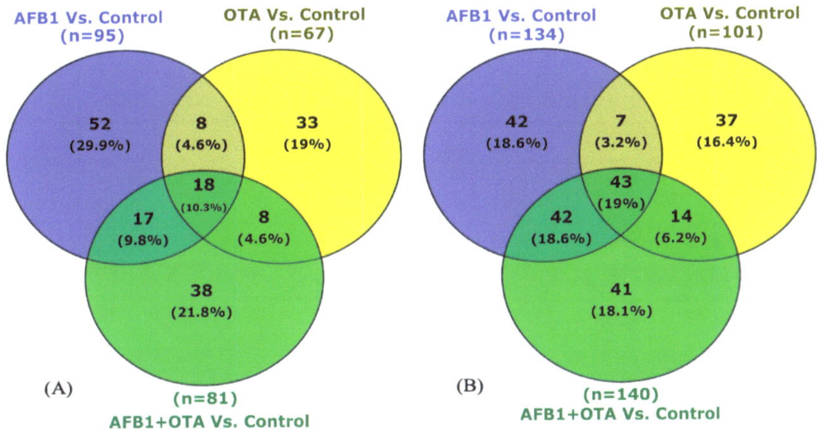

Figure 1. Venn diagram representation of common DEPs for male (**A**) and female (**B**) rats exposed to mycotoxins versus the control. $p < 0.05$ were significantly different from the control.

In male rats exposed to FW (Figure 2A), 116 proteins were differentially expressed in the FW + AFB1 group with respect to the one with only AFB1, 71 with OTA versus OTA group, and 122 with the combination (FW + AFB1 + OTA vs. AFB1 + OTA). In females (Figure 2B), a similar scenario is observed: 104 were identified with AFB1, 77 with OTA, and 115 with the combination.

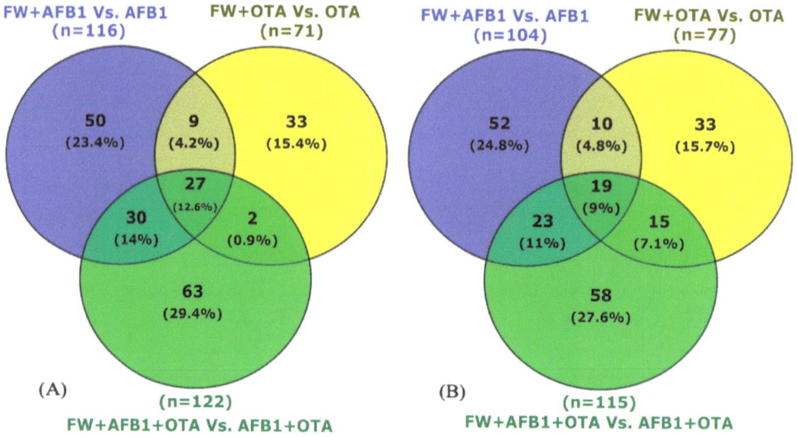

Figure 2. Venn diagram representation of common DEPs for male (**A**) and female (**B**) rats exposed to FW and mycotoxins versus the corresponding mycotoxin. $p < 0.05$ were significantly different from mycotoxins group.

In the presence of FW and P (Figure 3), the DEPs figure was higher than single FW when mycotoxins were administered individually, reporting a number of 127 proteins for males (Figure 3A) and 137 for females (Figure 3B) with AFB1 compared to 158 and 190 for males and females exposed to OTA, respectively. However, in the presence of both mycotoxins, the number decreased to 145 for males and 162 for females.

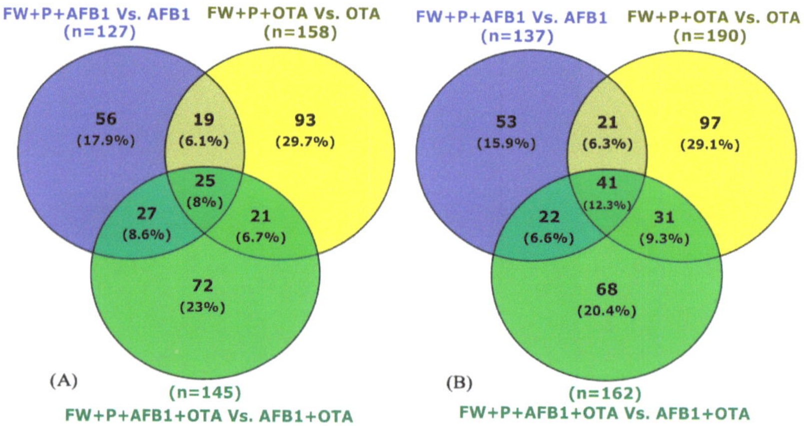

Figure 3. Venn diagram representation of common DEPs for male (**A**) and female (**B**) rats exposed to FW + P and mycotoxins versus the corresponding mycotoxin. $p < 0.05$ were significantly different from mycotoxins group.

2.2. Gene Ontology of Differentially Expressed Proteins

Functional annotation of the differentially expressed proteins (DEPs) was performed using the DAVID database [20] in order to identify the most significant biological processes (BPs) and molecular functions (MFs) involved in DEPs found in each comparison. The feed exposure to AFB1 affected hepatic metabolism. Compared to control, it altered the expression of urea cycle, glycolysis and gluconeogenesis, and amino acid biosynthesis proteins, which is in line with the results found by Sun et al. (2019) that reported an upregulation of proteins involved in cancer-related pathways of metabolism, amino acid biosynthesis, and chemical carcinogenesis [21]. Furthermore, it caused oxidative stress. These results were mostly observed in females in which, besides Hsp70 overregulation, Gpx1 and Sod1 downregulated expression was identified. Exposure to OTA feed causing an altered response to oxidative stress was evident in both sexes. The effects were similar, but the proteins involved were different, except for Prdx1 which resulted downregulation in both sexes. Its downregulation has been associated with the activation of the PI3K/AKT pathway and therefore the promotion of cancer [22]. In addition, ATP synthase F1 subunit beta (Atp5f1b), an important protein for hepatic mitochondrial function, was significantly downregulated as much in males as in females. Several studies demonstrated that the reduction in its expression exacerbates mitochondrial dysfunction and oxidative stress [23]. When rats were fed with AFB1 + OTA, changes occurred in the expression of proteins involved in metabolism such as the urea cycle, glycolysis/gluconeogenesis, and amino acid biosynthesis as in the AFB1 case. In addition, oxidative effects were observed. Specifically, in females, several antioxidant enzymes (Gta1, Gstm1, Sod1, and Cat) were downregulated. Along with these findings, the downregulation of these enzymes has been associated with the onset of diverse cancers [24,25]. Only after exposure to both mycotoxins did a reduction in the expression of structural chromatin constituents occur. Reduced expression of these components could have a negative effect on the maintenance of genome integrity.

After exposure to mycotoxins and the individual functional ingredient (FW), the response to xenobiotic stimulus emerged as the most significant BP in both males (n = 18 to 20) and females (n = 9 to 17) (Figure 4A,B).

Likewise, similar findings were observed when rats were exposed to both functional ingredients (Figure 5). The response to xenobiotic stimulus was the most common biological process in male (n = 18–20) and female rats (n = 9), though it was more pronounced in males.

According to that, it is well known that liver plays a crucial role in metabolizing and detoxifying xenobiotics via phase I and phase II enzymes, and many of the proteins involved are key players in the detoxification pathways activated by AFB1 and OTA exposure. In fact, the upregulation of oxidative proteins suggests higher oxidative stress in the liver, which could indeed be part of a positive feedback mechanism that the liver cells use to maintain homeostasis. While these parameters may not directly reflect hepatotoxicity, they offer valuable insight into the liver's adaptive response to stress [26]. Among the proteins affected, the most significantly altered were the mitochondrial enzyme involved in ketogenesis Hmgcs2, glutathione S-transferases (Gsta1, Gstm1, Mgst1) which conjugate toxic metabolites with glutathione to facilitate their excretion, and oxidative stress biomarkers superoxide dismutase 1 (Sod1) and catalase (Cat). Moreover, heat shock proteins such as Hspa8 and Hspd1 which protect cells from stress-induced damage were upregulated with the combination of toxins (Log Fold Change (FC) > 2) but not with the individual exposure (LogFC < −2). Additionally, enzymes implicated in energy metabolism and cellular repair (Aldh9a1, Adcy1) were strongly downregulated in the combined exposure (LogFC < −1.80) but not in the single ones, suggesting a synergistic effect of the toxins. According to that, previous studies confirmed the hepatotoxic modulation of xenobiotics metabolizing enzymes in the presence of AFB1, but at the same time, the capacity of coffee extracts to activate detoxifying enzymes for its degradation was demonstrated [27]. Likewise, the degradation of AFB1 was recently proven by employing diverse bacteria species, as well as different waste products containing high amounts of phenolic compounds [28,29]. Moreover, the modulation of the xenobiotic transformation system induced by OTA was involved in hepatic metabolism processes in vitro and in vivo [30,31]. However, as in this case, it has been demonstrated that plant extracts and their bioactive compounds may act by inducing xenobiotic detoxification and biotransformation pathways [32].

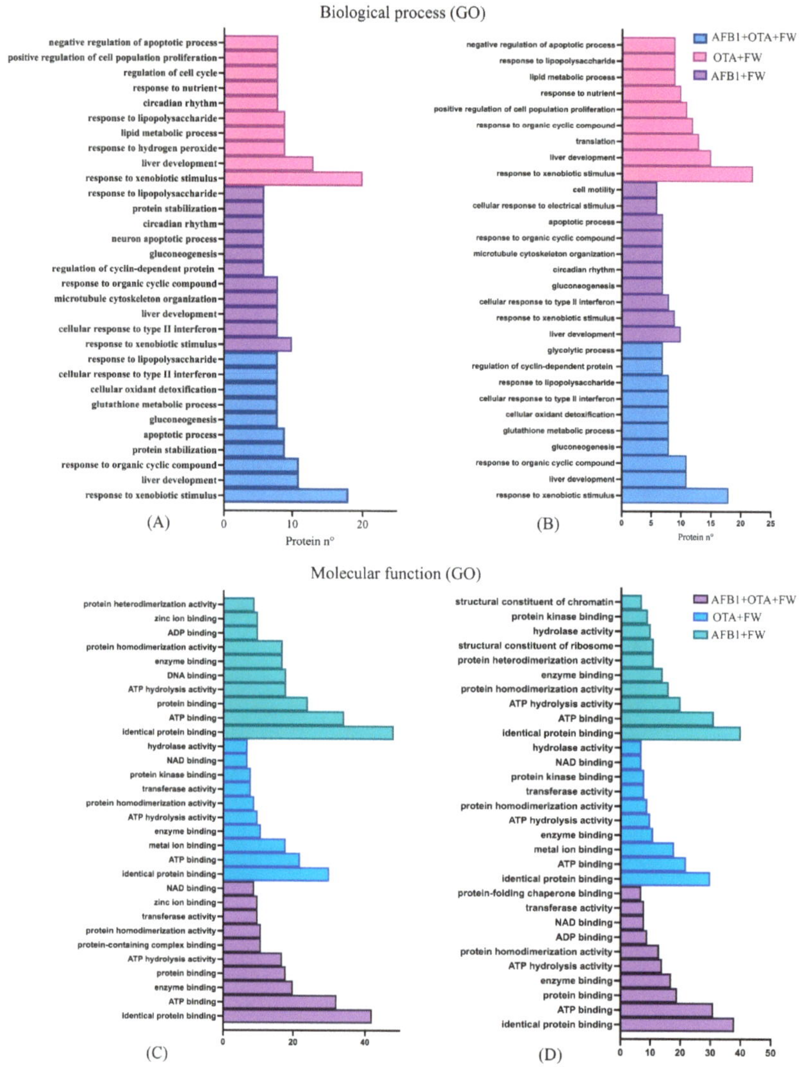

Figure 4. Gene ontology (GO) functional annotation of differentially expressed proteins for biological processes and molecular functions of male (**A**,**C**) and female (**B**,**D**) rats exposed to FW + AFB1, FW + OTA, and FW + AFB1 + OTA compared with respective mycotoxins without functional ingredient.

The following foremost BP was related to liver development with both functional ingredients and in the two sexes, reporting a number of 10 to 13 findings for males (Figures 4A and 5A) and 8 to 10 for females (Figures 4B and 5B). This biological mechanism is essential for growth, differentiation, and maturation of the liver and is tightly regulated by various signaling pathways and proteins that control cellular functions such as proliferation, differentiation, and metabolic adaptation. In the context of mycotoxin exposure, proteins such as Atp5f1b, UDP glucuronosyltransferase family 1 member A6 (Ugt1a6), adenylate kinase 2 (Ak2), and aldehyde dehydrogenase 9 family member A1 (Aldh9a1) and ornithine transcarbamylase (Otc) were downregulated when the rats were exposed to mycotoxins individually. Accordingly, diverse studies have reported the healthful effect of bioactive components contained in food in the increase in cellular antioxidant defense systems at the hepatic level [33]. Therefore, these results indicate that the combined action of these bioactive ingredients may actively participate in favorable mitigation processes. However, when both AFB1 and OTA were administered together, the expression of these proteins was notably increased, suggesting an adaptive response by the liver to counteract the toxic effects and promote recovery. Additionally, BPs related to glutathione metabolism, apoptotic processes, gluconeogenesis, response to nutrients, and circadian rhythm were also affected, but in a lower manner. In terms of MFs (Figures 4C,D and 5C,D), identical protein binding was the most enriched function in both sexes (n = 30 to 48), followed by ATP binding (n = 18 to 34), enzyme binding (n = 11 to 20), and ATP hydrolysis activity (n = 10 to 20).

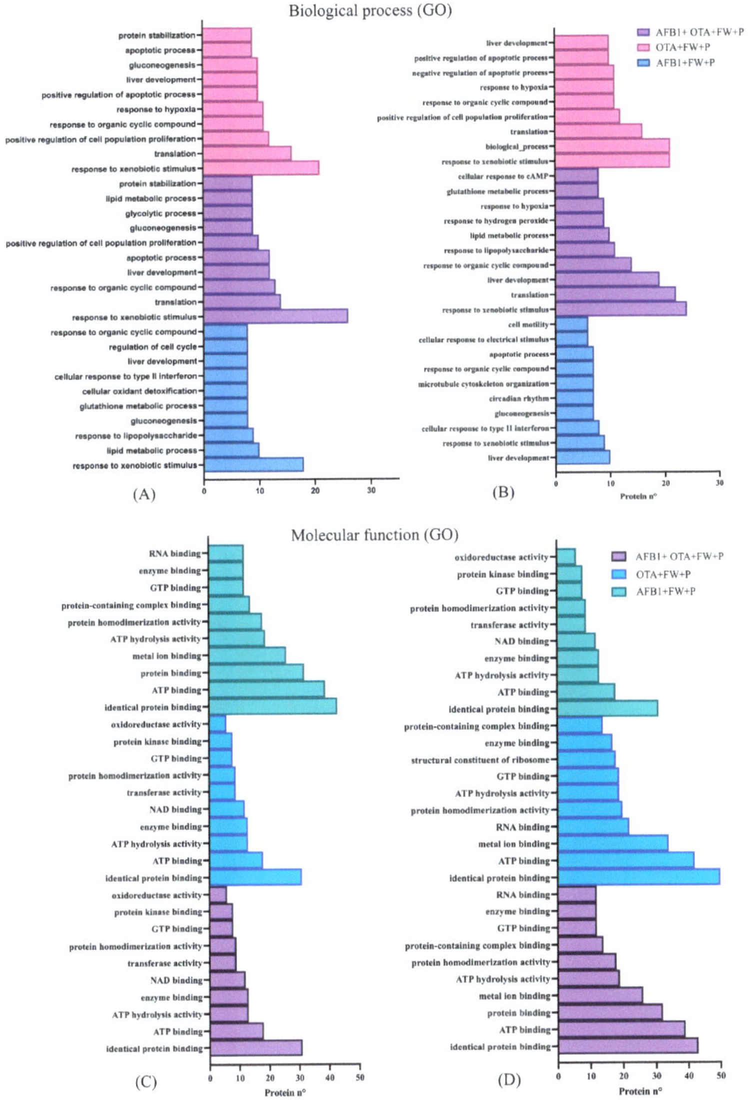

Figure 5. Gene ontology (GO) functional annotation of differentially expressed proteins for biological processes and molecular functions of male (**A**,**C**) and female (**B**,**D**) rats exposed to FW + P + AFB1, FW + P + OTA, and FW + P + AFB1 + OTA compared with respective mycotoxins without functional ingredients.

To deepen these results, a heatmap was generated from the proteomic data to visually represent the general changes in protein expression following exposure to AFB1, OTA, and their combination (AFB1 + OTA) in the presence of FW or FW + P in both male and female rats compared to control feed (Figure 6). The heatmap displays downregulated proteins (green) and upregulated proteins (red), providing a clear overview of the proteomic response to mycotoxin exposure across the different conditions. The detailed list of DEPs altered in BPs is included in Supplementary Material Table S1 for male and Table S2 for female.

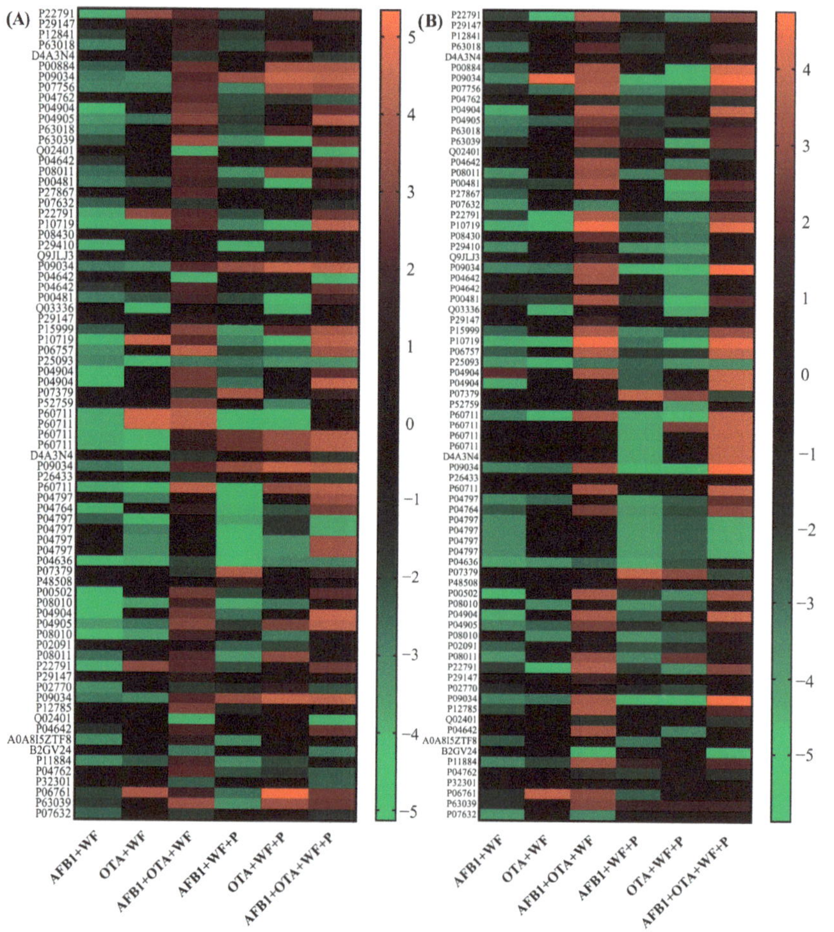

Figure 6. Heatmap representation on the expression of DEPs involved in the main biological processes after AFB1, OTA, and the combination (AFB1 + OTA) exposure in presence of FW or FW + P in male (**A**) and female (**B**) rats compared to control. The red-to-green gradient represents the logarithmic fold change value for upregulation (Log2FC > 0) and downregulation (Log2FC < 0), respectively. Black box is log2FC = 0. $p < 0.05$ significantly different from the mycotoxin groups.

Steady outcomes were perceived in both sexes, revealing matching trends in protein expression. When exposing rats to each mycotoxin separately supplemented with FW or FW + P, the preponderant part of proteins displayed a moderate downregulation (LogFC < −2), particularly in the AFB1 group, hinting at the positive action of bioactive compounds against the toxin. However, occasionally with OTA, a few proteins were upregulated, especially in males and with both ingredients. Differently, when rats were exposed to the mycotoxin mixture, the expression profile outlined a significant upregulation (LogFC > 2), particularly in the combined group (AFB1 + OTA + FW + P). This condition exhibited a potential synergistic consequence of the two toxins, where the simultaneous exposure may exacerbate the biological response compared to a single one. In line with that, numerous investigations have previously reported an AFB1 and OTA additive effect in vivo and in vitro, emphasizing the potential risk of their co-occurrence [34]. A recent metabolomic study, for instance, reported a synergistic effect of AFM1 and OTA in mice livers, displaying the alteration of metabolites related to oxidative stress [35].

2.3. Metabolic Pathways Analysis

Understanding the mechanism of action of pathways involved in the primary functioning of the liver has helped to clarify the metabolic alterations that occur in the presence of AFB1 and OTA and, notably, verify the beneficial role of the functional ingredients. For this purpose, the KEGG visualization tool related to DEPs in this study allowed the identification of the main processes altered in rats exposed to feed tainted with mycotoxins and combined with FW or FW + P, reporting that the most significant signaling pathways affected were predominantly linked to metabolic responses (Figure 7). In fact, these routes showed the highest number of modified features, higher in males exposed to FW and AFB1 alone (n = 54) or in combination with OTA (n = 52) whereas, in females, they were lower with single mycotoxins (n = 34) than combined (n = 52). In the presence of pumpkin, the situation was reversed between the genders.

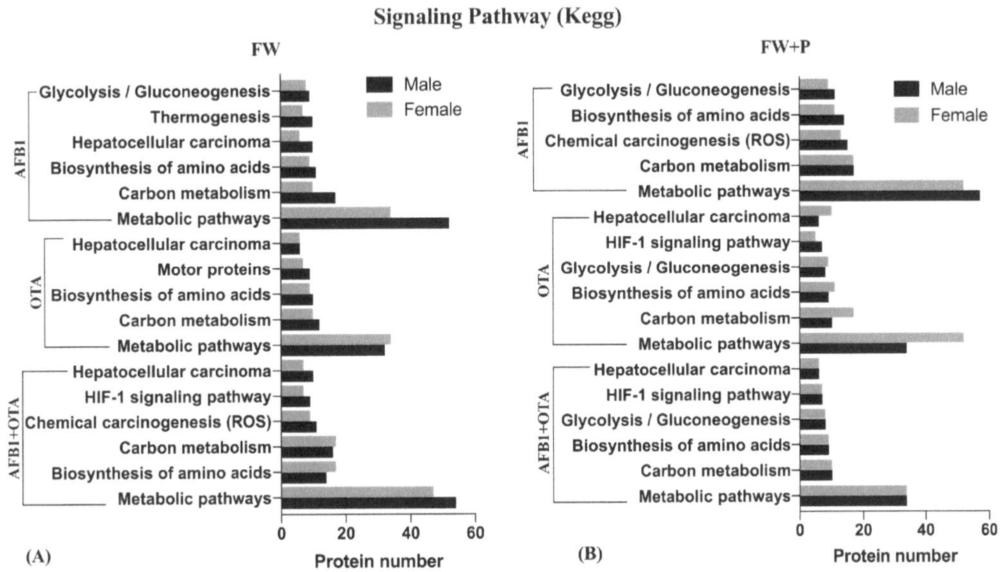

Figure 7. KEGG pathway visualization of significant signaling pathways in rats exposed to mycotoxins in combination with fermented whey (FW) (**A**) or fermented whey + pumpkin (FW + P) (**B**) feed related to the number of proteins involved compared with the exposure without functional ingredients.

However, it is well known that the liver is the largest metabolic organ which plays a specific role in digestion, metabolism, absorption, and transport of nutrients, biodegradation of toxic compounds, and processing of various hormones and cytokines secreted by the viscera [36]. Moreover, it is essential for the biosynthesis of amino acids (AAs) that serve as the building blocks for several key proteins as well as being the center of glucose metabolism and fatty acid diverting [37]. In the present investigation, carbon metabolism emerged among the most commonly altered pathways (Figure 7), along with biosynthesis of AA, suggesting consequent disruptions in energy production besides cellular processes. Indeed, one-carbon metabolic pathways aim to activate serine metabolism to glycine, the glycine cleavage system (GCS), and the metabolism of choline and other amino acids. For that reason, recent studies indicated that cancer cells may modify or become increasingly dependent on these pathways in order to maintain the supply of carbon units which are necessary for their proliferation [38]. Additionally, chemical carcinogenesis of ROS and hepatocellular carcinoma were also impacted ($n > 10$), particularly relevant given the liver's central role in metabolizing AFB1 and OTA.

Focusing once more on the overall expression of DEPs, the heatmap revealed a distinct pattern in protein expression across the multiple groups. The detailed list of DEPs altered in MPs is included in Supplementary Material Table S3 for male and Table S4 for femal.

In this case, when exposing rats to AFB1 + FW and AFB1 + FW + P, an extend downregulation can be observed compared with AFB1 (Log < −1.6). Very small differences between sexes were found in the downregulation trend observed with AFB1 exposure with the inclusion of functional ingredients compared with the mycotoxins only. With OTA, the downregulation trend can be observed more clearly when adding FW + P, especially in females (Figure 8). Nonetheless, in the latter, an upregulation of certain specimens is also displayed, reporting higher values in females (LogFC > 2.0) than in males (LogFC > 1.7). Conversely, when both AFB1 and OTA were administered together, a clear upregulation of proteins was observed, suggesting once again a synergistic effect between the two toxins. Nevertheless, the increase in protein expression in response to combined toxin exposure was slightly more pronounced in females (LogFC up to 4.7) (Figure 8A) than in males (LogFC up to 4.5) (Figure 8B), further supporting the hypothesis of a stronger synergistic effect in females. Thus, the contribution of FW or FW + P in the diet modulated the toxic effects of AFB1 and OTA when they were administered singularly, highlighting the potential for combined exposures to exert stronger effects than individual toxins.

Remarkably, several proteins identified through the proteomic analysis were linked to the hepatocellular carcinoma (HCC) pathway, a key area of concern following exposure to mycotoxins. In fact, HCC is a primary form of liver cancer which often results from chronic exposure to various toxic agents and is the sixth most common malignancy worldwide [39]. In this study, several proteins involved in liver function and cancer development were significantly affected by AFB1 and OTA exposure (Figure 9).

Among them, important members of the actin family such as beta actin (Actb), beta-actin 2 (Actbl2), actin gamma 1–1 (Actg1l1), and actin gamma 1 (Actg1) often dysregulated in cancer were downregulated under exposure to individual mycotoxin and bioactive ingredients, hinting at their helpful role. In fact, these actin monomers are fundamental for cytoskeletal polymerization and integrity and are directly implicated in the maintenance of assembly and turnover of diverse cellular processes [40] and were upregulated in male and female rats after exposure to mycotoxins individually. Among them, Actb, Actg1, and Actin 5 (Act5) were strongly upregulated after single administration (LogFC > 14).

Additionally, antioxidant response proteins belonging to the glutathione S-transferase (GST) and NAD(P)H quinone oxidoreductase 1 (NQO) families were significantly downregulated in male and female rats, with LogFC < −2.6 for FW + AFB1 and LogFC < 1.8 with both functional ingredients. On the contrary, in the combined exposure (AFB1 + OTA), expression was increased (LogFC > 2), as was the case in the single mycotoxin administration of AFB1 (LogFC > 10), highlighting a shift in cellular signaling that could favor tumorigenesis. In fact, GST is a key regulator of phase II enzymes which protect cells from oxidative stress in cancer [41], and herein, six types of GST-related proteins were broadly altered: GSTa1, GSTa2, GSTa3, GSTm1, GSTm2, and MGST1. Likewise, the upregulation of NPQ1, as in this case, has been associated with human liver injury [42]. Overall, the expression of the abovementioned proteins, particularly in combination with functional ingredients, could serve as potential biomarkers for liver carcinogenesis through the identification of important targets for therapeutic intervention.

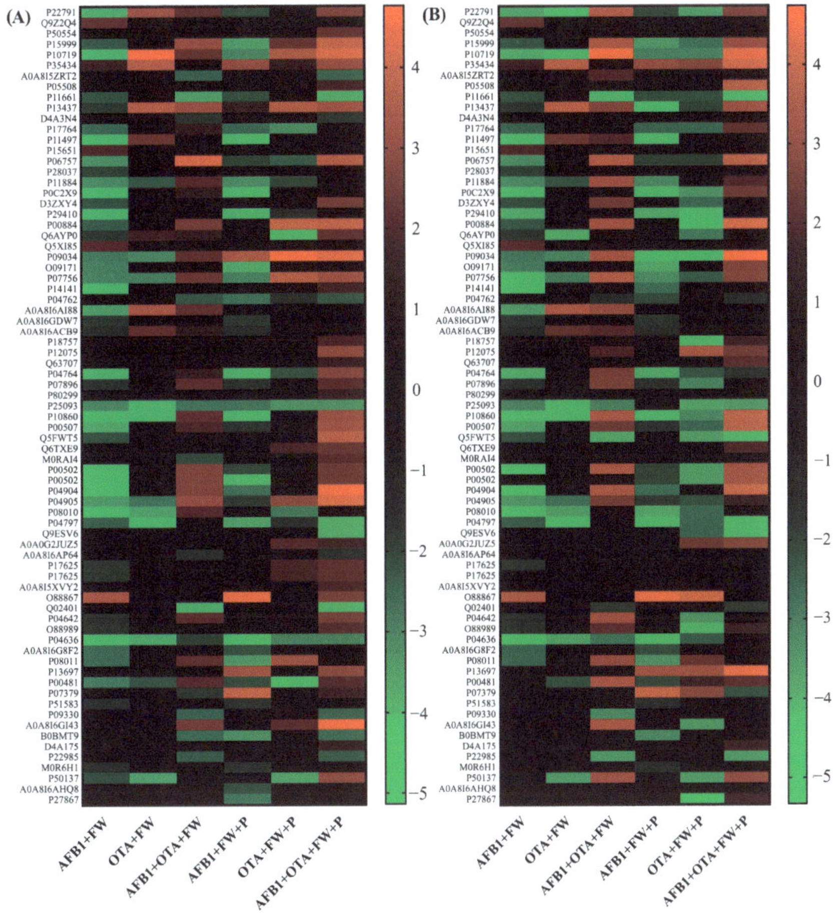

Figure 8. Heatmap representation of the expression of DEPs involved in the main signaling pathways after AFB1, OTA, and the combination (AFB1 + OTA) exposure in presence of FW or FW + P in male (**A**) and female (**B**) rat livers compared with the expression after exposure to mycotoxins without functional ingredients. The red-to-green gradient represents the logarithmic fold change value for upregulation (LogFC > 0) and downregulation (LogFC < 0), respectively. Black box is log2FC = 0. $p < 0.05$ significantly different from the mycotoxin's groups.

The present study highlights significant sex-specific differences in the hepatic response to mycotoxins (AFB1 and OTA) and their mitigation by bioactive compounds such as FW and P. These differences were evident in the number of differentially expressed proteins (DEPs), the biological processes (BPs) affected, and the pathways modulated under various experimental conditions.

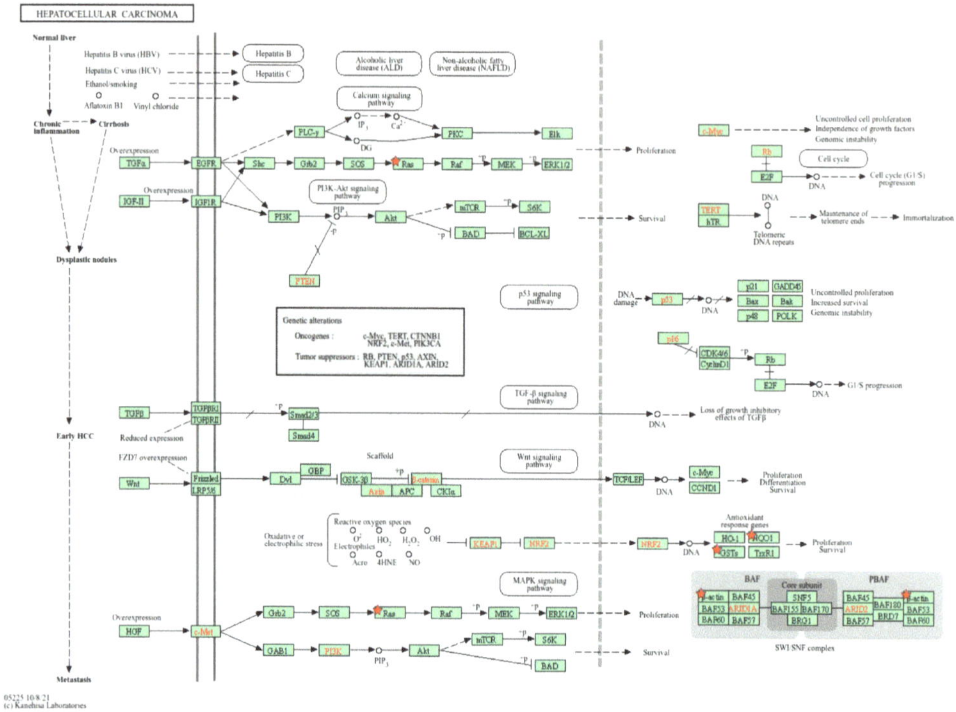

Figure 9. KEGG pathway visualization showing key molecular events involved in the development of hepatocellular carcinoma (HCC). The diagram highlights the critical signaling pathways, including those related to cell cycle regulation, apoptosis, and metabolic alterations, which contribute to the initiation and progression of liver cancer. Red stars indicate DEPs found in this study after AFB1 and OTA exposure and bioactive ingredients.

Female rats consistently exhibited a higher number of DEPs compared to males across all experimental groups. This disparity suggests a greater sensitivity of females to mycotoxin-induced hepatic changes. For instance, in response to AFB1 exposure, females exhibited 134 DEPs compared to 95 in males, while the combined AFB1 + OTA exposure amplified this effect further (140 DEPs in females vs. 81 in males). These findings align with previous reports suggesting that sex hormones may influence xenobiotic metabolism and the oxidative stress response, potentially rendering females more vulnerable to hepatotoxic effects [43,44]. The observed downregulation of key antioxidant enzymes such as Gpx1, Sod1, and Cat in females further corroborates this hypothesis, as it indicates a diminished capacity to counteract oxidative stress. In contrast, males displayed a more robust response to xenobiotic stimuli, suggesting a higher activation of detoxification pathways mediated by phase I and phase II enzymes.

Supplementation with FW or FW + P exhibited protective effects in both sexes, though the mechanisms and extent of mitigation differed. The response to xenobiotic stimuli emerged as a predominant biological process in males (18–20 proteins involved) compared to females (9–17 proteins), reflecting a sex-dependent variation in detoxification capacity. Conversely, females demonstrated a greater modulation of oxidative-stress-related pathways and metabolic processes, including amino acid biosynthesis and the urea cycle.

Interestingly, the combination of FW and P enhanced the mitigation effects, with a higher number of DEPs observed in both sexes compared to FW alone. For instance, FW + P supplementation in AFB1-exposed females resulted in 137 DEPs, compared to 127 in males.

These findings suggest a synergistic effect of FW and P in modulating hepatic responses to mycotoxins, particularly in pathways related to cellular repair and antioxidant defense.

Combined exposure to AFB1 and OTA exacerbated the hepatotoxic effects, particularly in females, as evidenced by a more pronounced upregulation of proteins (LogFC up to 4.7 in females vs. 4.5 in males). This suggests a synergistic interaction between the two mycotoxins that overwhelms the hepatic defense mechanisms, especially in females. The adaptive response observed in females, characterized by an increase in structural chromatin proteins and metabolic enzymes, may represent an effort to counteract the heightened toxic burden. However, this response appears to be less effective compared to the more stable protein expression profiles observed in males.

3. Conclusions

Proteomics studies of Wistar rats exposed to AFB1 and OTA with the addition of FW and P have highlighted their ability to counteract the negative effects of mycotoxins on hepatic responses, particularly in detoxification and development processes. Moreover, metabolic alterations induced by these toxins evidenced a significant variation in carbon metabolism and biosynthesis of AA, included in the liver's main functions. Interestingly, important biomarkers implicated in HCC were positively modulated by functional ingredients in both males and in females, but only with mycotoxins individually.

Based on these findings, the presence of FW or FW + P as functional ingredients in food may play a significant role in modulating toxic responses of mycotoxins, though further analysis is needed to fully elucidate the protective mechanisms.

4. Material and Methods

4.1. Reagents

For feed preparation, wheat flour, water, salt (NaCl), and sugar (sucrose) were acquired from a commercial market in Valencia, Spain. Aspergillus flavus ITEM 8111 was purchased from the Agro-Food Microbial Culture Collection of the Institute of Sciences and Food Production (ISPA, Bari, Italy) whereas Aspergillus steynii 20,510 was obtained from Spanish Type Culture Collection, CECT, Science Park of the University of Valencia (Paterna, Valencia, Spain). Goat milk whey coagulated by commercial rennet (starter culture R-604) was purchased from the ALCLIPOR society, S.A.L. (Benassal, Spain) while pumpkin used in this study was purchased from a supermarket (Valencia, Spain). It was peeled, the seeds removed, cut, and freeze-dried to then grind and obtain a homogeneous powder.

For protein precipitation, extraction, and digestion, ethanol was supplied by Sigma-Aldrich (St. Louis, USA), and dithiothreitol (DTT) with a purity of 99%, Trizma® hydrochloride, Tris-HCl with a purity of 99%, and trypsin were purchased from Sigma-Aldrich (St. Louis, MO, USA). Thiourea, purchased from Thermo Fisher Scientific (Kandel, Germany), and urea obtained from FEROSA (Barcelona, Spain) were used to prepare the lysis buffer used in protein digestion. Furthermore, iodoacetamide (IAA) with a purity of 98% was obtained from ACROS OrganicsTM, Thermo Fisher Scientific (Princeton, NJ, USA).

Finally, for proteomics analysis, methanol was supplied by Sigma-Aldrich. Acetonitrile (AcN) LC/MS-grade OPTIMA® ($\geq$99.9% purity) was supplied by Fisher Chemical (Geel, Belgium). Formic acid ($\geq$98%) was obtained from Sigma-Aldrich. Deionized water (<18, MΩcm resistivity) was obtained using a Milli-Q water purification system (Millipore, Bedford, MA, USA).

4.2. In Vivo Experimental Design

Male and female Wistar rats (weighing between 260–340 g) were obtained from the pharmacy animal facility at the University of Valencia, Spain. At the beginning of the study, rats were housed in polycarbonate cages in a windowless room with a 12 h light/dark cycle. The room conditions were carefully controlled to meet the species' requirements, with a temperature of 22 °C and relative humidity maintained between 45–65%. To ensure sterility during the procedures, nitrile gloves and FFP3 masks were worn when handling the animals or contaminated samples. This study was approved by the Animal Care and Use Committee of the University of Valencia (2021/VSC/PEA/0112).

After seven days of acclimatation, a total of 120 Wistar rats were divided into 12 groups, each consisting of 10 rats (5 males and 5 females) for the corresponding feeds. Among them, four test groups received mycotoxins individually or in combination, four were fed FW-contaminated feed, and the other four FW-P-containing contaminated feed. For the feeds containing functional ingredients, 35 g of FW and P were added to each during the preparation. This amount represents 1% (*w*/*w*). The control group was fed uncontaminated feed. The experimental conditions related to mycotoxin doses and their respective standard deviations are reported by [45].

The doses of aflatoxin B1 (AFB1) and ochratoxin A (OTA) used in the study were calculated based on the levels in the contaminated feed and the rats' daily intake: AFB1 dose varies from 176 to 387 μg/kg body weight per day, depending on the experimental group and sex of the rats. The dose of OTA ranged from 162 to 552 μg/kg body weight per day, with females generally receiving higher doses than males due to differences in feed intake relative to body weight. These doses were derived from feed containing AFB1 and OTA at concentrations of approximately 4.3–5.2 μg/g for AFB1 and 5.4–8.8 μg/g for OTA and were adjusted for body weight and feed consumption to reflect realistic exposure scenarios.

After 28 days, rats were sacrificed following isoflurane inhalation and organs were stored at −80 °C.

4.3. Protein Extraction, Reduction, Alkylation, and Digestion

Protein extraction was initiated using 50 mg of liver tissue which was homogenized in MilliQ-H_2O using an Ultra Turrax (IKA T10 standard). Afterwards, proteins were precipitated twice by adding 2 mL of cold ethanol to each sample, bringing the final volume to 2.5 mL. Samples were then centrifuged at 4.000 rpm, 4 °C for 15 min, the supernatant was discarded, and the pellets were resuspended in 500 μL of H_2O. Protein concentration was determined using a NeoDot UV/Vis Nano Spectrometer(ref) in order to standardize the concentration to 1 mg/mL to start the digestion. Subsequently, samples were resolved in 200 μL of lysis buffer (8 M urea/2 M thiourea/50 mM Tris-HCl) and underwent reduction and alkylation by adding solutions of DTT and IAA at a concentration of 200 mM and pH 7.8, prepared with MilliQ-H_2O and 0.4 M Tris stock buffer (pH 7.8, Tris base/MilliQ-H_2O). To break disulfide bonds, samples were incubated with 5 μL of DTT 200 mM for 1 h at 60 °C in a ThermoMixer C (Eppendorf). Samples were then incubated for 30 min at 37 °C to alkylate protein cysteine residues with 20 μL of IAA. Finally, trypsin enzyme (1 mg/mL) was added to start peptide digestion which was carried out overnight at 37 °C. After that, the reaction was stopped by adding acetic acid 5% (pH 5) and filtered prior to LC-MS/MS-Q-TOF injection.

4.4. Identification and Quantification of Proteins Through LC-MS/MS-Q-TOF

Two technical replicates of each biological sample (50 μg/mL) were injected into an LC system (Agilent 1200 LC) coupled to a triple quadrupole time-of-flight (Q-TOF) mass spectrometry (Agilent 6540 UHD) system using a C18 RP AdvanceBio capillary column for 2.7 μm, 120 Å, 2.1 × 150 mm peptide mapping. The method previously developed by [28] was followed. Briefly, a nonlinear gradient of 40 min at a flow rate of 0.2 mL/min was utilized. Two different phases were used in the process: phase A (H_2O in 0.1% formic acid) and phase B (acetonitrile in 0.1% formic acid). The elution gradient starts with 3% phase B for 1 min and increases to 40% at 21 min. In the next 3 min, it reaches 95% and was maintained during 1 min; Afterwards, it decreases to 3% for 6 min and maintained in the last 8 min. The experimental conditions were repeated three times independently.

4.5. Statistical Analysis and Bioinformatics

The software Spectrum Mill MS Proteomics Workbench Package Rev B.06.00.201 (Agilent Technologies) was used to process the chromatographic spectra. This software is capable of analyzing data from high-quality spectra, reducing false positives, and identifying proteins and peptides by matching them with the UniProt database. Entities were then sorted by their frequency of occurrence across all replicates within each experimental group following the MS/MS parameters previously retrieved and verified by [46]. Afterwards, the identified proteins were statistically filtered by using Mass Profiler Professional (MPP) software v15.0 (Agilent Technologies) and differences between the experimental mycotoxin and the control group were assessed using an unpaired t-test with Benjamini–Hochberg adjustment. Results with a FC $\geq$ 0.7 and a p-value < 0.05 were considered statistically significant and checked for the bioinformatics analysis, including the features which corresponded to the UniProt accession codes for *Rattus norvegicus*. Finally, the BPs, MFs, and metabolic pathways associated with these proteins were explored using the Database for Annotation, Visualization, and Integrated Discovery (DAVID). Graphical representations of the data were created with GraphPad Prism software version 8.0.0 (San Diego, CA, USA). The Venn diagram for DEPs was generated using the Venny 2.1 interactive tool [47].

Supplementary Materials: The following are available online at https://www.mdpi.com/article/10.3390/toxins17010029/s1, Table S1: DEPs obtained through statistical analysis comparing male rats exposed to mycotoxins and FW (AFB1+FW, OTA+FW, AFB1+OTA+FW) or FW+P (AFB1+FW+P, OTA+FW+P, AFB1+OTA+FW+P) versus the corresponding mycotoxin (AFB1, OTA, AFB1+OTA) related to biological processes. Values represent the Log2FC. Table S2: DEPs obtained through statistical analysis comparing female rats exposed to mycotoxins and FW (AFB1+FW, OTA+FW, AFB1+OTA+FW) or FW+P (AFB1+FW+P, OTA+FW+P, AFB1+OTA+FW+P) versus the corresponding mycotoxin (AFB1, OTA, AFB1+OTA) related to biological processes. Values represent the Log2FC. Table S3: DEPs obtained through statistical analysis comparing male rats exposed to mycotoxins and FW (AFB1+FW, OTA+FW, AFB1+OTA+FW) or FW+P (AFB1+FW+P, OTA+FW+P, AFB1+OTA+FW+P) versus the corresponding mycotoxin (AFB1, OTA, AFB1+OTA) related to metabolic pathways. Values represent the Log2FC. Table S4: DEPs obtained through statistical analysis comparing male rats exposed to mycotoxins and FW (AFB1+FW, OTA+FW, AFB1+OTA+FW) or FW+P (AFB1+FW+P, OTA+FW+P, AFB1+OTA+FW+P) versus the corresponding mycotoxin (AFB1, OTA, AFB1+OTA) related to metabolic pathways. Values represent the Log2FC.

Author Contributions: Methodology, S.T. and A.C.; conceptualization, S.T., A.C., and L.M.; software, S.T. and A.C.; validation, S.T. and A.C.; formal analysis, S.T., A.C., and P.V.-D.; investigation, S.T., A.C., and P.V.-D.; writing—original draft preparation, S.T. and A.C.; writing—review and editing, L.M., M.G., and J.M.; visualization, A.C., L.M., and P.V.-D.; supervision, L.M, M.G., and J.M.; project administration L.M. and J.M.; funding acquisition, L.M. and J.M. All authors have read and agreed to the published version of the manuscript.

Funding: This research was funded by Spanish Ministry of Science, Innovation and Universities for the project PID2022-140722OB-I00, MCIU/AEI/10.13039/501100011033/FEDER, EU and for the postdoctoral grant of Alessandra Cimbalo "Margarita Salas" (MS21-165). Silvia Trombetti was supported by post-graduated fellowship grant from the Department of Veterinary Medicine and Animal Productions, University of Naples Federico II (Italy), Project leader: Prof. Sante Roperto.

Institutional Review Board Statement: The animal study protocol was approved by the Ethics Committee of University of Valencia (protocol code 2021/VSC/PEA/0112—17 June 2021).

Informed Consent Statement: Not applicable.

Data Availability Statement: The original contributions presented in this study are included in this article and Supplementary Material. Further inquiries can be directed to the corresponding authors.

Conflicts of Interest: The authors declare no conflicts of interest.

References

1. Arce-López, B.; Coton, M.; Coton, E.; Hymery, N. Occurrence of the Two Major Regulated Mycotoxins, Ochratoxin A and Fumonisin B1, in Cereal and Cereal-Based Products in Europe and Toxicological Effects: A Review. *Environ. Toxicol. Pharmacol.* **2024**, *109*, 104489. [CrossRef] [PubMed]
2. IARC Working Group on the Evaluation of Carcinogenic Risks to Humans. Chemical Agents and Related Occupations. *IARC Monogr. Eval. Carcinog. Risks Hum.* **2012**, *100*, 9–562.
3. Suman, M. Last Decade Studies on Mycotoxins' Fate during Food Processing: An Overview. *Curr. Opin. Food Sci.* **2021**, *41*, 70–80. [CrossRef]
4. Persico, M.; Sessa, R.; Cesaro, E.; Dini, I.; Costanzo, P.; Ritieni, A.; Fattorusso, C.; Grosso, M. A Multidisciplinary Approach Disclosing Unexplored Aflatoxin B1 Roles in Severe Impairment of Vitamin D Mechanisms of Action. *Cell Biol. Toxicol.* **2023**, *39*, 1275–1295. [CrossRef]
5. Gupta, R.C.; Doss, R.B.; Lall, R.; Srivastava, A.; Sinha, A. Aflatoxins, Ochratoxins, and Citrinin. In *Reproductive and Developmental Toxicology*; Elsevier: Amsterdam, The Netherlands, 2022; pp. 983–1002. ISBN 978-0-323-89773-0.
6. Deng, J.; Zhao, L.; Zhang, N.-Y.; Karrow, N.A.; Krumm, C.S.; Qi, D.-S.; Sun, L.-H. Aflatoxin B1 Metabolism: Regulation by Phase I and II Metabolizing Enzymes and Chemoprotective Agents. *Mutat. Res. /Rev. Mutat. Res.* **2018**, *778*, 79–89. [CrossRef] [PubMed]
7. Heussner, A.; Bingle, L. Comparative Ochratoxin Toxicity: A Review of the Available Data. *Toxins* **2015**, *7*, 4253–4282. [CrossRef] [PubMed]
8. Escrivá, L.; Manyes, L.; Vila-Donat, P.; Font, G.; Meca, G.; Lozano, M. Bioaccessibility and Bioavailability of Bioactive Compounds from Yellow Mustard Flour and Milk Whey Fermented with Lactic Acid Bacteria. *Food Funct.* **2021**, *12*, 11250–11261. [CrossRef]
9. Peng, K.; Koubaa, M.; Bals, O.; Vorobiev, E. Recent Insights in the Impact of Emerging Technologies on Lactic Acid Bacteria: A Review. *Food Res. Int.* **2020**, *137*, 109544. [CrossRef]
10. Salehi, B.; Quispe, C.; Sharifi-Rad, J.; Giri, L.; Suyal, R.; Jugran, A.K.; Zucca, P.; Rescigno, A.; Peddio, S.; Bobiş, O.; et al. Antioxidant Potential of Family Cucurbitaceae with Special Emphasis on *Cucurbita* Genus: A Key to Alleviate Oxidative Stress-mediated Disorders. *Phytother. Res.* **2021**, *35*, 3533–3557. [CrossRef] [PubMed]
11. More, P.R.; Jambrak, A.R.; Arya, S.S. Green, Environment-Friendly and Sustainable Techniques for Extraction of Food Bioactive Compounds and Waste Valorization. *Trends Food Sci. Technol.* **2022**, *128*, 296–315. [CrossRef]
12. León-López, A.; Pérez-Marroquín, X.A.; Campos-Lozada, G.; Campos-Montiel, R.G.; Aguirre-Álvarez, G. Characterization of Whey-Based Fermented Beverages Supplemented with Hydrolyzed Collagen: Antioxidant Activity and Bioavailability. *Foods* **2020**, *9*, 1106. [CrossRef] [PubMed]
13. Rosa, L.S.; Santos, M.L.; Abreu, J.P.; Rocha, R.S.; Esmerino, E.A.; Freitas, M.Q.; Mársico, E.T.; Campelo, P.H.; Pimentel, T.C.; Cristina Silva, M.; et al. Probiotic Fermented Whey-Milk Beverages: Effect of Different Probiotic Strains on the Physicochemical Characteristics, Biological Activity, and Bioactive Peptides. *Food Res. Int.* **2023**, *164*, 112396. [CrossRef]
14. Kume, H.; Okazaki, K.; Yamaji, T.; Sasaki, H. A Newly Designed Enteral Formula Containing Whey Peptides and Fermented Milk Product Protects Mice against Concanavalin A-Induced Hepatitis by Suppressing Overproduction of Inflammatory Cytokines. *Clin. Nutr.* **2012**, *31*, 283–289. [CrossRef] [PubMed]
15. Zhao, Z.W.; Pan, D.D.; Wu, Z.; Sun, Y.Y.; Guo, Y.X.; Zeng, X.Q. Antialcoholic Liver Activity of Whey Fermented by Lactobacillus Casei Isolated from Koumiss. *J. Dairy Sci.* **2014**, *97*, 4062–4071. [CrossRef] [PubMed]
16. Radic, I.; Mijovic, M.; Tatalovic, N.; Mitic, M.; Lukic, V.; Joksimovic, B.; Petrovic, Z.; Ristic, S.; Velickovic, S.; Nestorovic, V.; et al. Protective Effects of Whey on Rat Liver Damage Induced by Chronic Alcohol Intake. *Hum. Exp. Toxicol.* **2019**, *38*, 632–645. [CrossRef] [PubMed]
17. Moawad, R.M.; Osman, A.H.; Hassanein, K.M.; Elkot, W.F.; Asar, A.M.; Alhag, S.K.; Al-Shuraym, L.A.; Alghamdi, O.A.; Al-Farga, A.; Zaidalkilani, A.T.; et al. Assessment of Sweet Whey Fortified with Bifidobacteria and Selenium on Reduction of Pesticide Liver Toxicity in Albino Rats. *Ital. J. Food Sci.* **2024**, *36*, 275–287. [CrossRef]
18. Van Summeren, A.; Renes, J.; Van Delft, J.H.M.; Kleinjans, J.C.S.; Mariman, E.C.M. Proteomics in the Search for Mechanisms and Biomarkers of Drug-Induced Hepatotoxicity. *Toxicology In Vitro* **2012**, *26*, 373–385. [CrossRef] [PubMed]

19. Wong, T.Y.; Yan, N.; Kwan, K.K.L.; Pan, Y.; Liu, J.; Xiao, Y.; Wu, L.; Lam, H. Comparative Proteomic Analysis Reveals the Different Hepatotoxic Mechanisms of Human Hepatocytes Exposed to Silver Nanoparticles. *J. Hazard. Mater.* **2023**, *445*, 130599. [CrossRef] [PubMed]
20. Huang, D.W.; Sherman, B.T.; Lempicki, R.A. Systematic and Integrative Analysis of Large Gene Lists Using DAVID Bioinformatics Resources. *Nat. Protoc.* **2009**, *4*, 44–57. [CrossRef]
21. Sun, Y.; Wen, J.; Chen, R.; Deng, Y. Variable Protein Homeostasis in Housekeeping and Non-Housekeeping Pathways under Mycotoxins Stress. *Sci. Rep.* **2019**, *9*, 7819. [CrossRef] [PubMed]
22. Xiao, H.; Yang, T.; Yan, L.; Feng, J.; Huang, B.; Jiang, Y. PRDX1 Is a Tumor Suppressor for Nasopharyngeal Carcinoma by Inhibiting PI3K/AKT/TRAF1 Signaling. *OncoTargets Ther.* **2020**, *13*, 9123–9133. [CrossRef] [PubMed]
23. Xiao, M.; Chi, X.; Zhu, X.; Xu, Z.; Zou, Y.; Peng, Y.; Luan, S.; Dong, J.; Dai, Y.; Yin, L. Proteomic Analysis of Laser Captured Tubular Tissues Reveals Complement Activation and Mitochondrial Dysfunction in Autoimmune Related Kidney Diseases. *Sci. Rep.* **2024**, *14*, 19311. [CrossRef] [PubMed]
24. Galasso, M.; Gambino, S.; Romanelli, M.G.; Donadelli, M.; Scupoli, M.T. Browsing the Oldest Antioxidant Enzyme: Catalase and Its Multiple Regulation in Cancer. *Free Radic. Biol. Med.* **2021**, *172*, 264–272. [CrossRef]
25. Asaduzzaman Khan, M.; Tania, M.; Zhang, D.; Chen, H. Antioxidant Enzymes and Cancer. *Chin. J. Cancer Res.* **2010**, *22*, 87–92. [CrossRef]
26. Temple, R. Guidance for Industry. *Drug Saf.* **2009**.
27. Cavin, C.; Mace, K.; Offord, E.A.; Schilter, B. Protective Effects of Coffee Diterpenes against Aflatoxin B1-Induced Genotoxicity: Mechanisms in Rat and Human Cells. *Food Chem. Toxicol.* **2001**, *39*, 549–556. [CrossRef] [PubMed]
28. Adegoke, T.V.; Yang, B.; Tian, X.; Yang, S.; Gao, Y.; Ma, J.; Wang, G.; Si, P.; Li, R.; Xing, F. Simultaneous Degradation of Aflatoxin B1 and Zearalenone by Porin and Peroxiredoxin Enzymes Cloned from Acinetobacter Nosocomialis Y1. *J. Hazard. Mater.* **2023**, *459*, 132105. [CrossRef]
29. Chariyakornkul, A.; Punvittayagul, C.; Taya, S.; Wongpoomchai, R. Inhibitory Effect of Purple Rice Husk Extract on AFB1-Induced Micronucleus Formation in Rat Liver through Modulation of Xenobiotic Metabolizing Enzymes. *BMC Complement. Altern. Med.* **2019**, *19*, 237. [CrossRef]
30. González-Arias, C.A.; Crespo-Sempere, A.; Marín, S.; Sanchis, V.; Ramos, A.J. Modulation of the Xenobiotic Transformation System and Inflammatory Response by Ochratoxin A Exposure Using a Co-Culture System of Caco-2 and HepG2 Cells. *Food Chem. Toxicol.* **2015**, *86*, 245–252. [CrossRef] [PubMed]
31. Gautier, J.-C.; Richoz, J.; Welti, D.H.; Markovic, J.; Gremaud, E.; Guengerich, F.P.; Turesky, R.J. Metabolism of Ochratoxin A: Absence of Formation of Genotoxic Derivatives by Human and Rat Enzymes. *Chem. Res. Toxicol.* **2001**, *14*, 34–45. [CrossRef]
32. Wu, J.-C.; Lai, C.-S.; Tsai, M.-L.; Ho, C.-T.; Wang, Y.-J.; Pan, M.-H. Chemopreventive Effect of Natural Dietary Compounds on Xenobiotic-Induced Toxicity. *J. Food Drug Anal.* **2017**, *25*, 176–186. [CrossRef] [PubMed]
33. Ganesan, K.; Jayachandran, M.; Xu, B. A Critical Review on Hepatoprotective Effects of Bioactive Food Components. *Crit. Rev. Food Sci. Nutr.* **2018**, *58*, 1165–1229. [CrossRef] [PubMed]
34. Frangiamone, M.; Lázaro, Á.; Cimbalo, A.; Font, G.; Manyes, L. In Vitro and in Vivo Assessment of AFB1 and OTA Toxic Effects and the Beneficial Role of Bioactive Compounds. A Systematic Review. *Food Chem.* **2024**, *447*, 138909. [CrossRef] [PubMed]
35. Gao, Y.-N.; Wu, C.-Q.; Wang, J.-Q.; Zheng, N. Metabolomic Analysis Reveals the Mechanisms of Hepatotoxicity Induced by Aflatoxin M1 and Ochratoxin A. *Toxins* **2022**, *14*, 141. [CrossRef] [PubMed]
36. Treyer, A.; Müsch, A. Hepatocyte Polarity. In *Comprehensive Physiology*; Prakash, Y.S., Ed.; Wiley: Hoboken, NJ, USA, 2013; pp. 243–287, ISBN 978-0-470-65071-4.
37. Liao, Y.; Chen, Q.; Liu, L.; Huang, H.; Sun, J.; Bai, X.; Jin, C.; Li, H.; Sun, F.; Xiao, X.; et al. Amino Acid Is a Major Carbon Source for Hepatic Lipogenesis. *Cell Metab.* **2024**, *36*, 2437–2448.e8. [CrossRef]
38. Newman, A.C.; Maddocks, O.D.K. One-Carbon Metabolism in Cancer. *Br. J. Cancer* **2017**, *116*, 1499–1504. [CrossRef] [PubMed]
39. Ferenci, P.; Fried, M.; Labrecque, D.; Bruix, J.; Sherman, M.; Omata, M.; Heathcote, J.; Piratsivuth, T.; Kew, M.; Otegbayo, J.A.; et al. Hepatocellular Carcinoma (HCC): A Global Perspective. *J. Clin. Gastroenterol.* **2010**, *44*, 239–245. [CrossRef] [PubMed]
40. Peng, J.; Bera, R.; Chiou, C.; Yu, M.; Chen, T.; Chen, C.; Wang, T.; Chiang, W.; Chai, S.; Wei, Y.; et al. Actin Cytoskeleton Remodeling Drives Epithelial-mesenchymal Transition for Hepatoma Invasion and Metastasis in Mice. *Hepatology* **2018**, *67*, 2226–2243. [CrossRef] [PubMed]
41. Li, T.; Zhao, X.-P.; Wang, L.-Y.; Gao, S.; Zhao, J.; Fan, Y.-C.; Wang, K. Glutathione S-Transferase P1 Correlated with Oxidative Stress in Hepatocellular Carcinoma. *Int. J. Med. Sci.* **2013**, *10*, 683–690. [CrossRef]
42. Aleksunes, L.M. Up-Regulation of NAD(P)H Quinone Oxidoreductase 1 during Human Liver Injury. *World J. Gastroenterol.* **2006**, *12*, 1937. [CrossRef] [PubMed]
43. Xu, L.; Yuan, Y.; Che, Z.; Tan, X.; Wu, B.; Wang, C.; Xu, C.; Xiao, J. The Hepatoprotective and Hepatotoxic Roles of Sex and Sex-Related Hormones. *Front. Immunol.* **2022**, *13*, 939631. [CrossRef] [PubMed]

44. Allegra, A.; Caserta, S.; Genovese, S.; Pioggia, G.; Gangemi, S. Gender Differences in Oxidative Stress in Relation to Cancer Susceptibility and Survival. *Antioxidants* **2023**, *12*, 1255. [CrossRef] [PubMed]
45. Vila-Donat, P.; Sánchez, D.; Cimbalo, A.; Mañes, J.; Manyes, L. Effect of Bioactive Ingredients on Urinary Excretion of Aflatoxin B1 and Ochratoxin A in Rats, as Measured by Liquid Chromatography with Fluorescence Detection. *Toxins* **2024**, *16*, 363. [CrossRef] [PubMed]
46. Cimbalo, A.; Frangiamone, M.; Juan, C.; Font, G.; Lozano, M.; Manyes, L. Proteomics Evaluation of Enniatins Acute Toxicity in Rat Liver. *Food Chem. Toxicol.* **2021**, *151*, 112130. [CrossRef] [PubMed]
47. Oliveros, J.C. Venny. An Interactive Tool for Comparing Lists with Venn's Diagrams. 2007–2015. Available online: https://bioinfogp.cnb.csic.es/tools/venny/index.html (accessed on 18 November 2024).

MDPI AG
Grosspeteranlage 5
4052 Basel
Switzerland
Tel.: +41 61 683 77 34

Toxins Editorial Office
E-mail: toxins@mdpi.com
www.mdpi.com/journal/toxins

www.ingramcontent.com/pod-product-compliance
Lightning Source LLC
Chambersburg PA
CBHW041355210926
51850CB00013B/3091
* 9 7 8 3 7 2 5 8 4 2 3 3 9 *